DEVELOPMENT, THE ENVIRONMENT AND FOOD: **TOWARDS AGRICULTURAL CHANGE?**

DEVELOPMENT, THE ENVIRONMENT AND FOOD: **TOWARDS AGRICULTURAL CHANGE?**

Pierre Jacquet
Rajendra K. Pachauri
Laurence Tubiana (Editors)

Raphaël Jozan, Sébastien Treyer, Viviane Gravey,
Sanjivi Sundar (Associate Editors)

AGENCE FRANÇAISE DE DÉVELOPPEMENT · IDDRI SciencesPo. · teri

CONTENTS

■ **Agence Française de Développement (AFD)** is a public development finance institution that has worked to fight poverty and support economic growth in developing countries and the French Overseas Communities for almost 70 years. AFD executes the French government's development aid policies. Through offices in more than fifty countries and nine French Overseas Communities, AFD provides financing and support for projects that improve people's living conditions, promote economic growth and protect the planet: schooling, maternal healthcare, help for farmers and small business, water supply, preservation of tropical forests, and fighting climate change, among other concerns. In 2010, AFD approved more than €6.8 billion for financing aid activities in developing countries and the French Overseas Provinces. The funds will help 13 million children go to school, improve drinking water access for 33 million people and provide €428 million in microloans benefiting more than 700,000 people. Energy efficiency projects financed by AFD in 2010 will save nearly 5 million tons of carbon dioxide emissions annually.
More information and publications available at www.afd.fr

■ **The Institute for Sustainable Development and International Relations (IDDRI)** is a think tank based in Paris (France) and Brussels (Belgium). Its mission is to bridge the gap between research and decision-making and investigate international policy issues on sustainable development and global governance. IDDRI focuses its activities on four major topics: climate change, biodiversity, global governance and urban fabric. It also seeks to facilitate discussions between the various stakeholders involved in global issues related to sustainable development.
Through its partnership with Sciences Po in Paris, France, IDDRI is also involved in teaching and participates in the development of research programmes with support from the Sustainable Development Centre of Sciences Po.
More information and publications available at www.iddri.org

■ **The Energy and Resources Institute (TERI)** was set up in 1974 to deal inter alia with issues relating to sustainable development, the environment, energy efficiency and the sustainable use of natural resources. Its goal is to develop innovative solutions for achieving sustainable development. Its activities range from the formulation of local and national strategies, to proposals for global solutions, to energy and environment-related issues. TERI is based in New Delhi, and also present in many other regions of India. It has 700 employees and is headed by Rajendra K. Pachauri who is also the Chairman of the Intergovernmental Panel on Climate Change (IPCC), which was awarded the 2007 Nobel Prize.
More information and publications available at www.teriin.org

Development, food and the environment: towards agricultural change?
Pierre Jacquet, Rajendra K. Pachauri and Laurence Tubiana (Editors)
Delhi: TERI Press, 2012
ISBN 978-81-7993-443-2

Editors. Pierre Jacquet, Chief Economist for the French Development Agency (AFD); Rajendra K. Pachauri, Director General of The Energy and Resources Institute (TERI) and Chairman of the Intergovernmental Panel on Climate Change (IPCC), which was awarded a Nobel Prize in 2007; and Laurence Tubiana, Director of the Institute for Sustainable Development and International Relations (IDDRI) and the Sustainable Development Centre at Sciences Po.

Associate editors: Raphaël Jozan (AFD), Viviane Gravey and Sebastien Treyer (IDDRI), Sanjivi Sundar (TERI).

Editorial coordinators: Raphaël Jozan (AFD), Isabelle Biagiotti (*Courrier de la planète*) and Viviane Gravey (IDDRI).

Translation and adaptation from French: Jim and Katell Johnson, and Gill Gladstone.

Text- and copy-editing: Jim and Katell Johnson.

Maps and figures: Légendes Cartographie, SARL.

Graphic design: Alain Chevallier, design and cover.

Administration: Raphaël Jozan (AFD)

Photo credits: Cover: Circular crop fields in Kansas, characteristic of center pivot irrigation, ©Nasa.

Pierre Jacquet
Chief Economist, AFD *(Agence Française de Développement)*, Paris, France

Rajendra K. Pachauri
Distinguished Director TERI, (The Energy and Resources Institute), and Professor, TERI University, New Delhi, India

Laurence Tubiana
Director, Institute for Sustainable Development and International Relations, (IDDRI), Paris, France

AGRICULTURE: A PREREQUISITE TO SUSTAINABLE DEVELOPMENT

A *Planet for Life* 2012, our annual book on sustainable development issues, focuses this year on agriculture and its relation to development and the environment. In 2011, agriculture has come back to the fore of the international agenda in the aftermath of the 2008-2010 price hikes in food and other farm commodities, as well as the various weather shocks leading to food crises in parts of Africa and to the humanitarian crisis in the Horn of Africa. Suddenly, a series of issues that were hardly mentioned gained prominence in the international public debate. Yet, little was really new. The 2008 and 2010 price shocks were indeed very large, but were not without historical precedent. Besides which, the challenge of food security was daunting long before the crisis: since the mid-1990s, the proportion of undernourished people in developing countries has stopped declining, remaining more or less stable at just above 15%. Given the population increase, this means that an increasing number of people, nowadays close to one billion, are suffering from malnutrition. However justified and understandable the post-crisis uproar may be, this crisis can thus be interpreted as a wake-up call rather than a fundamental cause of the problem.

This points once again to the role of crises in sustaining action and correcting unsustainable situations. As we have already underlined in former issues of *A Planet for Life*, what is "unsustainable" will by definition not be sustained, so that the central policy issue for sustainable development is when to act in order to minimize the costs of necessary adjustments. But if one wants to go beyond mere normative statements, a theory of action is needed, one of the central questions being when the adjustment can actually be decided and implemented – as opposed to what it should consist of. This is indeed the core challenge of sustainability. The minds of decision-makers and others tend to focus on issues at times of crisis rather than on an ongoing basis. There are many reasons for this: first, the human mind is able to manage complexity, but certainly cannot simultaneously focus on all issues, so that some issues will receive less attention, and it is of course understandable that short-term pressures define the priorities; second, when there is no perceived sense of urgency, there are always debates and oppositions to any course of action; third, policy makers are under great media pressure to show that they are in charge and know how to react to ongoing crises with concrete measures and solutions. As a result, they tend to look for quick fixes on current problems, and are in a reactive rather than a pro-active, crisis prevention mode. This is why it is so important, when a crisis concentrates attention on any issue, to use the political momentum not only to respond to the crisis, but to put in place measures and processes that will have a longer term effect and address the issues at their core rather than only the symptoms of their earlier mismanagement.

It was nonetheless encouraging that after so many years of neglect, agriculture and food security eventually received so much political attention in 2011. The French presidency of the G20 has succeeded in mainstreaming them as G20 priorities, and there is a good chance that these issues will still be addressed on priority in the following G20 exercises, starting with the 2012 Mexican presidency. The various chapters in this book shed light on the core underlying issues, focusing on the nexus of agriculture, food, development and the environment. They show that on virtually all aspects, there is hardly any consensus, and the debate goes on. As mentioned above, this may also explain why timely action is hard to come by without a crisis: crises and the perception of their potential impact act as powerful coordination incentives that help bridge disagreements and invite action.

Faithful to the general principles of this annual book, which aims at documenting sustainable development issues and actions, we have called on many highly specialized authors from different countries and perspectives, so that controversies may emerge. This diversity also shows how the formulation and reality of various stakes are both bound by time and geography. Our goal, in this book, is not to bridge gaps and reach conclusions: the challenge of sustainable development precisely amounts to finding the conditions for action, and this is a constant, domestic and international, negotiating process between the various stakeholders. As citizens, we are deeply concerned. But our book is an exercise in analytic documentation, rather than in advocacy.

The following, substantive synthesis by the book's scientific directors provides a substantial survey of the various chapters, highlighting major issues and controversies and questioning some of the conventional wisdom that shapes ongoing views on agriculture and sustainable development. This introduction highlights four main issues: production, ecology, the international dimension, and the role of public policies.

FEEDING THE PLANET

Feeding the planet has been presented as one of the crucial challenges ahead, given the expected population increase to 9 billion by 2050[1]. As will be suggested in this book, the main challenge, though, is not one of volume for global production. Prospective studies tend to show that enough land, human and water resources can be mobilized to provide the necessary calories. This suggests that production increases will be necessary and are possible in many parts of the world, notably in sub-Saharan Africa. However, there still remains a twofold challenge. First, there will be a need for complex and reliable logistics, so that food can be adequately and timely supplied to the populations that need it. It is, however, likely that the timing and geography of needs will not smoothly correspond to those of production. Hence, there is a major role for stockpiles and for delivery systems, and for all services that are needed in terms of quality control and maintenance over the supply chain. In developing countries, these issues remain crucial. Local commodity exchanges, like the Ethiopian commodity exchange in Addis Ababa, are useful building blocks for more effective production and logistical systems. As a first step, local exchanges provide a place to exchange information and settle disputes. They play a major role in helping de-personalize transactions, by providing collective guarantees (about delivery and quality) that allow producers and customers to transact without necessarily knowing each other. This may expand trade and production, and requires the building and maintenance of stockpiles as well as measuring and maintaining quality in a credible way. This is, however, vastly underdeveloped in the poorest countries and will require social and institutional, as well as financial, investments. Beyond the local dimension, there is also a logistical challenge in preparing for the next crisis, which points to the role of the World Food Programme (WFP) as a last resort food supplier and major contributor to crisis management. The G20, under French presidency in 2011, has given its support to a pilot project of the Economic community of Western African States (ECOWAS) to set up a targeted regional emergency humanitarian food reserve system to the creation of a strategic food reserve[2].

The second dimension is economic: the poorest and most vulnerable do not have access to food because they cannot afford it. Hence, economic availability is a more

1. For a discussion of these issues see Vindel, B. et P.Jacquet (2011), "Agriculture, développement et sécurité alimentaire", in P. Jacquet and J.H. Lorenzi (eds.), Les nouveaux équilibres agroalimentaires mondiaux, Paris: PUF et Descartes & Cie.

2. See G20 (2011), Final Declaration of the Cannes Summit, http://www.g20.utoronto.ca/

serious problem than physical availability. This leads some[3] to the conclusion that an increase in food production is not necessary because there is enough of it already, especially if one is able to deal with the considerable waste of resources in that area. Pervasive waste is indeed a serious issue in developed and developing countries alike, and may represent about one third of production, but reducing it calls for heavy investments and a substantial and sustainable evolution of consumption and production behaviours, so that it is not amenable to quick fixes. Besides, in the poorest developing countries, farming still remains key to improving the standards of living. FAO studies show that the potential is there, especially in Latin America (Brazil, Argentina, Colombia and Bolivia) and sub-Saharan Africa (Soudan, Angola, Democratic Republic of Congo) in terms of land and irrigation. The considerable lag in farm productivity in sub-Saharan Africa also suggests that the productive potential is very high. Surely, economic development is not only about farm production, but it is hard to see how to engineer a path toward a diversified economy without first building on the assets and know-how existing in the rural areas and which can be expressed through farming. This is also why the focus promoted by the G20 on agricultural investment and production is welcome.

HOW TO PROMOTE SUSTAINABLE AGRICULTURE?

However, climate and environmental risks prevent the increases in food production and productivity from taking place along the classical, intensification route and invite the development of modes of intensification that will respect the land, soil and environment and will save energy. Techniques in agroecology seem very promising in that respect, but they require an increased effort in agricultural research, the development of adapted seeds, the involvement of local farmers in the experimentation of new farming practices and the evaluation of these experiments, and the interaction between scientific and technical progress with local know-how and knowledge.

Farming also produces many environmental services such as land maintenance, preservation of landscape, or in some instances carbon and energy savings. There has been much discussion already of the multifunctional dimension of agriculture, and this dimension should increasingly be recognised and valued. The focus of a multifunctional agriculture would still be agricultural production, but a production that fully takes account of all its externalities, both positive and negative. This is particularly important in Europe, where the strengthening of a multifunctional, productive agriculture is at the core of the reform of the Common Agricultural Policy currently under discussion.

In developing countries, farm development requires the involvement of small farmers. There is a tendency to set up export-oriented farming in opposition to family farming that essentially produces staple foods. This is overly simplistic. The modernization of agriculture is not a change of state that takes place overnight,

3. Like the Green European Parliamentarian Franziska Keller, during a hearing organized by the European Parliament Commission on Development on 4th October 2011

but a dynamic that requires the involvement of local actors, namely small farmers and producer groups. This is why a focus on smallholders is essential to develop an agricultural system that will be able to meet the food requirements and to anchor the modernization and sustainable development of agriculture.

Land grabbing has been very much in the news in 2011 and is further examined in this book[4]. There are several issues that need to be differentiated. The debate is sometimes cast in Manichaean terms between a view that celebrates foreign technology and capital inflows and the development of an export potential through modern farming and those who object to the capture of land by foreign investors who don't care for local food needs but count on profitability through exports. Here again, the well-being of local farmers is a crucial factor to monitor. Behind this whole issue of land grabbing, there is the question of land rights. The specific problem does not come as much from the foreign investor as it comes from the capacity of local governments to deprive local farmers from land use and to arbitrarily contract with foreign investors. Beyond this, if such contracts exist, they must be transparent and clearly specify the respective obligations of contracting parties, as would be necessary for all forms of foreign direct investment.

Finally, there is no single approach to sustainable agriculture: rather than a state to reach, it is also a process through which a mutually satisfactory trade-off is obtained between conflicting productive or non-productive land use choices, such as food production, exports, industry, energy production, housing, reforestation... There was heated debate in 2011 about the role of biofuel production as a cause of the recent food crises. Certainly, the use of land to produce biofuels implies that there will be less production of other crops, which may ex post prove detrimental to the evolution of the prices of these crops. But the solution is neither to forbid biofuel production nor to ignore the potential problem. It will rather rest in a twofold combination: the formulation of integrated and consistent farm policies, clearly setting the objectives and taking into account the challenges of food security, in order to provide a framework in which to define the necessary trade-offs and manage them over time; and the development of innovative approaches, such as the experimentation of various combinations of food crops and biofuels. One of the guiding principles is to let farmers decide the best way to use their own land. These choices can usually be altered across farming seasons, so that such decisions are not irreversible.

THINK INTERNATIONAL

In the aftermath of the recent food crises, there have been some calls in favour of restoring food self-sufficiency at the national level. However, without the occurrence of massive movements of population, this is unpractical, as the aforementioned developments suggest. On the contrary, international trade is bound to be a crucial component of food security. Our energies should therefore be devoted to

4. For a thorough discussion of landgrabbing see Afrique Contemporaine (2011) issue on Land grabbing, No. 237, October.

the strengthening of the multilateral system for world trade. What is now at stake is not a rather sterile debate between free trade and protectionism, but the stability and credibility of trading rules, especially when crises occur. We need a reliable trading framework to which countries may be willing to entrust their food security. Obviously, we are not there yet.

One of the most conflicting debates has been on how to deal with export bans in times of crisis. These bans have a clear rationale: allow local consumers to access cheaper food by using domestic production for domestic consumption. Such measures disrupt world markets and world prices and may compound crises, both through their direct effect on world supply and through the uncertainty they generate on future supply conditions. Yet, the prohibition of export bans does not seem very realistic: at times of crises, national interests always prevail. However, two complementary approaches may help: first, the negotiation and introduction of collective disciplines, a sort of good practices, on the use of such measures; second, a decision to address crises immediately through collective discussions, rather than through unilateral, uncooperative moves. G20 discussions in 2011 have been useful in these respects. In Cannes, according to the June 2011 Action Plan on Food Price Volatility and Agriculture that was submitted by the G20 agriculture ministers, G20 governments have agreed to remove food export restrictions for food purchased for non-commercial humanitarian purposes by the World Food Programme. They have also launched a "Rapid Response Forum" to coordinate responses in times of market crises.

REVISITING FARM POLICIES

In everything that has been said so far, there is a recognition of the need for active farm policies to help mobilize the productive potential, to promote investments towards sustainable agriculture and to put in place a set of international rules conducive to food security. The development of agriculture has suffered, in industrial as well as developing countries, from the demise of such policies. Indeed, there is a belief that the contribution of the markets has led to the demise of the state Moreover, the very notion of liberalism was associated with the illusion that free markets would on their own be better contributors to the common good than public policies. Yet, neither politics nor good economics is compatible with this view, even though it has become conventional wisdom. The fact is that public spending went on increasing in most countries (cases in point are the current debt crisis in the Eurozone, but also the situation of public debt unsustainability in other countries such as the US and the UK), but it did so in reaction rather than proactively, and without an agreed framework on the role of public policies. In Europe, for years the main objective of CAP reform was the need to cut support for budgetary and international reasons; instead there should have been a debate on what the CAP should be trying to achieve and with what instruments. Encouragingly enough, this debate now seems to have started.

Situations and policy needs differ between advanced and developing countries. In the poorest countries, agricultural development should be a clear policy priority. Farm policies should simultaneously address land rights, training and capacity building, infrastructure, research, data collection, access to credit, organization of markets, among others. The modernization of agriculture is an acute requirement in a context of: rapid population increase; migration pressures toward cities with rates of growth that are greater than the capacity to equip them with facilities and basic services; and the need to promote job creation in rural areas. This situation differs from the one in which European countries found themselves at the start of their agricultural modernization, therefore the policies and trajectories for this modernization will have to be reinvented, notably by engaging smallholder farming.

We wish to highlight two further dimensions: risk management and the relationship with the private sector. In 2011, there was much discussion on food and agricultural price volatility. Indeed, this volatility substantially increased over the course of the 2000s. Yet, there were many successions of high and low volatility sequences across the last century, with notable periods of high volatility for example after the World Wars and in the 1970s in the aftermath of the oil shock and of the demise of the Bretton Woods system[5]. While there is therefore a need to better understand what has led to the recent crises, there is also a risk of myopia by doing so, especially by focusing on the detrimental role of speculation. As mentioned above, food security was a problem before the recent crises, even at times when there was much less "financialisation" of agriculture than today. Speculation may be excessive and costly, and there is a good case to rethink and strengthen regulations. But this work also needs to take into account the useful role played by speculators: they make risk hedging possible, and they act as signals of potential imbalances between supply and demand that policies have ignored for too long and should consider. More generally, it is likely that speculation has increased the amplitude of volatility and the impact of changing expectations on prices. But price volatility also signals fundamental causes linked to the low elasticity of the supply and demand of farm products.

Volatility matters in two ways. First, the level of prices impacts on production and on poverty. Prices have increased recently in response to the expected evolution of demand (demographic and economic growth, change in diets) and of supply (weather uncertainties, natural catastrophes, energy prices, etc.), at a time when the level of food stocks was too low to act as a buffer. Price increases may sustain production, but they are costly in the short term for vulnerable populations. According to the FAO, for example, the number of people suffering from malnutrition in Africa increased by 20 million between 2007 and 2008, reaching a level of 240 million where it stabilized. Second, price uncertainty matters because it affects the expected income of farmers and may lead to underinvestment, and also because it penalizes fiscal management.

Farm policies should address volatility and help move from a mode of crisis management, which is costly in human and financial terms, to a mode of risk management. This is not only linked to a proper use of market-based, risk hedging instruments such as futures and options or other financial contracts. These instruments may

5. See for example Roache, S.K. (2010), What Explains the Rise in Food Price Volatility, IMF Working Paper WP/10/129

prove very useful, notably in industrial and emerging countries. However, they are less adapted to the needs of poorer countries, given their costs, the need to transact on high volumes, the need to understand financial and derivative markets and also the fact that they deal with international risks that often do not reflect the local risks facing producers. Local prices may differ from international prices for a number of reasons: the specific character of local production, exchange rate variations, uncertainty on transaction costs, etc. Risk management invites a local approach, beyond the sole price risk. As NEPAD recognised in a letter sent to the G20 in September 2011, farm policies should integrate a dimension of risk management, and donors should help in doing so. The G20 has also supported an initiative to create a joint, multi-donor platform for agricultural risk management that would act as an advisory body for developing countries in assessing risks and choosing the best options to address them, through policy measures as well as market instruments.

The second dimension we wish to highlight relates to the involvement of the private sector. Agricultural development essentially rests on private firms of various sizes, domestic and foreign. But it also engages the common interest, through food security and through the various environmental services (or negative externalities) it exerts. There is thus a de facto interaction that should lead to a deeper public-private partnership, which underlines one of the crucial dimensions of a renewal of public policies, in agriculture and elsewhere: how to more effectively engage private firms in the production of common goods? There are obvious complementarities, which still need to be exploited. The G20, in 2011, has given support to reinvented public-private partnerships along such lines, and this is also one of the initiatives taken by the World Economic Forum with the creation of a New Vision for Agriculture, through a taskforce which groups the CEOs of major international food and food processing firms. This taskforce has concluded country partnership agreements with a number of countries, including Vietnam, Indonesia, Mexico, Tanzania, and with NEPAD.

In 2011, the debate and mobilization seem to have progressed in an encouraging direction. The G20 has mainstreamed food security and agricultural development as a clear priority and has validated the Action Plan adopted by the G20 agricultural ministers in June. This Plan insists on production and investment, and reaches meaningful agreement on various issues: the need for more data collection and transparency, notably through the AMIS (Agricultural Markets Information Systems) initiative, through the Rapid Response Forum mentioned above, through the implementation of a system of emergency food reserves, through new disciplines on export bans, through the promotion of risk management and risk management instruments within consistent, national farm policies, and through public-private partnerships. These recommendations were delivered under short-term pressure, and many issues have clearly been left untouched, exposing ongoing disagreements. However, fundamental issues have been addressed which may open the way for better coordination and more effective policies toward agricultural development, if the collective momentum can be sustained. ■

Viviane Gravey,
Institute for Sustainable Development and International Relations, Paris, France

Raphaël Jozan,
Agence Française de Développement, Paris, France

Sébastien Treyer,
Institute for Sustainable Development and International Relations, Paris, France

DEVELOPMENT, THE ENVIRONMENT AND FOOD: TOWARDS AGRICULTURAL CHANGE?

At the end of the 2000s, the theme of agriculture reappeared in international debates, taking front of stage in numerous workshops, technical publications, and programs launched by NGOs, governments and international organizations. A consensus has emerged and points to the urgent need for massive investment in the sector, which is (once again) viewed as one of the prime engines for development and food security, as well as for poverty reduction.

But what exactly does this consensus cover? International debates encompass widely varying points of view, and food security itself is a protean notion. As Laurence Tubiana and Noura Bakkour explain in the first chapter, it can be understood as an issue linked with poverty and access to food, the quantity and quality of foodstuffs, security – particularly in regions in the throes of major political crises – etc. Today more than ever, it has become a challenge for international governance, on an equal footing with widespread environmental changes and the economic crisis, as evidenced by the interest the G20 has shown in the subject since 2010.

Agriculture and food security challenges should not be reduced to simply a matter of food production (Chapter 2). This book thus invites the reader to discover these

COUNTRIES AND REGIONS COVERED BY *A PLANET FOR LIFE* 2012

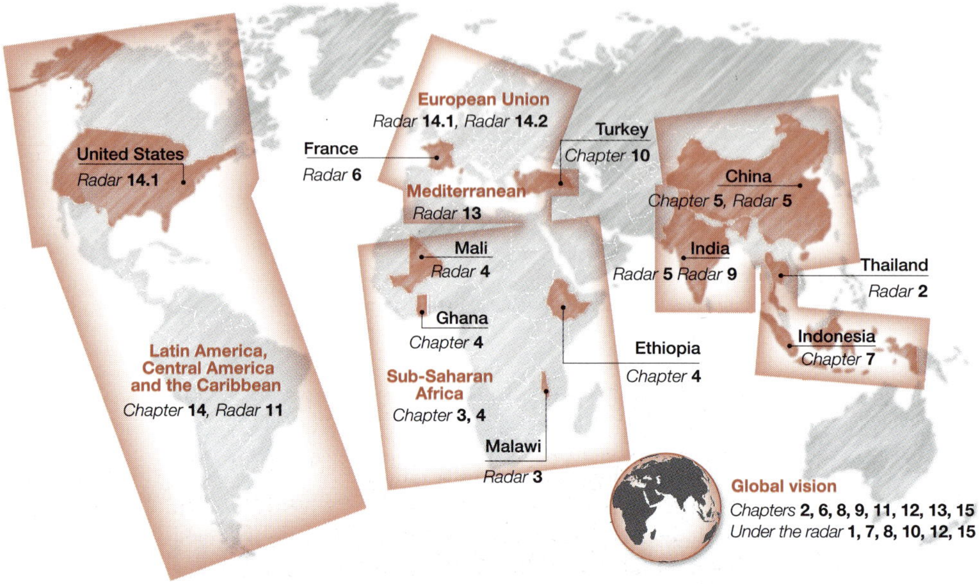

challenges in all their complexity. It takes into account international agricultural research (Chapter 9), issues of access to, and use of, natural resources (Chapters 6 and 7), the agri-food industries (Chapter 11), our consumption patterns[1] and agricultural employment (Chapters 3 and 4). Nonetheless, the question of agricultural production remains pivotal. While the idea of investing in agriculture is gaining ground and although several countries or regions appear to be offering opportunities for investment in agricultural land[2] debates are still raging on as to which agricultural models to choose and how agricultural policies should be implemented.

Issues pertaining to food and agriculture widely differ from one region to another. This edition of *A Planet for Life* takes us to all five continents, to the USA and the European Union (Radar 14.1 and 14.2), to India and China (Radars 5 and 9, and Chapter 5), to Latin America (Chapter 4 and Radar 11) and Africa (Chapters 3 and 4). The book also sheds light on international issues: the growing influence of private stakeholders (Chapters 13 and 15) and the changes in global governance (Chapters 12 and 15, and Radar 12).

Certainly, the current effervescence surrounding the topic is due as much to the urgency of the issues as to their appropriation by a proliferation of players with conflicting interests and values. At the same time, the theme covers complex problems linking global and local scales, as well as short- and long-term stakes that spark controversy and debate, and this book sets out to elucidate their terms and meanings.

1. For consumption patterns, notably meat and organic products consumption see Radars 2 and 5.

2. For a discussion of the scarcity of cultivable land see Radar 7.

AGRICULTURE AS AN ENGINE FOR DEVELOPMENT?

The late 2000s witnessed the emergence of an agriculture-for-development agenda. Developing and emerging countries were encouraged to integrate into their national policies the idea of agriculture as a major development engine, and Developed countries encouraged to do the same with respect to their development policies. This agenda is particularly foregrounded in the World Bank's *World Development Report 2008*[3]. This volume emphasizes the role of agriculture in the different developmental "stages", which are modelled on the historical sequence seen in the developed world: the gradual transition from an agriculture-based economy to one underpinned by industry and then services, with a rural-urban shift. Agriculture provides an initial momentum through its productivity gains, which enable the accumulation then transfer of labour and capital from one sector to the other, hand in hand with better standards of living and an increase and diversification of demand: only a minority of the population can remain and specialize in agriculture (generally those with the best production factors), the majority being absorbed into non-farm and industrial activities, most often through rural-to-urban migration.

These three trends – agricultural specialization, rural diversification, migration – are presented as possible exit routes from rural poverty, but "the question of the viability of these exit strategies today is not really debated" according to Bruno Losch (Chapter 3), and "previous transitions have clearly taken place at other historical moments". The current phase of globalization characterized by the State's withdrawal, trade liberalization and breakthroughs in communications technology has created a totally different playing field. Those countries with a lag in structural change and a high dependency on agriculture, as is the case of many African countries, are facing the constraints of fierce competition, an unstable international business environment and the consequences of global change, which have drastically reduced their margin for manoeuver.

While the agricultural sector in developed and emerging economies have a concentration of production units, increasingly motorized production processes and a large share of research and technological innovations, most of the agricultural labour force still work on smallholdings using manual and barely or non-motorized production methods, which are unable to ensure livelihoods. Although some workers migrate to centres of economic activity, they do not necessarily transfer their land to more productive systems. A core of family members remains in the village and engages in multiple jobbing, leading to subsistence farming and providing cheap labour for the large agricultural production chains, as is described for cotton-growing in a small Turkish village (Chapter 10). There is no automatic transfer of farm labour to industry, but rather the risk of a downward spiral that traps a good part of the population in poverty.

In Africa, but also in parts of Central Asia, Eastern Europe and Latin America the

3. World Bank, 2007, World Development Report 2008: Agriculture for Development. Washington D.C.: World Bank.

lack of productivity of existing farms creates opportunities for development and a possible answer to the food challenges of tomorrow highlighted Radar 0. Increasing production by 70% by 2050 necessitates a modernisation of agriculture: more concentrated units of production, motorisation, greater reliance on research and development etc. But even though transforming farming units of production and increasing productivity is a pivotal issue, focusing only on certain farms producing a limited set of commodities would be a major mistake, turning today's opportunities into yesterday's errors by replicating a non-inclusive development scenario.

This book also highlights the importance of a widely dispersed web of very small-scale economic actors in the processing and distribution chains, who could well hold certain keys to development and the future of food security (notably Chapter 2), as is the case for urban and peri-urban agriculture which contributes to provide African cities with food products, yet who remain ignored by most agriculture and food policies[4].

Sustainable agriculture can only be built on economic growth that encompasses both the agricultural and industrial sectors, rural and urban areas, and stimulates productivity for these two different spaces taken as a whole. The development equation, explains Xinshen Diao and his co-authors in Chapter 4, remains specific to each country, to its spatial economy and to the interrelations between sectors and territories. These interacting changes may disadvantage a particular area, as illustrated by Chinese industrial growth, which has only been achieved through a certain abandonment of rural areas (Chapter 5): the challenge is then to rethink development policy to make it geographically more balanced. Moreover, while transforming production structures and enhancing their productivity remain a major challenge, this must not be limited to the farms linked into global industries.

Not least, it would be a mistake to narrow down the importance of agriculture as a development engine to developing countries alone. Agriculture also plays a prime role in many emerging economies (cf. the case of Latin America in Chapter 14) and is at the heart of the global sustainable development stakes that are challenging agricultural development models and the associated framework policies.

ENVIRONMENTAL CONSTRAINTS ON LONG-TERM AGRICULTURAL TRAJECTORIES

Apart from having to step up its global production, agriculture had also been saddled with the role of development engine during the years of the green revolution, and there are lessons to be drawn from this experience. Today, replicating this model is not on the cards as its negative environmental and social impacts are now known (see Radar 5). Beyond the inherent limits of this model, the problems of resource availability, degradation of natural capital and other environmental threats such as climate change are all game changers and strongly constrain future agricultural trajectories.

Among these different environmental challenges, resource availability is gaining

4. See Radar 3 for a discussion of urban agriculture in Malawi.

ground in policy processes around the world (as illustrated by the EU strategy *a resource efficient Europe*). Contemporary foresight exercises (Radar 0) illustrates the dangers of a response solely targeting resource scarcities, namely a risk to focus solely on increasing production, without integrating social and environmental, but also mid to long term economic dimensions.

To respond to agricultural and food challenges, States often massively open up "available" land to a growing number of investors who take positions on agricultural assets, lured by the opportunities offered by Africa, Ukraine, Russia, Kazakhstan or Brazil, to mention only a few favourite targets. According to the Land Matrix report[5] (Anseeuw et al., 2012) over 70 million hectares have been involved in transactions since 2006.

Some commend the virtuous spiral, when corporations, States and donors respond positively to the challenges. Others denounce the fact that multinationals or powerful economic players (national or foreign) seize control of agricultural and natural assets, which often results in accelerated withdrawal of water for irrigation or in continued deforestation. This raises worries about strong impacts on global climate balances and biodiversity, as well as on local natural resources that other stakeholders rely on. The environmental issue is even more directly felt as a conflict of access to resources when "available" lands are used by expropriated local communities.

Faced with this race for land, the question of its scarcity is particularly relevant. Laurence Roudart shows (Radar 7) that agricultural areas can still be extended and that there is no global land scarcity, though admittedly soil degradation could in the long run become a cause for concern. There are however many situations where regional and local land scarcity may be chronic. States that have land available need to make the most of this margin of manoeuver in order to develop sustainable agriculture practices that fully integrate social and environmental constraints. The participation of peasant and professional organizations in discussions should be guaranteed, and investment commitments need to be well defined so that it is not simply a question of transferring land rights, but rather of ensuring that transactions materialize well and truly as investments.

Would increasing yields be enough to check the expansion of cultivated areas and attenuate environmental impacts? In the case of deforestation in Indonesia, Anne Booth and her co-authors (Chapter 7) show that this proposal is not robust to economic analysis. Certainly, efforts to increase the productivity of food and cash crops are proving successful, but this has not led to the containment of rampant deforestation. According to the authors, this seems to confirm the need for supportive public policies in order to dovetail the two objectives of enhancing agricultural productivity and controlling the expansion of cropland.

Land "scarcity" is by no means the only environmental issue that needs to be overcome. In addition to the scarcity of water and fossil energy sources, a whole raft

5. Anseeuw, W., Alden Wily, L., Cotula, L., and TAYLOR, M., 2012, "Land rights and the Rush for Land. Findings of the Global Commercial Pressures on Land Research Project", Rome: International Land Coalition.

of environmental problems combine to produce a real systemic challenge for agriculture, implying a need for innovative solutions. Uno Svedin (Chapter 6) points out that in our Anthropocene era, where human-related environmental changes (climate change, modified cycles for nitrogen, phosphorous, etc.) are already of a magnitude comparable to geological changes, changes in agriculture have substantially contributed to the global impact of human activity on all natural cycles. Today we are no longer in what some scientists such as Johan Rockström and Uno Svedin (Chapter 6) refer to as a "safe operating space for humanity", while the pursuit of our current agricultural and food model is constantly moving us even further out of this safe space. Climate change in particular poses a threat to the resilience of farming systems and, as Jean-François Soussana shows (Chapter 8), this impels us to a total rethink of our approach to agriculture. It will take a great deal of innovation if we are to invent smart, knowledge-intensive and resilient agricultural practices capable of increasing productivity and also environment-friendly. Yet, getting what Michel Griffon (2006) calls a "doubly green revolution" off the ground will not be simple, given that how to put it into practice is subject to controversy and debate. Benoit Labbouz and Sébastien Treyer (Chapter 9) point out that implementing sustainable and resilient agricultural systems inevitably hinges on innovation. However, the on-going debate on the new paradigm of agricultural innovation (new green revolution, agro-ecology, etc.) shows that this will require a sea change in the current functioning and priorities of international agricultural research. What is needed is a radically new approach to innovation that gives farmers a true role, since it is they who produce knowledge specific to local contexts – knowledge vital for ensuring optimal respect of environmental constraints and enhancing a territory's cultural and natural capital. While the authorities steering international agronomic research are the strategic hubs for far-reaching socio-technical choices, they remain reluctant to engage in a genuine transition.

AT THE CROSSROADS OF THE CHALLENGES POSED BY DEVELOPMENT, FOOD SECURITY AND THE ENVIRONMENT, THE TRANSFORMATION OF THE AGRICULTURAL SECTOR IS AT THE HEART OF THE GLOBAL STAKES OF SUSTAINABLE DEVELOPMENT.

A COMPLEX AND GLOBALIZED ECONOMIC FABRIC

Agricultural trajectories in different regions of the world are interlinked by international markets, which crystallize criticisms and hopes. By analysing various agro-industrial chains, this book provides a more effective assessment of the complex economic fabric underlying the concept of markets, power relations and the multiple arenas for the transactions needing to be factored in if the sustainable agriculture equation is to be solved.

In the agro-industrial system, which is now increasingly internationalized along all its supply chains, industrial processes operate on a just-in-time basis and rely on logistics chains where supplies must be secured and traceable (Chapter 11). The technical characteristics of raw and processed food products must be appropriately defined and calibrated and comply with a raft of standards covering health rules, as well as industrial standards for mass production and supply chains. Markets are multifarious and materialize much more as technical-economic networks that are highly segmented and specialized on a product-by-product basis. They are endlessly diversifying within the 'quality economy' through labelling systems and protected designations of origin (Radar 13), which do not jeopardise other developments. Consumer demand for non-standard products is in tune with current industrial trends (standardization, innovation, etc.).

This food economy has two major consequences. First of all, it translates into a landscape built around oligopolies, where a handful of corporations control nearly three-quarters of the market. Ten countries concentrate 85% of global agribusiness and 70% of its workforce. Certainly, in the early 2000s, the supremacy enjoyed by major rich countries (United States, France, Germany and Japan) evolved into a leadership shared with the emerging world (China, Russia, Brazil and India). These developments, however, have not changed the structure of the markets. The second consequence is that the price of gaining a foothold in these networks is constantly rising and value has been redistributed throughout these networks. The agri-food industry is now becoming more service-oriented and currently spends more on marketing than on agricultural commodities, thus disadvantaging small producers.

This dynamic is an indicator of how difficult it has become for non-industrialized countries to integrate the global agricultural economy. Opportunities for market access depend on their scale of production, on the availability of logistics and on their access to information. Developing countries producers are rarely endowed with such means. A company with a firm foothold in international markets can more easily invest in processing and commercialization, as well as in direct production, whereas local producers outside of the value chains find it very hard to do so. This analysis gives deeper insight into why the governments of developing countries strive to attract the industry's key players onto their territories. As the levers for public action have dwindled greatly since the 1980s, governments often have no other feasible option in the face of international competition than to offer access to natural resources in order to interest potential investors in their territory.

Yet, other schemes to connect up local production with regional and international markets also exist, and do not exclude small local producers. In many countries, we can see initiatives based on quality signals for local value-added products. The creation of protected designations of origin to develop the market for olive oil from the southern shores of the Mediterranean shows that these experiences can succeed if they are part of an overall framework (Radar 13) calling on the participation of all operators in the value chain and on intervention from powerful market players. In Latin America, mechanisms have been created to articulate agri-businesses and micro-suppliers. This is the basis of the "supplier programs" set up in Mexico and Chile, where large agri-food companies are able to give technical assistance to small and medium-sized firms operating in all productive sectors. These mechanisms are particularly well-adapted to the reality of a highly fragmented agricultural sector and the agro-industry's supply requirements (Radar 11). A win-win relationship is thus established between distributors and producers, without distorting the market. Finally, Jean-Louis Rastoin suggests in chapter 11 that our current agro-industrial model in its increasingly service-oriented format may be economically efficient, but its sustainability is far from certain, and will only be achieved by its cohabitation with an alternative model based on proximity, on a network of small enterprises, valorising idiosyncratic natural and cultural patrimonies.

MANAGING THE EFFERVESCENCE AROUND THE AGRICULTURAL SECTOR

The current crisis, characterised by an increase in agricultural commodities' price and their volatility (Chapter 12) has in fact opened a window of opportunity that could trigger the dynamics of more sustainable and more balanced development, without sacrificing agricultural and rural development. Many countries have increased the budgets of the ministries of agriculture (Chapter 4). For these countries, the "comeback" of agriculture is synonymous with a more endogenous development strategy, geared to leveraging the under-employed and developing natural resources, whose value had been obscured by pre-crisis ready access to financial capital. For developed countries, the time for reform has also come, heralded by the upcoming American *Farm Bill* (2012) and future European CAP (2014). Alongside these national initiatives, international mobilization – the G20 and the reform of the FAO (Radar 15) – betokens the heightened importance given to this challenge by the donor community.

This window of opportunity is not limited to putting agriculture back onto the agenda of development agencies. In fact, its "comeback", and its previous "abandonment" referred to in the World Bank's World Development Report 2008, often seem overstated. The statistical analysis proposed by François Pacquement (Radar 15) shows that agriculture has always been high on the donors' agenda. It is important to note that this window of opportunity in fact stems from a combination of constraints: rising prices, stronger constraints and climate change, continuing demographic pressures, questions of agricultural employment, etc., which together all challenge our current agricultural and food models.

This window of opportunity thus invites us to ask the question of change in the agricultural world: not the question of whether agriculture should indeed change, as today's challenges make this seem inevitable, but of which direction this change should take, what definition should be given to sustainable agriculture and food? The different sections in this book reveal some conflicting tensions: on one side, there is the effervescence that has surrounded agricultural and food issues since the mid-2000s, and on the other side, the difficulty of reforming agricultural policies and changing not only production methods but also processing and consumption patterns.

Today's policies are seemingly stuck in the rut of projects geared to a conventional modernization of agriculture, Despite the criticism levelled at its environmental and social impacts, support for the conventional model remains dominant, be it in India

(Radar 9) or in Europe and the United States, a fact that Antonin Vergez and Simon Liu suitably recall in their analysis of how sustainability criteria integrate into the public policymaking of the planet's two largest agricultural players. The interview with Henri Nallet (Radar 14.2) highlights not only the European Union's unique opportunity to reconstruct its Common Agricultural Policy on sustainable foundations, but also the difficulties of an exercise that will be carried out under unprecedented budget constraints. This opportunity, and need, for reform, is fully felt in Latin America, where policies are undergoing deep changes, becoming more participative and inclusive of the whole agro-food chain since the beginning of the 2000s.

Despite this come-back of agricultural policies, the task of defining sustainable agriculture could indeed slip out of the public actors' control, as Eve Fouilleux shows in Chapter 13. Private sector standards are increasingly shaping the contours of sustainability in the models of international agricultural production. The growing importance of private stakeholders in defining guidelines for global agriculture does however carry risks, warns Olivier De Schutter, (Chapter 15) and needs to be controlled both at the national level – particularly in the developing countries that today lack strong policies in this regard – and at international level by a reform and deepening of global governance.

However, the current state of global governance when it comes to agricultural matters is far from encouraging. Apart from the setting up of a reformed Committee on Food Security (Radar 15), WTO negotiations are at a standstill. As Michel Petit underlines, the main components of an agreement are on the table – the abolition of export subsidies, substantial reductions in tariffs and other barriers to accessing internal markets, cuts in internal supports for farmers in rich countries. Yet today, those who stand to lose by such an agreement are eminently better mobilized that the potential winners, which is undermining the possibility of concluding the Doha Round. A further example is the G20 agricultural negotiations. These have given a topmost place on the agenda to agricultural issues, which would have been unthinkable only a few years ago. It is true that some interesting advances have been made (cf. general introduction) and that the G20 decisions have successfully provided the means to extinguish the present blaze. However, in view of the challenges examined here, it is not sure that the tools, instruments and commitments to agricultural development negotiated and agreed on by the different States will really be able to help prevent further crises (Chapter 12).

At the crossroads of the challenges posed by development, food security and the environment, the transformation of the agricultural sector is at the heart of the global stakes of sustainable development. To help steer these changes towards greater sustainability, this book makes us aware of how crucial it is to also change our representations of agriculture, change the visions that guide projects for change and the policies regulating this sector. ■

WILL HUMANITY BE ABLE TO FEED ITSELF IN 2050? THE RECENT EXPLOSION OF FORESIGHT STUDIES

Sébastien Treyer,
Institute for Sustainable Development and International Relations, Paris, France

Since the early 2000s, an increasing number of foresight reports have sought to explore the possible futures of the global food system, including the overall balance between the growth in demand for food (and other uses of agricultural biomass for energy, animal feed or fibres such as cotton) and the potential for increased agricultural production. At the origin of this profusion is the diagnosis that the era of the Green Revolution is coming to an end - a 40 year period that saw spectacular yield increases and that is now showing signs of reaching its limits (indications of the stagnation of yields that had previously been increasing significantly, the impacts of agriculture on the environment, the risks associated with climate change impacts on future production, unequal access to technologies...). In addition to these production uncertainties, the demographic challenge remains very important (average projections are for the global population to increase from six to nine billion people between 2000 and 2050); especially since nutritional transitions tend to result in an increase in total calorie intake and in the proportion of it that is derived from animal products per capita, given that every calorie of animal origin requires between three and ten calories of plant feed for its production. The reasons for concern seem clear, but the main objective is to explore, in a rigorous and structured way, the different possible scenarios and the magnitude of the challenges ahead, to avoid jumping to conclusions too quickly. The profusion of foresight scenarios runs the risk of being accused of Malthusian bias, disregarding the regional and local priorities or other dimensions that are not taken into account, due to its focus on the limits of the global agricultural supply. What do these scenarios really tell us?

Firstly, the most oft-quoted scenario in the media is the FAO's baseline scenario, which focuses on a 70% increase in global food supply between 2005

▸ In addition to trend-based scenarios, converging around the fairly standard FAO scenario that sets the scale of the challenge at around a 70% increase in global food supply between 2005 and 2050, recent foresight exercises have explored several trend-breaking scenarios, indicating possible changes, manoeuvre room and risks (loss reduction, demand management, promising technology models...).

FIGURE 1 WHAT SCENARIOS FOR THE GLOBAL BALANCE BETWEEN FOOD DEMAND AND SUPPLY IN 2050?

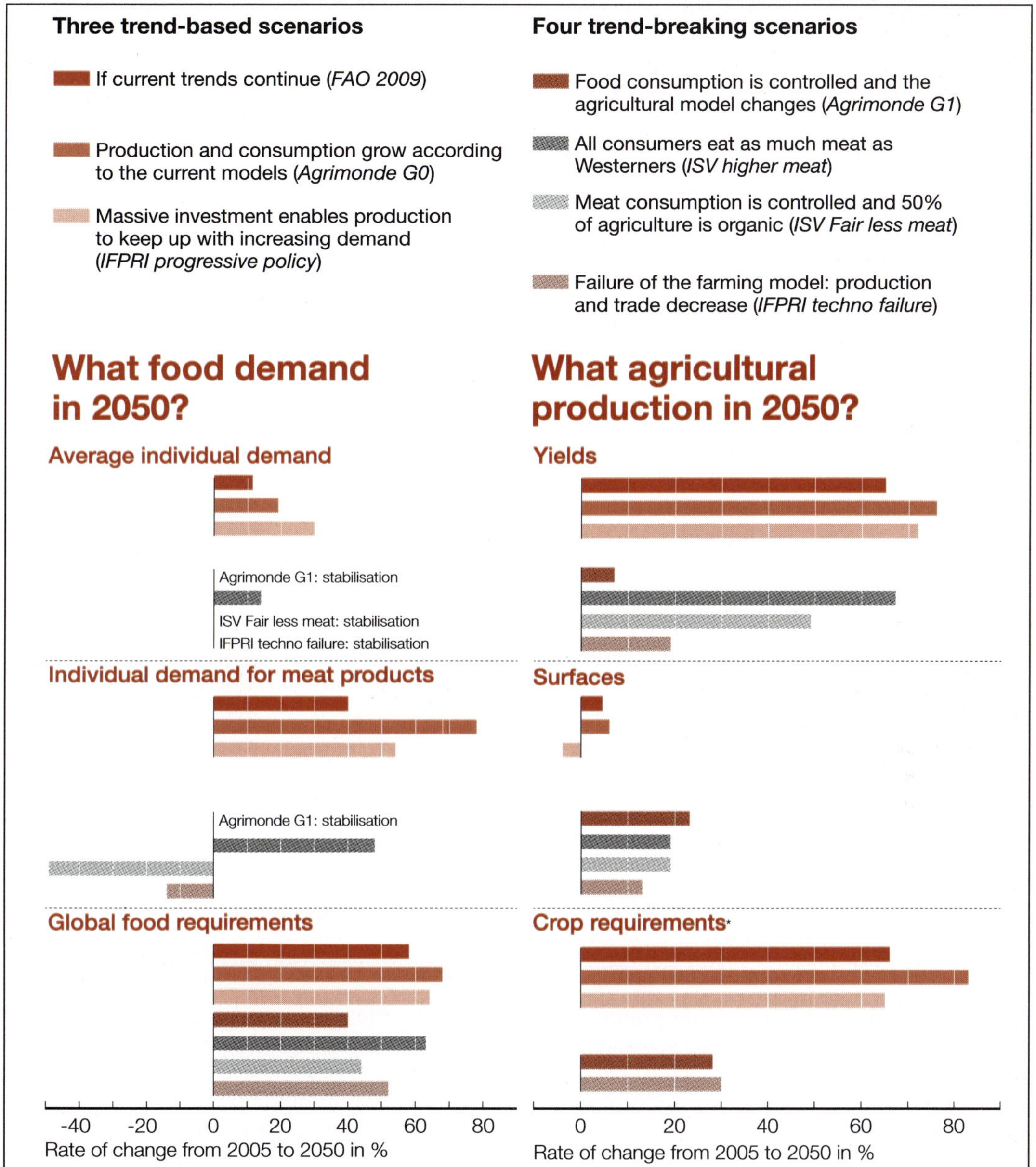

Sources: M.-A. Even and C. Laisney, February 2011, *La demande alimentaire en 2050: chiffres, incertitudes et marges de manoeuvre*, Analytical note from the Centre d'Etudes et de Prospective, Analyse n°27 and M.-A. Even and J. Vert, June 2011, *La production agricole mondiale à l'horizon 2050: comparaison de quatre prospectives*, Analytical note from the Centre d'Etudes et de Prospective, Analyse n°28.

* Plant needs for scenarios *ISV higher meat* and *Fair less meat* vary under different assumptions regarding plant and animal productivity and area under cultivation.

FIGURE 2 AGRICULTURE AND FOOD CHALLENGES IN MAJOR WORLD REGIONS IN 2050: LESSONS FROM THE AGRIMONDE EXERCISE

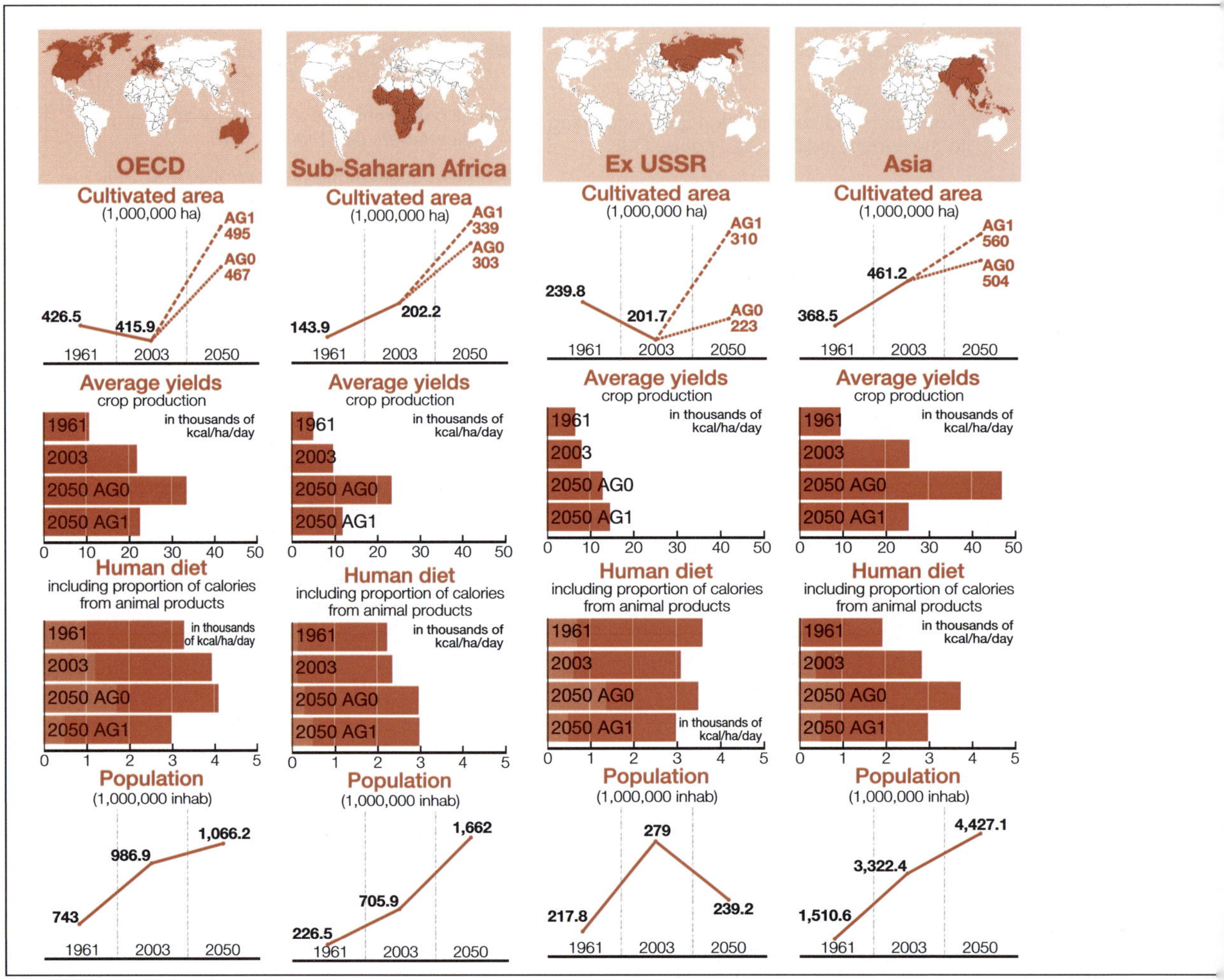

and 2050, which corresponds to the continuation of food and demographic trends and to assumptions that are deemed credible in terms of the production increase in different regions of the world, provided that a major investment effort is made during this period. This scenario is consistent with other trend following scenarios. However, the message is subtle: the scope of the challenge is major, but it remains lesser than the achievements of the last forty years (during which global production increased by 150%). Above all, most of this increase in production has to take place in developing countries, because the issue of food security cannot be reduced to a global balance between food supply and demand: to ensure households have access to food products, increasing agricultural production in the least advanced countries is an essential contribution to the improvement of incomes and to the resources of food production. These scenarios present such an increase in production as a necessary condition, but it is obviously not sufficient, since the impact on incomes from increasing a country's production depends on a set of other

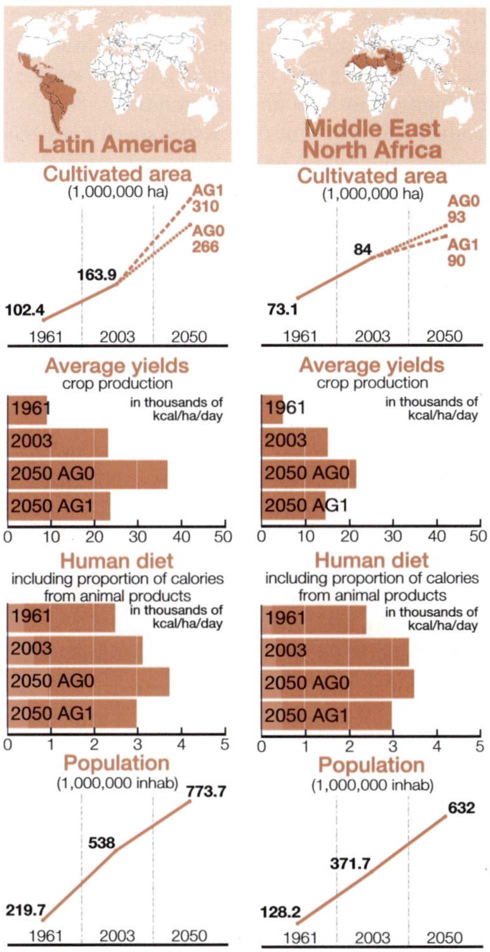

For large parts of the world, two contrasting scenarios (the AGO trend-based scenario and the AG1 scenario of a transition to agro-ecology and the control of food demand) make consistent assumptions on the magnitude of food demand and agricultural production. In Asia, and in North Africa and the Middle East, the scenarios show the limits of the production potential at the continuing rate of population growth (although AGO forecasts the continuation of an explosion in crop yields in Asia, while AG1 highlights what could happen if diets do not reach the peaks of Western ones). In the OECD, Latin America and former USSR countries, the potential for increased production is still significant (increase in cultivated area and higher yields) when compared to a slow growing or controlled demand. Despite its potential for increased production, the rate of population growth and therefore demand in sub-Saharan Africa will represent a major challenge, even to obtain a minimum level of food availability per capita, which could make this region dependent on imports if improvement in agricultural productivity is not fast enough.

factors, including inequality and other sectors of the economy.

In turn, the trend-breaking scenarios raise very useful questions on the possible deviations from the trend scenarios. The first question concerns the idea that the increase in food demand may not be a foregone conclusion, particularly the convergence of diets towards the current Western model. Several scenarios explore the stabilization of the average food availability at around 3,000 kilocalories per day per person, and reductions of varying extents in the proportion of animal products in diets: to justify such a break from the trend, these scenarios mention not only the reduction of consumer waste and the potential effects of policies and nutritional recommendations, but also more profound transformations of food chains that would divert the entirety of humankind from following Western consumption patterns. While some point to the need for a 70% increase in crop production, other scenarios indicate that a growth of only 30% might be sufficient if demand increase is mitigated. Conversely, one trend-breaking scenario seeks to maximise the demand by assuming that all regions of the world will consume as much meat as Western countries, which leads to a result that is very similar to the trend scenarios, even though the latter do not suppose a maximum demand. Therefore, there is room for discussion in relation to what is commonly seen as an irrepressible demand trend.

Another important question can be raised regarding the doubts over the assumptions of increasing yields that are made in the trend-based scenarios, which are sometimes very optimistic: are these assumptions robust in light of the potential impacts of climate change, environmental degradation (water, biodiversity), the depletion of certain minerals, such as phosphorus, and of fossil fuels? Underlying these figures is a major controversy on agricultural technical models (green or doubly-green revolution, see Chapter 9), and the very different predictions on the potential of various future technologies. Besides the very optimistic trend-based scenarios, which we can question in terms of the sustainability of the assumptions they make, even with the support of biotechnology, there are also scenarios that seem overly pessimistic (Agrimonde 1), which make assumptions that are overly conservative, due to climate constraints, regarding yield growth, as they are even lower than the assumptions made in the scenario that anticipates the failure of agricultural policies. It is worth noting that all scenarios, even those making conservative assumptions on yields, converge on a low or moderate increase in the area of land cultivated: such an increase therefore seems certain, given the pressure of the demand, but all the scenarios make similar assumptions about the problems of infrastructure and organization, in particular regarding land and social issues, associated with bringing new land under cultivation.

The different scenarios raise issues specific to each major world region: on one side are regions such as Asia, North Africa and the Middle East, which have a limited production potential and run the risk of becoming dependent on food imports, and for whom the limits of agricultural development could pose problems both for employment and access to food for the poor. Despite its potential to increase agricultural production, sub-Saharan Africa could belong to this category if a rapid increase in agricultural productivity is not guaranteed. On the other side are regions (such as the OECD, Latin America and the former USSR) where there is potential for a growth in production, which could lead to a race to feed the world, but where some issues will be strongly debated regarding the availability and remuneration of the agricultural workforce and also particularly the environmental impact of production growth.

Finally, several of these scenarios highlight the importance of factors for change or for inertia that are as yet unquantified: power relationships, particularly among supply chains stakeholders, issues of employment in rural areas and of migration... ■

Food security, along with climate change and the current economic downturn, is a systemic crisis calling for a global coordination. Ensuring food security for all necessitates investment in agriculture, but it is vital that such investment places agriculture at the heart of sustainable development and good governance.

AGRICULTURE AND FOOD SECURITY: TAKING THE MEASURE OF A GLOBAL CHALLENGE

Laurence Tubiana and Noura Bakkour,
Institute for Sustainable Development and International Relations IDDRI, France

Rising commodity prices of 2008/2009 have brought the issue of food security back to the top of the international agenda.

At the end of the 1990s, food security was mainly presented as a domestic issue, while the global market was ensuring the overall adjustment of supply and demand. Since then this domestic issue has become an international problem requiring coordinated solutions. When world food commodity markets rocketed in 2008 driven primarily by increased demand from rapidly developing countries, as well as by competition for resources from first-generation biofuel production (Godfray, 2010), some experts suggested that the spike in food prices would herald a period of rising and more volatile food prices as agricultural commodities markets were increasingly interconnected with financial markets. This crisis was not for them a cyclical crisis but the foreshadowing of major new trends. Indeed contrary to events in previous market crises, particularly those of 1951 and 1974, the drop in prices, that was sharp due to the global economic crisis, has been short-lived. Commodity prices (especially agricultural ones) started to rise again in late 2010 and stayed subject to significant volatility. These price increases caused "hunger riots" in a number of countries around the world, opening a period of political instability and increasing the number of people suffering from hunger and malnutrition at the world level.

AGRICULTURE IN A GLOBAL CONTEXT

THE RETURN OF AGRICULTURE AS A GLOBAL ISSUE Countries reacted to this increase in prices by trying to insulate the local markets from international instability, sometimes restricting exports to give preference to local consumers, reinforcing tensions in global markets. Food security was no longer a domestic issue but a global food crisis on every government agenda.

This *de facto* globalization of the food crisis has been translated in the political field: the G8 and G20 agendas have included this issue since 2008, the French Presidency even made it an important point of the November 2011 G20 resolutions by creating a "Global partnership for agriculture, food and nutrition". The Secretary-General of the United Nations has taken several initiatives including the creation of the "High Level Task Force on Food Security" and the appointment of a special representative. Finally, the FAO has initiated a reform and revived its food security committee[1].

This crisis added to other more structural concerns such as on the responsiveness of global agriculture to medium and long-term demand, the decline in public support for development and of national funds dedicated to agriculture, the impact of climate change, the situation of the "bottom billion" people excluded from gains generated by globalization, and the unsustainability of food consumption patterns.

These trends converge into making food security a global problem, in addition to environmental challenges and the difficulties generated by financial market instability.

The transformation of agriculture in developed, emerging and developing countries is one of the prerequisites to escape from this systemic crisis. It can play a key role in reducing poverty and inequality through the development of rural areas and the support for smaller producers. Many studies have shown the positive impact of the reduction of inequality on economic growth. But the development of agriculture and rural areas is effective in the reduction of inequalities, not only because it is in itself central for livelihoods and provides a source of food security for a considerable proportion of the world population, but also because it produces positive structural impacts on human capital, health and education. It remains an engine of economic growth even in developed countries, as shown in Ireland, which experienced a financial crisis and is currently recovering through its agricultural exports. The revival of agricultural production, however, that has historically largely benefited consumers through lower prices must now be thought of in a new context, outside of the frame of the Green Revolution: that of the total availability of natural resources and the control of the impact of production on the environment.

FOOD SECURITY IS NO LONGER A DOMESTIC ISSUE BUT A GLOBAL FOOD CRISIS ON EVERY GOVERNMENT AGENDA.

These resources – water, arable land, minerals and more generally soil fertility – are not universally rare or depleted and are even renewable, so long as human activities make reasonable use of them. But the overall findings remain unchanged, a more intensive use of natural resources is not a sustainable way to meet the growing food demand, even though many uncertainties remain about the complex ecosystem changes and the sustainability of alternative production patterns.

1. For more details see p. 258.

FIGURE 1 AGRICULTURE AND ECONOMY

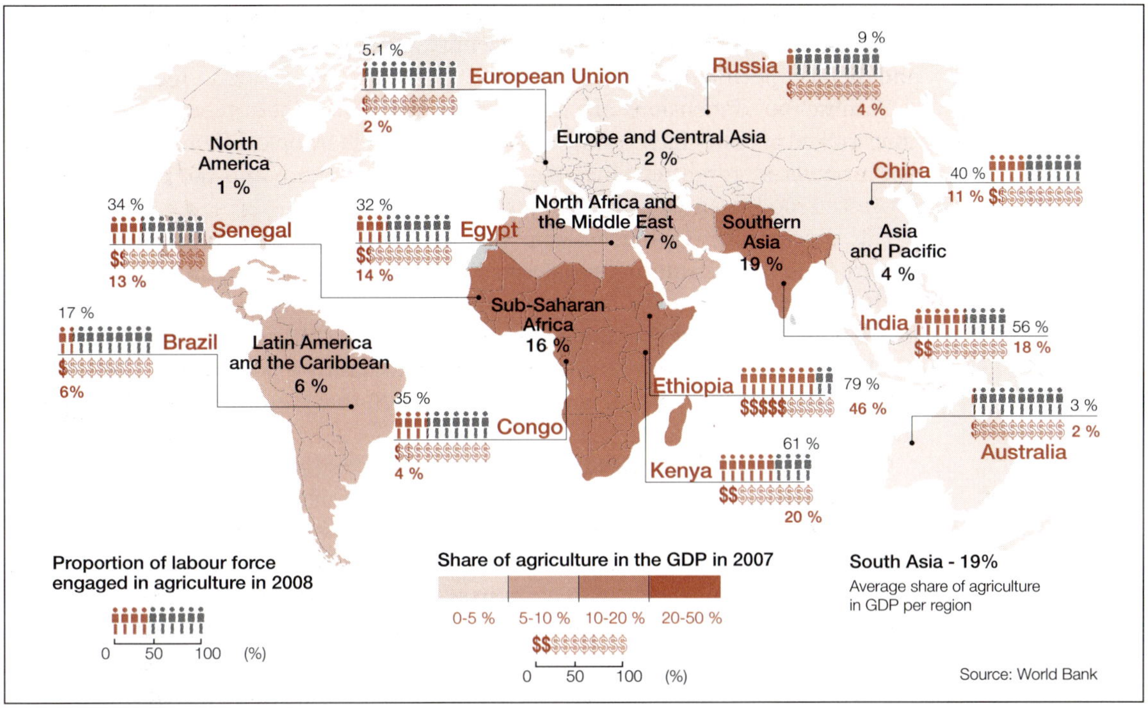

The proportion of agriculture in the economy and in the workforce varies widely by country. The area remains economically important especially in Africa, even if the comparison between Ethiopia and Congo, for example, reveals very different profiles in terms of employment created. Even in countries that still have a strong rural agricultural workforce such as Ethiopia and Kenya, the agricultural structure itself and the links with other sectors explain the varying contributions of the sector to the creation of national wealth. Finally, large economically important countries such as India or China, where agriculture represents a small proportion of the GDP, retain a significant portion of their labour force in agriculture. All of these factors reflect the difficulty of addressing the overall issue of agriculture and of proposing strategies for the future.

WHAT DOES THE CONCEPT OF FOOD SECURITY ENCOMPASS? The fact that food security is becoming a global issue does not mean that coordinated international action can be easily implemented, because views on food security do not necessarily overlap.

Food security has a "canonical" definition adopted at the 1996 World Food Summit and confirmed in 2002: "*Food security exists when all people, at all times, have physical and economic access to sufficient, safe and nutritious food to meet their dietary needs and food preferences for an active and healthy life*".

Actually food security is a multifaceted notion. At least four dimensions, that are sometimes complementary and sometimes discrete, are raised in international discussions: the issue of self-sufficiency in basic commodities. Countries that are heavily dependent on commodity imports may face food insecurity in times of global scarcity; the issue of malnutrition and hunger, both structural (the most disadvantaged urban and rural populations in poor, emerging or developed countries) and cyclical or specific (failed states, civil wars, climatic and production disasters); the dimension of food safety; and finally the nutritional quality of foods and the cultural and dietary specificities of given regions.

Of these four dimensions, the first two are more often the subject of international concern and of policy intervention; the last two, which involve different actors, should not however be forgotten since the issues of malnutrition and the poor quality of processed food are now gaining importance.

This diversity of approaches explains the focus on different modes of action according to the institutions and actors involved.

The vision based on the issue of availability and therefore agricultural production – which is a condition that is necessary but not sufficient, since access to this production must also be guaranteed – is predominantly a position held by the FAO and the institutions of global agricultural research that were created in the wake of the Green Revolution.

An alternative vision based on the concept of the right to food, related to the Universal Declaration of 1948, and the development of economic and social rights, is supported by the United Nations and a growing number of NGOs. Some countries, including India and Brazil, have translated this vision into national legislation.[2] Food security is also seen as a fight against hunger and malnutrition, one of the Millennium Development Goals, and is an integral part of poverty reduction policies. This point of view prevails within the development agencies of the United Nations, and first and foremost the World Bank, a strong correlation being found between hunger and development. The 2008 World Development Report (World Bank, 2007), the first to focus on agriculture for 20 years, identified a 'decades-long neglect of agriculture'.

But food security has also been part of the debate on agricultural trade liberalization and public policy interventions in markets. This debate is not over, even if it has been muted by the paralysis of trade negotiations in the WTO. The issue resurfaced, however, with the wave of export controls at the time of the 2008 crisis, reflecting the fragility of the trading system when such crises occur. Finally food security is also the objective pursued by intergovernmental humanitarian organizations and major NGOs in situations of crisis, and food insecurity is seen as a political destabilizing factor in terms of migrations, political upheavals and regime changes, subversion and terrorism. In this it represents a security issue in the traditional sense.

The diversity of approaches, and the fragmentation of the institutions and stakeholders that advocate these approaches, does not facilitate a comprehensive response at a time when we need to rethink the main pillars: agricultural production, environmental issues, issues of equity and stability. It must be recognized, however, that the issue of interactions between the different facets of the problem is the focus of much work and proposals for solutions, in research as well as in public policy. Even if each institution or group of actors gives priority to one dimension, the complexity of the interactions is acknowledged by every one: one such example is how development

2. This vision is further developed in Chapter 15 by Olivier De Schutter

institutions, even the most specialized ones like the FAO, are considering environmental concerns along with production objectives.

Similarly, the *vertical inter-relations* are subject to specific reflections. The issue of food security is deployed at different levels: international (e.g. the trade regime and the coordinated efforts of development assistance for agriculture in particular, etc.); regional (regions have specific vulnerability situations and their own institutional devices); national (with specific policies regarding both legislation and availability or accessibility); and finally, local (food insecurity is always linked to specific territories, resources and populations). These levels interact with the increasing integration of economies into the global economy: for example, the increasing international investment in the purchase of agricultural land.

We must also take into account the complexity, or even the "complexification", of the relationships between the stakeholders of the food chain: the late twentieth century food revolution has been characterized by dramatic shifts throughout the entire food system. Important drivers and obstacles for changes to agricultural systems lie beyond primary production, the downstream supply chain and the upstream research and extension services.

AGRICULTURE AT THE HEART OF SUSTAINABLE DEVELOPMENT Agriculture is high on the global agenda because it is both a sector deeply impacted by global changes (climate change, water resources degradation, invasive species…) and an important driver of these global changes, due to its environmental impacts. As an example,

BOX 1 FOOD AVAILABILITY AND HISTORICAL TRENDS IN UNDERNUTRITION

Twentieth century food policy discourse mostly revolved around hunger and food availability issues in the developing world. Real advances were made in terms of research, development and technology transfer between the 1940s and 1970s that increased agriculture production around the world. These advances were measured in terms of delivering more food and for more people affordably. The increases in agricultural production fostered by the "Green Revolution" are often credited with having helped to avoid widespread famine, and for feeding billions of people. While the Green Revolution has had significant impact in Asia (mainly smallholders and consumers) and Latin America (mainly larger scale farming), the impact in Africa was relatively small, not because the rate of production did not increase, but because it could not keep up with the population growth rate. So while the Green Revolution has certainly provided helpful insight and certainly deserves special attention in drawing lessons learned, it has not solved the problem of how to feed and nourish an expected 9 billion by 2050.

While the number of undernourished people dropped consistently from the 1970s, it began to rise again from the mid-1990s. Today, the statistics in hunger and malnutrition are not only worsening but devastating. According to the United Nations Food and Agriculture Organization, which measures "undernutrition" to estimate food security in the world, 925 million people are undernourished in 2010, this number represents 13.6 percent of the estimated world population of 6.8 billion, and nearly all the undernourished are in developing countries (FAO, 2011a). This figure of 'undernutrition' has increased since the mid-1990s, in part due to the current worldwide economic crisis; in part due to the neglect of agriculture by governments and international agencies, particularly for the very poor; but most notably, this increase is due to the significant increase of food prices in the last several years.

FIGURE 2 WHAT DO WE KNOW ABOUT HUNGER

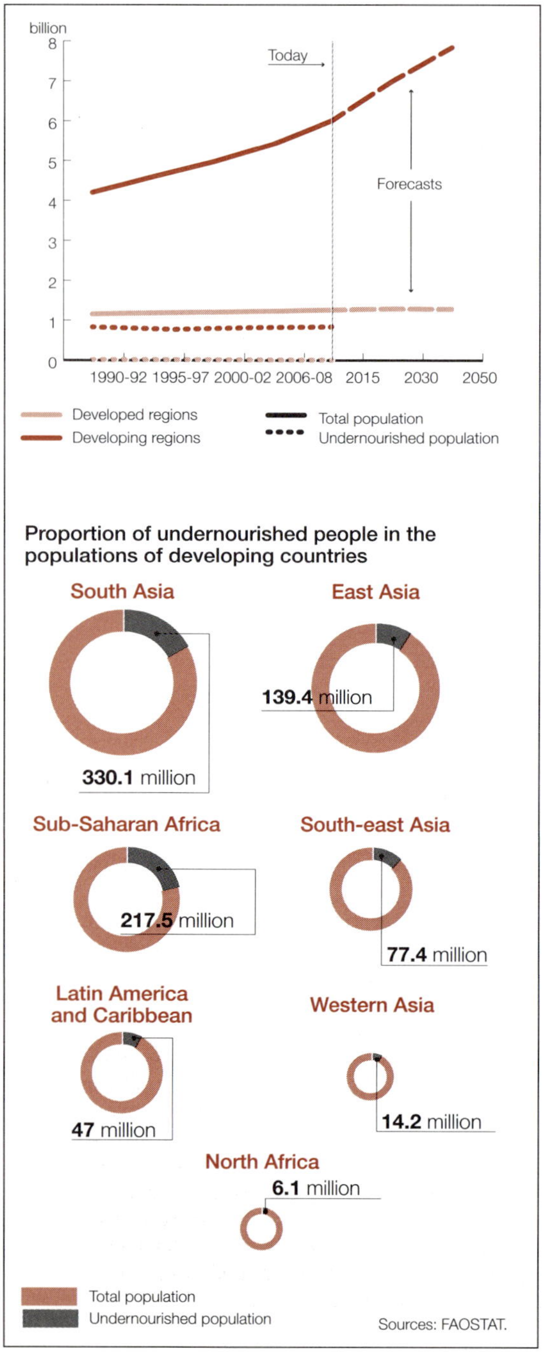

stakeholders from the agricultural sector themselves are advocating that the Climate negotiations should take the sector into account, particularly for what concerns the funding of adaptation, but also for mitigation actions (as has been the case during the Agriculture and Rural Development Day in Durban, COP17, 3rd December 2011)[3].

There is a growing consensus that with rising resource scarcity, climate change, and concerns about environmental costs, business as usual in the way agriculture uses natural resources is not an option (MCINTYRE et al, 2009). Possible options for the future of agriculture therefore need to be assessed with respect to fundamental requirements.

Increasing the productivity of resource use is central. This can translate into optimising the productivity of land, the efficiency of water use, but also into reducing the use of external inputs, therefore minimising the dependence on non-renewable resources like fossil energy, and also reducing the environmental impacts. Increasing the productivity also has to do with better closing natural cycles (of carbon, water, nitrogen, phosphorus, organic matter for soil fertility...) at the scale of the ecosystem, which would at the same time reduce the need for

3. See page 117 for more details.

The media generally state the symbolic figure of one billion people as suffering from hunger. While the method used by the FAO to calculate this figure is currently being debated, we know that the absolute number of undernourished people remains greater than the target set by the first of the eight Millennium Development Goals, which aims to half the proportion of the population suffering from hunger between 1990 and 2015. We can nevertheless acknowledge the efforts made in preventing the proportion of the undernourished population increasing at the same rate as the total world population, with issues specific to each region: despite the green revolution, many poor people in Asia still suffer from hunger; in Latin America, specific programmes seem to have borne fruit; in Africa, the challenge remains primarily demographic; in developed countries, the growth in inequality generates situations of malnutrition that are poorly referenced in the overall statistics.

external inputs and the impacts on the environment. Resilience to climate change and in general to environmental risks and shocks is also a key feature that has to be ensured for the sustainable future of agriculture, particularly for the most vulnerable farmers.

In general, the access of poor smallholders to technologies ensuring the improvement of productivity and their resilience is a central challenge: accessible technologies, and public support to invest in improving productivity, will also play a key role.

Agriculture is back on the agenda, and new investments might be expected: it is therefore important to take into account that current and emerging challenges make it necessary to get the basics right, in order to make the best use of these investments.

RETHINKING GLOBAL AGRICULTURE

Now that agriculture is regaining its rightful place high on the international agenda, the key question is to define what kind of agricultural system we should be investing in. The critical importance of food to human well-being and social and political stability makes it likely that governments and other organizations will want to encourage food production beyond that driven by simple market mechanisms.

The long-term nature of returns on investment for many aspects of food production and the importance of policies that promote sustainability and equity also argue against purely relying on market solutions. So how can more food be produced sustainably?

Pursuing today's dominant modern model of large-scale highly-mechanized and input-intensive agriculture, and expanding it to new areas where food production will need to rise in the future, is simply no longer an option, as we are hitting natural boundaries (Buck & Sherr 2011). Harnessing agriculture's potential to create sustainable development will require changing our current models towards ones that can produce positive social, economic, and environmental outcomes at multiple scales.

Agricultural policies are in dire need of reform, and this applies both to countries in the North and in the South. There are many challenges and constraints but there are also opportunities and objectives that must be seized and achieved.

THE CHALLENGES On both the demand and supply side, agriculture is facing a multitude of challenges. On the demand side, these include food and nutritional security, population growth, changing pattern of demand driven by increased income and competition for food crops for biofuel production, in addition to the emergence of agriculture (food, land) as a globally traded investment commodity.

According to the FAO, food production will need to increase by at least 70 percent by 2050 in order to meet the needs of an expanding population and changing diets (FAO, 2011b). This fragile global food security situation is in part due to the various pressures of rising prices, speculation on agricultural commodity markets or the growing pressure from biofuel, but it is also, and perhaps more fundamentally, the reflection of decades of underinvestment and misinvestment in the agricultural sector, both by governments and donors. Governments of low-income countries with agriculture-based economies have only been devoting around 4 percent of their national budgets to agriculture (World Bank, 2007) and donors decreased the share of official development assistance (ODA) for agriculture from a high of 18% in 1979 to a low of 3.5% in 2004 (World Bank, 2007).

On the supply side, the challenges faced by agriculture include limited availability of land, water, mineral inputs and rural labour as well as the increasing vulnerability of agriculture to climate change. Some would argue that globally there is plenty of land and water but that we are just not using resources efficiently. The World Bank estimates that there are over 200 million hectares of degraded and underused land in Africa that can be brought into production. Yields of major cereal crops in Africa can easily be doubled, even tripled.

THE OPPORTUNITIES Many new opportunities are emerging for agriculture. They include increased awareness by governments, donor interest in supporting and investing in agriculture development in low income countries, growing interest of private investors in sustainable agriculture and increasing consumer demand for sustainably produced food. Dynamic new markets, far-reaching technological and institutional innovations, and new roles for the state, the private sector, and civil society all characterize the new context for agriculture.

In order to seize these many opportunities, further agricultural research and a better application of the research results already obtained are needed. Currently, neither the international research centres of the CGIAR nor national research centres are adequately funded. On the positive side, new players like the Gates Foundation are stepping in, and financially supporting agriculture research.

There are multiple possibilities for innovation in agricultural technologies and practices, ranging from innovations at the seed level due to genetic improvements (marker aided selection, GMOs), innovations at the farm level with a better inclusion of ecological principles in agronomy, and innovations in connecting farmers to their markets (such as IT applications). As of now, there is no global consensus on which level or type of innovation will yield the best results for low implementation costs. [4]

TRANSFORMING AGRICULTURE TOWARDS A SUSTAINABLE, LOW CARBON AND RESOURCE CONSERVING MODEL FOCUSED ON SMALLHOLDERS WILL NOT HAPPEN BY CHANCE, BUT ONLY BY DESIGN.

Yet, as Olivier De Schutter (2011) notes, transforming agriculture towards a sustainable, low carbon and resource conserving model focused on smallholders will not happen by chance, but only by design. Achieving this transition will require strong political will at national levels and strong international cooperation, and an array of supportive policies and public investment in infrastructure, agricultural research and extension services and the creation of conducive economic conditions, including fairer markets.

4. This discussion is further developed in Chapter 9

FIGURE 3 WHICH FERTILIZER FOR WHICH TYPE OF AGRICULTURE?

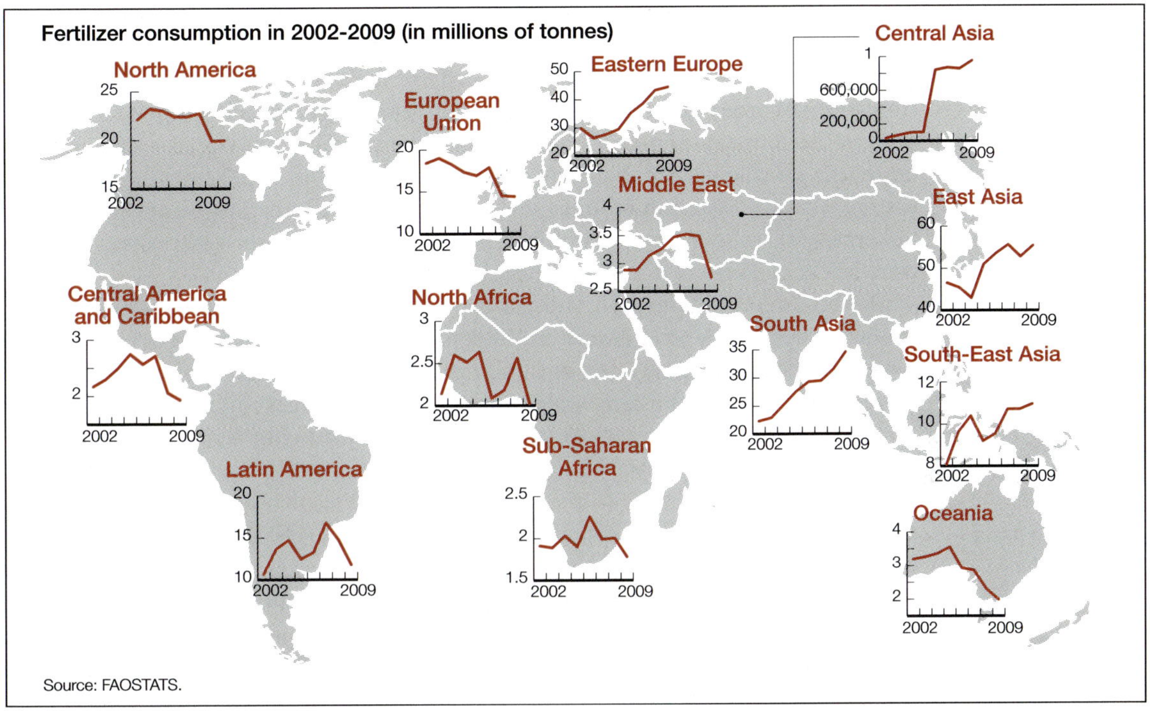

Fertilizer consumption is increasing worldwide (+14% between 2002 and 2009), a figure that encompasses contrasting situations: the reduction of fertilizer use in most developed countries does not offset the sharp increase in their use in Asia or Eastern Europe. Input use in sub-Saharan Africa remains very low, reflecting the limited financial resources of agricultural actors more than an environmentally-motivated choice.

NEW FUNDAMENTALS AND POLICY OBJECTIVES Making agriculture more effective in supporting sustainable growth and reducing poverty starts with a favourable socio-political climate, adequate governance, and sound macroeconomic fundamentals. It then requires the definition of an agenda for a transition toward sustainable agriculture, an agenda and transition that may differ radically from one country to another. This is a global challenge which concerns all types of agriculture, but it confronts differently countries who have already undergone an industrialisation of farming, after World War II or through the Green Revolutions of the 1960s to 1980s (OECD countries, Asia, Latin America), and countries whose agriculture remains mostly "traditional", that is low input, high labour, extensive models. These different agricultures face, in each region, diverging conditions of supply and demand.

But in the face of this global challenge, four new fundamentals need to be considered. The first of these is the need to impact not solely agriculture but the entire food chain: input manufacturers, researchers, farmers, agro-industries, retailers and consumers are all concerned and sustainability of agriculture cannot be attained without taking into account the sustainability of the broader food system. The connections between the different steps of the value chain need to be enhanced, with better market entry and access, new kinds of links between producers and consumers, and overall more efficient value chains that minimise losses. The second new fundamental is the role of farmers. There is a need to empower farmers as key actors of this transition: in order to ascertain that research reaches fields, farmers need to participate in the innovation process; choices of agricultural technologies and transition pathways need to ensure that they offer decent employment. Farmer organizations can be a precious tool to mobilise farmers and as such need to be developed. But new methods to identify research priorities integrate farmers' competences and to transmit knowledge and innovations are essential.

The third new fundamental concerns the interaction between agriculture and other economic sectors. A programme of "Agriculture for development" will only work if we foster the linkages between agriculture and other economic sectors. Finally, the fourth new fundamental is to encourage private investments alongside public intervention in agriculture, and to oversee these new investments to make sure that they are inclusive of the needs of smallholders.

REBUILDING A GREEN ECONOMY SCENARIO WITH SUSTAINABLE AGRICULTURE

There is no single solution, but a key component is an appropriately scaled framework for making policy and governance decisions. This in turn will help ensure that any economic development initiative based on agriculture is a truly sustainable effort that has been discussed and validated in a democratic and transparent manner avoiding situations of unfair rents. Public intervention will be necessary to correct market failures (in particular environmental externalities, positive and negative), and to build safety nets and invest in infrastructure. Pursuing an "agriculture for development" agenda for a country implies defining what to do and how to do it. What to do requires a policy framework anchored on the behaviour of agents – producers and their organizations, the private sector in value chains, and the state. How to do it requires effective governance to muster political support and implementation capacity, again based on the behaviour of agents – the state, civil society, the private sector, donors and global institutions.

To use agriculture as a development engine and a provider of public goods such as environmental protection and the sustainable management of natural resources, and food of high quality, a country would need a coherent and comprehensive programme.

The ability to deliver social peace and a legitimate governance system within a sound macro economic environment is a condition for any successful plan for agriculture. This basic premise was all too often missing in agriculture-based countries until the mid 1990s, particularly in sub-Saharan Africa. The agricultural programme will have to be environmentally sustainable both to reduce the environmental footprint of agriculture as well as to sustain future agricultural growth. In this perspective, taking advantage of the local knowledge as well as the advances in science and technology, biosciences, ICT, ecology, is key.

The design has to be specific to each country and its natural conditions and fitted to local stakeholders' projects and capabilities through broad participation. Finally, monitoring the progress of implementation and ensuring adequate administrative capacity is as important as the policy design.

The new interest generated by agricultural matters is not or should not be limited to developing countries and to an "agriculture for development" agenda. Agriculture – and all stages of our food systems – has a key role to play towards achieving sustainable development in both North and South countries (as illustrated by the FAO Greening the Economy with Agriculture initiative, or the OECD Green Growth strategy).

Northern countries have to rethink their agricultural policies through the spectrum of sustainability (also the efficiency of public policies), as markets signals and public support are still an incentive for unsustainable agricultural practices through subsidies harmful to the environment and employment. We need to discuss what public goods agriculture provides and how much as a society we are willing to pay for their provision. This needs to be done in coherence with our international commitments (trade, climate change, etc). In the implementation of the major changes necessary, countries are sovereign and will choose their own options: but the sustainability challenges of the three dimensions (economic development, social and environmental) make it important to assess the strategies chosen, and to realize that the path to sustainable development may be narrow, given the challenges. Whatever the choices, the fundamentals have to be right.

Furthermore, consumers and companies also have to act alongside governments and farmers. Companies in particular have to apply responsible corporate governance to their core activities, both in the upstream and downstream of food production. They can play a crucial role as input providers for agriculture: sustainability in agricultural production means a sustainable use of inputs. Similarly, as processors of agricultural products, companies are responsible for the quality and nutrition components, as well as the content of their communication campaigns as they play an increasing role in food consumption. They are part of the solution even more so when global food producer and distributor groups are shaping consumer choices. In a world where obesity as well as malnutrition affects a greater number of poor people, companies cannot ignore the impact of their policies.

Finally there is a great need for a global debate on future consumption models. Even if the diversity of food consumption is wide and will hopefully remain so, unsustainable trends have to be assessed. As is the case in energy consumption patterns, a global population of 9 billion cannot follow the modern western path of high meat consumption based on a very inefficient use of energy and natural resources. A global discussion on this subject is unavoidable.

Globalization has played a major role, both positive and negative, for food security: from now onwards, no solution can be deployed only at the national level. The different visions of the role of agriculture and food will have to find some common ground, away from the traditional trade divide. Reintroducing into the discussion the different components, including the local and global public goods that sustainable agriculture can produce, is certainly a positive step forward.

Moreover governments will have to face a paradox. Ensuring food security is, and has always been throughout history within any political regime, a responsibility of governments and an element of their legitimacy. As such it is an important component of their sovereignty. Indeed, there are tragic examples, like the famine in North Korea, that demonstrate how much some governments consider food security to be part of their exclusive domain. But because of the global interactions from climate change to market instability, the full exercise of sovereignty, in this case and in many others like disease control or global warming, requires global action and coordination. ■

REFERENCES

Buck, L. and Scherr, S. (2011) Moving Ecoagriculture into the Mainstream *in* Nierenberg,D. and Halweil, B. (ed) 2011 State of the World, innovations that Nourish the Planet, a Worlwatch Insitute Report on Progress Toward a Sustainable Society http://www.gcsu.edu/adp/docs/State_20of_20the_20World_20Innovations_20that_20Nourish_20the_20Planet_1_.pdf (accessed 19/01/2012)

De Schutter, Olivier (2010) Agroecology and the Right to Food *Report presented at the 16th Session of the United Nations Human Rights Council [A/HRC/16/49]*, 8 March 2011

FAO 2011a, The State of Food and Agriculture 2010-2011: Women in Agriculture, Closing the gender gap for development, http://www.fao.org/docrep/013/i2050e/i2050e00.htm (accessed 19/01/2012)

FAO 2011b, The State of Food Insecurity in the World: how does international price volatility affect domestic economies and food security, http://www.fao.org/publications/sofi/en/ (accessed 19/01/2012)

Godfray, H.; Beddington, J.; Crute, I., Haddad, L.; Lawrence, D.; Muir, J., Pretty, J., Robinson, S.; Thomas, S., Toulmin, C.; (2010) Food Security: The Challenge of Feeding 9 Billion People, *Science 327, 812*

Lerin F, Petit M : volatilité des prix internationaux agricoles et alimentaires et liberalisation en Afrique du Nord, Options Méditerranéennes Paris 2010

McIntyre, B., Herren, H., Wakhungu, J., Watson, B., 2009, "International assessment of agricultural knowledge, science and technology for development (IAASTD): synthesis report", Island Press, Washington

Word Bank, 2007, *World Development Report 2008: Agriculture for development,* World Bank, New York

AID AND THE ISSUE OF AGRICULTURE

François Pacquement,
Agence française de développement, Paris, France

Originally responsible for ensuring the continuity of colonial improvements and providing an auxiliary function in the cold war, official development assistance (ODA) underwent significant changes during the 1990s, when it had to acquire a new raison d'être in a period of extended globalization. International solidarity was reaffirmed in the objectives set jointly by donor countries, and in the growing mobilization of aid to enable developing countries to contribute to global policies that are in the process of implementation (infectious diseases, climate, etc.). Aid for agriculture is emblematic of these changes: it now presents itself as a contribution, sometimes to the global challenge of sustainable development, and sometimes towards solidarity (the fight against poverty, the reduction of hunger). Despite fluctuating opinions in the underlying discourse, agriculture occupies a prominent position in donor activities. While the general perception is that, regrettably, the last twenty years has seen a decline in aid, how has the situation actually evolved? From statistical data and a summary analysis of the strategies of major donors, this focus brings together some elements of analysis that nuance these perceptions.

THE UNDERLYING BASIS OF AGRICULTURAL AID

Agriculture is a sector apart. This is due, firstly, to the place occupied by agricultural production in multilateral trade negotiations, which immediately poses a problem of coherence: can we provide support to an activity with one hand, while imposing diverse restrictions on the distribution of its products with the other? Secondly, particularly since the 1970s, food crises and their media coverage have stirred up public emotions, while simultaneously highlighting the limits of markets, both delocalized and abstract, and of public intervention.

Overall, while aid has been distributed since the 1970s according to a ratio of one third multilateral to two thirds bilateral, multilateral aid is proportionately more engaged in the agricultural sector, where its share goes up to 40% in a relatively stable ratio, despite the fluctuations (due to the fact that these are commitments and not disbursements). The main bilateral donors are Japan, the United States and France, which together have accounted for at least 50% of such aid over the last 15 years. While French support has remained stable at around 10% of total aid to agriculture since 1995, US aid has risen sharply from just over 5% to nearly 30%, while Japanese aid has gradually decreased from 40% to less than 15%.

THE PERCEPTION OF A DECLINE IN AID TO AGRICULTURE IS NOT UNFOUNDED

The analysis of ODA[1] statistics, based on the refer-

1. The ODA statistical analysis encounters several difficulties. Firstly, it only provides a partial vision of aid: the data do not include contributions from development banks, such as the World Bank, which play a prominent role. Secondly, the data are very general and do not provide details of the contribution forms for the identified sectors. Thirdly, sectoral data are based solely on commitments, which lead to disbursements that may be late, in amounts that are sometimes lower than expected.

ence data of the OECD's Development Assistance Committee (DAC), enables some major trends to be highlighted.

The share of agriculture in total ODA has been falling steadily to a low point in 2007. This trend must be considered in comparison to those of other sectors: agriculture is a sub-sector of the productive sectors but it can benefit from infrastructure programmes, which are accounted for separately.

The evolution of these major sectors allows three periods to be distinguished. Until 1973, the sectoral distribution was unstable: initially high, the share of the productive sectors diminished in favour of investments in social services and infrastructure, and in "other" areas, a category including diverse activities that are difficult to allocate by sector. From 1973 to 1989, the dynamics were stabilized: investments in economic services and infrastructure increased. From the 1990s there was an increase in multi-sectoral and transversal operations and in social sectors, which preceded the formulation and then the adoption of the Millennium Development Goals in 2000, at the expense of the productive sectors and, to a lesser extent, of economic infrastructure and services, in accordance with the restrictions introduced by the Helsinki Agreement on aid related to these sectors (Pacquement, 2009).

The importance of the share of the "multi-sectoral/transversal" sector clearly demonstrates the uncertainty attached to sectoral distribution and leads to a major relativization of the decrease in aid to agriculture. Within the productive sectors alone, which may not encompass the entirety of aid to agriculture, the share of agriculture has been dominant since 1975, over 60% from 1983, increasing further until 1999 and then fluctuating to reach more than 70% at the end of the period. An analysis of the productive sectors that have benefited from aid directed towards trade policies, which seem to have partially gained from the sectoral reallocation, enables these fluctuations to be understood.

FIGURE 1 THE POOR RELATION OF BILATERAL AID?

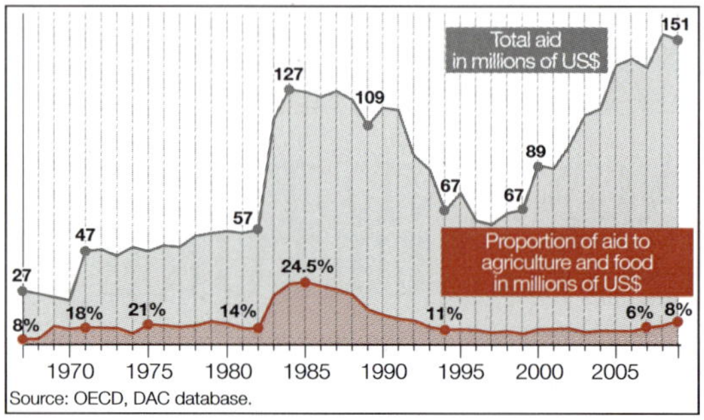

Agricultural and food aid have never represented more than 20% of the total bilateral aid allocated to developing countries. Since the mid-1990s, this proportion has stagnated at a particularly low level (less than 10% of the total). Since 2006, the slight increase that occurred in response to the food crises has not brought this proportion to over 8% of the total.

BOX 1 WHAT IS AID TO AGRICULTURE?

Aid to agriculture has two complementary forms:

– Aid to the agricultural sector and to rural development, according to the usual methods of investment aid and project aid, for the financing of more general sectoral policies;

– As an insurance mechanism to support the export incomes of agricultural products (to cope with price fluctuations and quantitative harvest deficits, resulting from a climate shock for example) with European aid (STABEX for the ACP countries and COMPEX for the others).

Added to this are:

– Contributions to international agronomical research either through national research centres (France, United Kingdom) or through the funding of international CGIAR centres, or through institutions such as the FAO and the World Bank, even if the latter's emblematic reports on global development have only addressed the issue on three occasions (1982, 1986 and 2007).

– Contributions to market conditions, which are unaccounted for in aid, mainly from development banks (at the market conditions).

FIGURE 2 AID TO AGRICULTURAL PRODUCTION HAS REMAINED RELATIVELY SUSTAINED

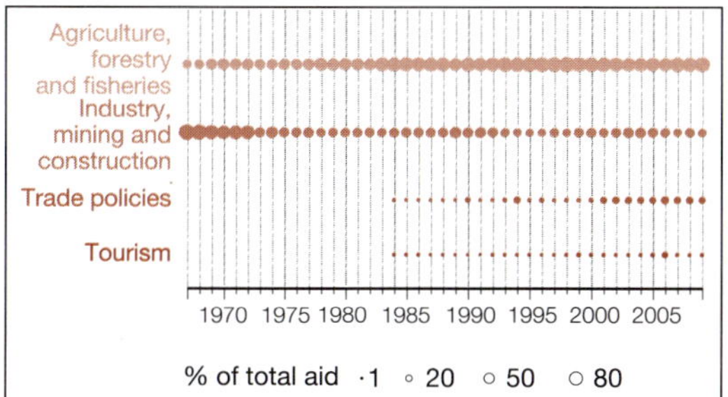

Source: OECD, DAC database.

As a productive sector in which aid is in decline in favour of infrastructure and services, the agricultural sector has been relatively well sustained. Cooperation in terms of trade policy or tourism development, more recent subjects, remains in the minority.

FIGURE 3 AGRICULTURE IS NOT A STRATEGIC SECTOR

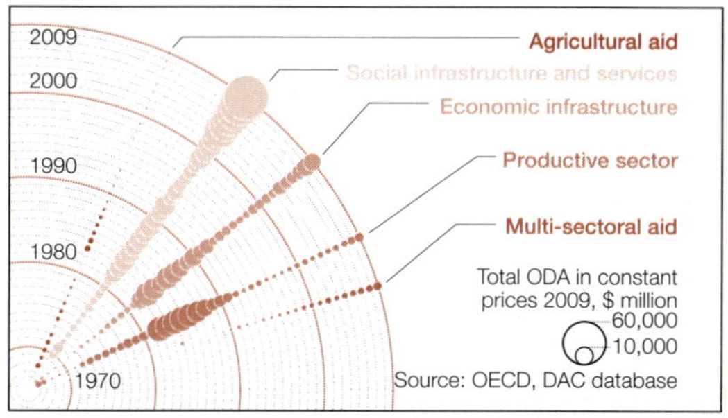

Aid to agriculture has never been a major and strategic investment for donors of bilateral funds; while aid for infrastructure, social and economic services has predominated for decades. The food crises of the mid-2000s have done little to change this situation.

CONCLUSION

In total, while the statistical analysis struggles to show the sectoral components of aid over a long period, it does however reflect that agricultural aid is an extremely significant component of donor contributions: against a backdrop of decreasing aid to productive sectors, aid to agriculture resists this trend. In addition, its various paradigm shifts can help form links to different sectors. To hold the opinion, based on a cursory look at the statistics, that agriculture has returned to the agenda of aid agencies, is to ignore the inherent vagaries in the construction of the data; agriculture has always maintained favour with donors, for whom the rural world remains both an area of poverty and one that holds potential for development, potential that is even more interesting when results are more easily discernible in terms of foreign trade, for example. ■

REFERENCES

OECD, 2009, Creditor Reporting System 2009: Aid Activities in Support of Agriculture, 2002-2007.

Pacquement François, 2009, Building global policies : development assistance, a source of inspiration ? IDDRI, *Idées pour le débat* n°5/2010 October pp1-22.

World Bank, 2007, Independent Evaluation Group, World Bank global support and its assessment, Washington.

CHAPTER 2

The food crisis is above all a crisis of poverty, not a crisis of global agricultural production. Some of the traditional actors in the agricultural sector raise Malthusian fears in an attempt to obtain a new legitimacy in a changing environment of food security governance, where new actors are entering onto the stage.

FOOD CRISIS: A RESHUFFLING OF STAKEHOLDERS

Nicolas Bricas and Benoit Daviron,
Centre International de Recherche Agronomique pour le Développement (CIRAD), UMR Moisa, Montpellier, France

In Chinese, the word 'crisis' is expressed by the combination of two signs: danger (wei) and opportunity (ji). In accordance with this definition, the current food crisis that many observers have identified as ongoing since 2008, comprises 'danger', specifically the suffering, both acknowledged and silent, of vulnerable populations affected by rising prices; alongside 'opportunity', which in this case is the chance for numerous stakeholders to voice their opinions regarding the handling of the crisis and to claim rights or defend legitimate positions in the institutional game. There have been countless summits (at high and low levels), seminars and roundtables on the topic of food security; and a multitude of TV and radio programmes on the subject. Since 2008, in France alone there have been no fewer than ten books published on the topic of feeding the world, humankind or humanity[1].

In many ways, the issues at stake since 2008 with regard to food security are not completely new. Indeed, to many observers the crisis is purely the manifestation of the end of a cycle of agricultural disinvestment, as occurred in the 1950s and the 1970s (HLPE, 2011). It is therefore unsurprising to see that history has repeated itself. The only difference is that the world has changed in the meantime, so that the current crisis engages a much broader range of stakeholders.

This new scene of discussions and actions provides the focus for this article. Firstly we examine the way in which the current price crisis is not related to a reduction in supply, and that the food crisis appears to be above all a crisis of poverty. We then

1. Books by Barnier, Brunel, Charvet, Dorin et al., Ghersi, Gleizes et al., Griffon, Parmentier, Fumey, Petit, Rastoin, Thinard, Worldwach Institute, Guillou and Matheron.

look at how new stakeholders, both international and local, have emerged into the field of food security. The article concludes by highlighting the growing importance of non-state stakeholders, civil society, NGOs, the media and the private sector in areas of governance for food security.

FOOD SECURITY AND AGRICULTURAL PRODUCTION

There is a paradox between the way in which the understanding of food security has evolved over 30 years to encompass the recognition of its multi-factorial and multi-sectoral nature, and how it is now often reduced to the single dimension of food availability.

The definition of food security following the 1976 World Food Conference focused on food availability and the risk of failing to meet increased demand. This rising demand was primarily seen as a demographic question, at a time when the stabilization of world population growth seemed a distant or even hypothetical prospect. This definition arose in a particular context, similar in some respects to that which has existed since 2008: climatic aberrations affected food production while oil and grain prices soared on the international markets. With thoughts dominated by Malthusian population growth curves and price spikes, the two major responses to emerge were to increase agricultural production and market regulation, through the storage of surplus supplies. Forty years on, demand is still accelerating, especially given the new prospects for biomass energy, and we are once again amid climatic aberrations and soaring prices; it is not therefore surprising that yesterday's solutions are making a reappearance today.

As in similar moments in history, future food supply scenarios have been envisaged to draw attention to the importance of the agricultural sector. Indeed, one of the advantages of such instruments is to highlight, as Malthus did, the rapidly rising demand for food relative to agricultural supply. The conclusions are therefore always the same: all necessary efforts must be made to increase supply through investment in agricultural research and infrastructure, such as irrigation for example.

Historically, this equation has been solved partly through the acceleration of the growth in supply, particularly by the Green Revolution, and partly through the moderation of demand through population control. Significant dietary changes were considered to accompany population growth, economic development and urbanization. Today, the majority of recent works remain convinced of the need to accelerate the growth in supply. Only a few directly address the question of diet, particularly the consumption trends for animal products and envisage 'unfashionable' demand-reduction scenarios (Paillard et al., 2010; Foresight. The Future of Food and Farming, 2011).

However, the definition of food security has changed dramatically, especially with the decisive contributions of Amartya Sen who put forward the concept of access: the objective is not to produce more, the priority must be to give each individual the ability to access the means to produce and/or to consume. From a purely agricultural issue, today's perspective has shifted the focus so that the question is now one of capabilities, purchasing power and thus poverty. The definition adopted at the 1996 World Food Summit, and which continues to hold sway today, reflects this emphasis on access and not just availability. And it is this shift that explains the emergence of institutions onto the stage, such as the World Bank, that are no longer solely dedicated to agricultural production but are more broadly concerned with economic development.

FIGURE 1 CEREALS AND VEGETABLE OILS, NOT JUST FOR FOOD

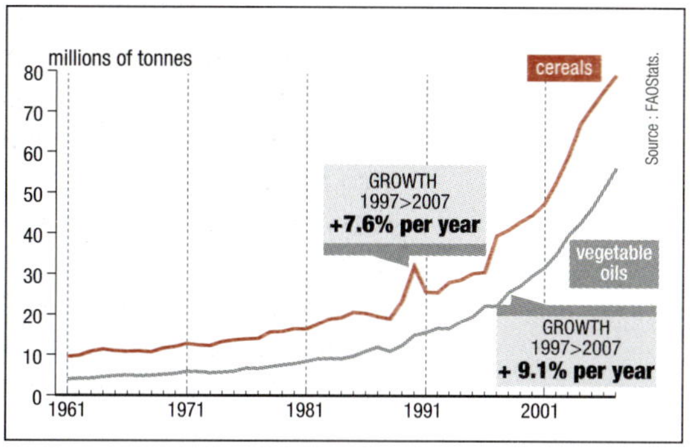

Non-food uses for cereals and vegetable oils have been increasing since the 1960s, but have undergone an explosion in the last decade: +7.6% for cereals, +9.1% for oil per year. This represents real competition for the supply of food markets.

However, due to the price crisis and the conclusions of the projective scenarios that look ahead to 2050, there has been a return to a reduced vision of food security as a balance between supply and demand. The spectre of Malthus is apparent in the media and political rhetoric, despite the fact that all the data show there is currently no global agricultural production crisis. Worldwide agricultural production continues to accelerate – in fact it has never been stronger – and the growth in food availability per capita has not slowed in recent years.

Sub-Saharan Africa, for example, which is considered food insecure at the regional level, has seen an increase in the consumption of cereals even during the recent crisis years, due to increased production.

If there is a food crisis then it exists because international oil, cereal and food oil prices have increased rapidly, which has had an impact on impoverished populations, particularly those that have accumulated in cities. And if these prices have risen, it is not so much because of a lack of production, but rather due to an increased demand, especially for non-food (and non-feed) uses and in particular the production of agro-fuels (Figure 1).

In an increasingly tense market with reduced global stocks, climatic aberrations have resulted in higher prices. Meanwhile, speculation on the derivative markets of agricultural raw materials has exacerbated the situation, along with the imposition of export restrictions in some countries. Over time, these conditions have combined to change a price increase into a price explosion.

In this context it is clear that the stakeholders of the crisis are not only farmers; in reality a much broader range of agents is involved.

DIVERSIFIED STAKEHOLDERS

The first stakeholders to be directly affected by price increases are international market operators, traders, brokers, financial analysts, banks, etc. Some of these agents are linked to the agricultural sector, through the use of futures markets to hedge against risks related to prices. But these agents derive more from the world of trade than from that of finance, evolving in a sphere that is relatively autonomous from the agricultural sector, of which they have little knowledge. Their tools are complex and poorly understood in the food industry, which considers financialization as an obscuration of market functioning. This partially explains the complexity of the debate on the controversial role of speculation in these markets; information is difficult to obtain and relationships with professionals have become uncommon.

Biofuel producers now join the ranks of buyers of agricultural commodities, contributing to the ratcheting up of market tension. Not only has their demand become significant and contributes directly to raising oil and maize prices, and also in consequence to increasing the prices of the other substitutable cereals, but they also buy or rent agricultural land for energy crops. These stakeholders also strengthen the connections between the agricultural and energy commodity markets, since their competitiveness depends on oil prices. This interconnection, which is a factor of instability in agricultural markets, is a new phenomenon that broadens the circle of stakeholders involved in food security to include energy sector stakeholders.

National states, which are also directly concerned with food prices, continue to play a role in regulating flows and prices in the interfaces with the international market. During the 2008 crisis, they fulfilled an important function in mitigating the price spikes for their domestic consumers. Many governments have thus decided to reduce taxes on food imports, although this approach is not in accordance with WTO agreements that prohibit the use of variable taxes. The countries that have adopted this tactic have incurred a major financial cost: in 2008 tax losses for the eight countries of the West African Economic and Monetary Union (WAEMU) amounted to between 480 and 880 million euros (Soulé et al, 2008). In parallel, many countries have also banned or restricted the export of food, fearing the loss of too much of their own produce onto the international market and the consequent risk of sudden price rises on their domestic markets. These measures, when widely applied to countries that export a significant amount of their agricultural produce, have made supplies scarce and pushed up international prices.

Through either tax cuts or export restrictions, what is clear is that states are asserting their sovereignty to manage the crisis, even if the measures that they implement appear contradictory with those developed at the international level and are subject to criticism from the international community. Although, as discussed later, a more global scale of governance has emerged from the crisis, the role of states, that a quarter of a century of liberalization had challenged, is making a comeback to the centre stage and resuming an important role.

States are not the main protagonists in the regulation of domestic markets in developing countries, particularly Africa, a fact that was true for the 1970s price crisis and even more so today. Private operators, traders and processors have for a long time been accused of existing only to make a profit on the backs of consumers and small producers, but have finally been recognized for their ability to supply rapidly growing consumer markets.

LOCAL STAKEHOLDERS IN THE AGRI-FOOD SECTOR

Africa provides an interesting case study, as a continent that continues to attract attention due to the important share of its population that is still considered to be

FIGURE 2 THE SPECIFICITIES OF THE AFRICAN DIET

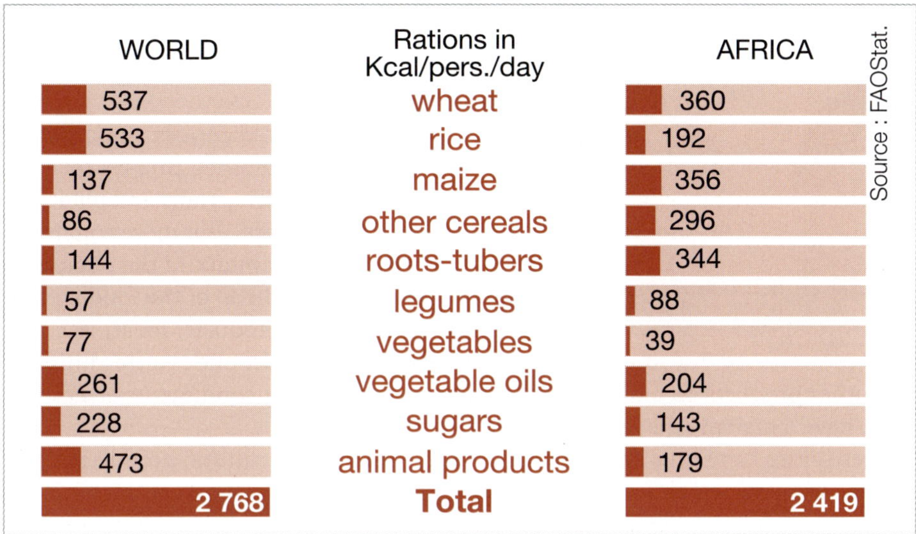

Average food intake per person in Africa is slightly below the world average. The proportion of local cereals and tubers in the average diet, compared to maize and rice also show the resistance of traditional eating habits.

food insecure. Indeed its recent development has gone against some preconceived ideas. Thus, contrary to popular belief, food in Africa remains largely dominated by local products. Although often described as highly dependent on cereal imports, on the basis of FAO data it can be calculated that wheat and rice imports for the entire continent only provide 12.2% of the available calories. While imports represent 61.7% of the available wheat food supply on the continent, the proportion is much lower for rice (39.5%)[2]. Such figures put into perspective the dominant, if not exclusive, position occupied by cereals in the food security debate. The average calorie intake for the whole African continent shows a great diversity (Figure 2). Compared to world data (Figure 3), Africa derives a much lower proportion of its calorie intake from calories of animal origin (8% compared to 19% for the global average).

Starch crops (cereals, roots, tubers and plantains) represent 70% of the available calories in Africa, while they only account for 52% in the world data, with product distributions that vary from one country to another: some are dominated by cereals, some by roots and tubers (cassava, yam, sweet potato, etc.). As a proportion of calorific intake, these roots and tubers are thus three times higher in Africa than in the world as a whole.

The situation is certainly different in cities, where imported products such as Asian rice and European or American wheat have gained significant market shares. However, the share of these products throughout the urban food market is limited, except in special cases such as Dakar, which is frequently cited as an example despite its unrepresentativeness of the continent as a whole. Surveys conducted in 2008

2. FAOSTAT data for the average calorie intake in Africa 1997-2007

FIGURE 3 AFRICA DEPENDS ON CEREAL IMPORTS, BUT NOT ONLY

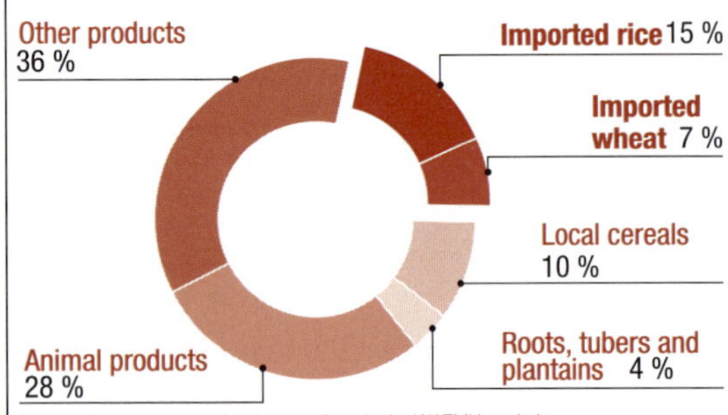

Source: Survey on household expenditure in the WAEMU capitals.

Cereal imports represent about 20% of urban household expenditure in West Africa. This amount is significant but must nevertheless be put into perspective when considering the factors of food insecurity: local products remain the staple diet.

on urban household expenditure in the capitals of the WAEMU countries (Abidjan, Bamako, Bissau, Cotonou, Dakar, Lome, Niamey and Ouagadougou) shattered the myth that these cities were being fed on imports (Figure 3).

In economic terms, imported cereals amounted to 22% of the food market, compared to 14% for rice, maize, millet, sorghum, cassava, yams and plantains and other local starchy products. But these commodities that support the calorific intake of the poor, only constitute just over a third (36%) of the food market of the eight cities. Animal products, meat, fish and dairy products account for just under a third (28%), and other products including oils, fruits, vegetables, sugar, condiments and beverages, have a market equivalent to that of starch (36%).

These cities have thus become a real outlet for local agricultural production, the market value of which now exceeds all agricultural export markets. Many foods formerly regarded as "subsistence" and designated for auto-consumption are now grown as cash crops. In addition, an intermediary sector has developed, involving collection, packaging, transportation, processing and distribution, which is providing cities with much more varied produce. Trends in urban food demand are increasingly moving towards products (such as flour, semolina, granules, fermented pastes, dried cassava chips, oil, etc.) that require more packaging, production, transportation and trade, thus providing employment for millions of people. In the eight capital cities of the WAEMU countries, the urban restaurant sector alone accounts for 24% of the food market, i.e. one billion euros per year. Businesses in this sector, both upstream and downstream in the supply chain, are poorly recognized by public authorities due to the diffuse and "informal" nature of the sector, and because it largely involves women and is often considered as a sign of underdevelopment. Abandoned by vocational training policies, excluded from bank credit systems and from the development of sanitary regulations and often persecuted by municipal authorities, the industry nevertheless feeds the vast majority of the population with low-cost foods that correspond to the population's eating habits. Far from simply reproducing rural culinary traditions, the sector is constantly innovating and contributing towards the invention of modern urban foods, while borrowing very little from Western models. Indeed, the sector is becoming one of major job creation, particularly for vulnerable populations, and has a strategic link with food security (Broutin & Brice, 2006) (World Bank, 2007). Over the past half-century, African cities have experienced population growth at a particularly rapid pace: in 1950 there were 33 million people, today this number stands at 420 million. However, despite this extraordinary growth rate, these cities have never experienced a food supply failure, except as a consequence of war, due to the expansion and organization of this intermediary food sector.

It is clear that in the larger cities this sector is changing. The emergence of a middle class is beginning to lead to the generation of a new type of business model, often more formal and mechanized, that offers products with higher added value, a higher level of quality control and with packaging that enables such products to compete in self-service shops alongside standard industrial foods. It is new entrepreneurs that are behind these businesses, often women, who sell not only on urban markets but also export to markets that cater to the diaspora throughout Europe and North America. Traders, who are gaining recognition as economic stakeholders, organize themselves to benefit from warehouse receipts and bank loans. In East Africa, they have established financial exchanges where futures contracts are traded. The largest traders have now reached a stage where they can negotiate with public authorities. Their role during the crisis is, however, controversial. Such agents have probably contributed to the mitigation of price rises by reducing their margins. However, a widely held belief persists that these traders have maintained their higher charges even after the price spikes began to return to lower levels.

AS HAS BEEN THE CASE IN LATIN AMERICA AND ASIA FOR SOME TIME, AFRICA IS A CONTINENT WHERE MULTINATIONAL AGRI-FOOD COMPANIES ARE NOW ACTIVELY PROSPECTING.

These developments, especially the growing awareness of the demographic importance of urban markets on a continent with strong economic growth, are drawing the attention of foreign investors. As has been the case in Latin America and Asia for some time, Africa is a continent where multinational agri-food companies are now actively prospecting. Such companies are attracted to countries with a developing middle class and they invest in industrial processing and supermarket distribution sectors. Indeed, supermarkets are multiplying across Africa, more so in Eastern and Southern Africa, and a little more gradually in Western and Central Africa. Europe, which has long dominated in these investment types, now has to compete with China, India, Brazil and South Africa.

Consumers, the final agents in the food chain, have for the first time strongly registered their protest. Population growth, which is still rapid in countries where demographic transition is incomplete, and urbanization, which is also rapid, mean that about 20,000 to 30,000 young people arrive on the labour market each year, per million inhabitants. This pace will continue for the next 20 years (Beaujeu et al., 2011). Employment has lagged a long way behind this pace of population growth and millions of young people (63% of the African population is under 25) have entered into crowded cities which, despite the difficulties, continue to appear as places with greater opportunities than rural areas. Only countries that have invested heavily in rural infrastructure such as transportation, communication, health, education, access to water, etc. have been able to reduce migration into cities.

In 2008, 2010 and 2011, riots occurred, that were dubbed by the media as "hunger riots", in response to the high costs of living. The rioters were members of a desperate urban population that felt it lacked any future, which had exhausted its resistance capacity after enduring successive years of crises, and who were exasperated by the political regimes in place, accusing them of not doing enough to create employment and to guarantee a more equitable distribution of resources. In most cases, the riots took place in cities with high levels of social tension; first oil prices began to rise, followed by a few notable foods, which together provided the spark to light the tinderbox.

The media have also played a role in the simultaneous nature of the riots and the creation of what was then called the "food crisis". The diffusion of television and the internet has indeed drawn every citizen closer to one another, and instantly

BOX 1 DIVERGENT INTERPRETATIONS AND ACCOUNTS OF THE IMPACTS OF PRICE CRISES

Curiously, beyond the so-called "hunger" riots, we actually know very little about the transmission of higher international prices on domestic markets and their effects on household behaviour. There are few examples where statistical systems actually track consumers. Systems to prevent food crises are predominantly rural and monitor precipitation, pest damage and production volumes and prices. The FAO also calculates the number of food insecure people in the world, based on an estimate of food availability, which roughly takes into account the income and purchasing power of rural and urban populations. Through this method, since 2008 the FAO has been able to declare a sharp increase in the number of food insecure people, and announced the symbolic crossing of the one billion threshold for the number of people classified as "hungry", with the intention of influencing decision-makers in order to obtain a mandate to manage the crisis. However, the choice of indicator is not neutral: by basing the calculation on food availability, crisis resolution is viewed in terms of moving towards the promotion of production, such as the greater distribution of fertilizers and seeds and incentives for new agricultural investment, all in the name of the challenge to feed nine billion people by 2050. Curiously, the 2008 high-level summit organized by the FAO in Rome did not even mention poverty and called only for a boost to agricultural production (Bricas & Daviron, 2008).

In response to the simplification of food insecurity to a problem of availability, other approaches have developed. These new approaches do not seek to objectively measure this insecurity, but in line with the proposals of S. Maxwell, they are based on how this insecurity is felt by the population. The calculation is no longer based on production statistics, indicators of income distribution or price data, but on extensive population surveys. Another new aspect to these methods is the use of data that is no longer public but private: on the local level, NGOs and development or humanitarian projects produce their own data that is usually independent of national statistical tools, for which they have no use (Dury et al, 2011). On a national level, some NGOs cover such a large geographic area that their data are considered as reference data, and are used by the FAO. Internationally, some studies have emerged that were carried out in several countries either by private polling institutes or by NGOs. For example, since 2007 a Gallup poll has been conducted in over 100 countries. On the basis of a large sample number, these surveys have concluded that in terms of the experience of food insecurity, the situation is improving at the global scale (Headey, 2011). Due to rapidly rising incomes in major emerging countries and because of rising agricultural commodities prices for producers, the number of people lacking money to buy food has decreased significantly. Oxfam conducted a survey in dozens of countries involving more than 16,000 people, which showed that the price rise may have changed eating habits (Oxfam, 2011). While all these data must, just like those of FAO, be used with caution, they reveal the dual vision of food security, which is either seen in terms of availability or poverty. In particular, they also reveal a game where the players vie for the choice of indicator to be used in arguments, since it is critical to steering the direction of the debate. Beyond technical and methodological issues regarding indicators, issues that are largely debated in national, regional or international forums involve a power struggle between agents that want the responsibility of managing the crisis.

kept them informed of world events. The fall of the Haitian government in April 2008, against a backdrop of rising prices was instantly known throughout Africa and generated protests that targeted the same objective. These riots related to the high cost of living are not new. The world has seen this before, often as localized events that were then considered as primarily internal phenomena. What is new is the simultaneity of the riots, about thirty occurred within a few weeks in Africa, Asia and America. And it is because this looks like an epidemic that concerns have been raised in international forums, including the United Nations, in case there is a risk of global destabilization.

A MULTITUDE OF INNOVATIVE INITIATIVES

It is this fear of destabilization that the international community has mobilized itself to address. Successive summits involving heads of state or governments have thus been organized in an attempt to alleviate the crisis. While the authorities in charge of food security had become less visible and were lacking the necessary authority to place this issue high on the agenda, the 2008 crisis has had the effect of reshuffling the stakeholder pack and providing an opportunity to reform these institutions. Two simultaneous and apparently contradictory movements can be read in the multiplication of institutional initiatives.

The first, as mentioned earlier, is a reaffirmation of the sovereign power of states to manage the crisis. This takes the form of trade policies (reducing import taxes and blocking exports) and, for some countries with poor agricultural resources, also entails large-scale land acquisitions in other countries to secure domestic supplies. The price surge caused a crisis of confidence in international markets as a means of securing stocks. In fact, some countries encountered difficulties when trying to buy wheat or rice on the international market and are consequently trying to diversify their supply modes, for example by boosting domestic production or by investing in production in other countries in addition to market purchases. The Doha Round was interrupted at the end of 2008 by the stance taken by some countries that refused to make any trade compromises to ensure food security for their populations. In 2011, the breakdown of WTO talks means that the prospect of an international agreement is very uncertain in the short term.

WHAT IS NEW IS THE SIMULTANEITY OF THE RIOTS, ABOUT THIRTY OCCURRED WITHIN A FEW WEEKS IN AFRICA, ASIA AND AMERICA.

While these agreements and international institutions are weakening, there is a parallel emergence of new global institutions and a reformation of old ones, such as: the High Level Task Force (HLTF), the Reform of the Committee on World Food Security, the Scaling Up Nutrition (SUN) movement, etc. One feature of the evolution of these institutions is the emergence of non-state stakeholders: NGOs and multinational private corporations. The limitations of public funding that the financial crisis has exposed means that hopes are now resting on private funding. The SUN, which works in the field of nutrition, has a substantial budget at its disposal for financing actions. The Gates Foundation is also very active in this area, and so are companies that provide food supplements.

CONCLUSION: WHAT CAN WE LEARN FROM THIS EMERGENCE OF NEW STAKEHOLDERS?

Since 2008, the traditional agents in the agricultural sector (trade unions, ministries, the FAO, agro-supply companies) have used the food crisis as a means to re-establish themselves with a new legitimacy, once again by citing Malthusian scenarios. This approach still has the ability to re-mobilize, as it puts the spotlight onto the future supply and demand equation. A semantic shift betrays such approaches: these stakeholders do not talk about individual food security, but about feeding the world or the supply/demand balance at the world level. Agricultural production clearly has an important role to play, especially in areas where it is barely sufficient to cover food and monetary needs. But the recognition of the multi-sectoral nature of food security calls for a reconsideration of the role of agriculture by expanding its mandate well beyond that of providing sufficient biomass. It plays a role in job creation, in health and nutrition, in the stability of prices and incomes, and in culture, and it must therefore be evaluated on these functions. But since these functions are important, all the other sectors contributing to them are important too.

SINCE 2008, THE TRADITIONAL AGENTS IN THE AGRICULTURAL SECTOR HAVE USED THE FOOD CRISIS AS A MEANS TO RE-ESTABLISH THEMSELVES WITH A NEW LEGITIMACY, ONCE AGAIN BY CITING MALTHUSIAN SCENARIOS.

The food crisis has provided a window of opportunity for the reconstruction of governance that is based on more than a simple reflection of the increasing importance of the global level, but instead on a dual movement where states are regaining a level of influence that liberalization policies had tended to strip away from them. The interplay between stakeholders is therefore more complex, especially as new agents have gained considerable importance: firstly private companies, ranging from upstream agricultural ones – seeds, fertilizers and pesticides – to agri-food processers and supermarkets; and secondly NGOs. But the significance of these non-state agents is not limited to agricultural supply. It also extends to information systems – such as those that focus on the experience of food insecurity, on prices or on satellite imagery – including global ones, with consequences that are still poorly understood.

Ultimately, it is the people themselves, which we will not term as 'consumers' here, that are notably absent from food security governance, a large part of whom auto-consume their own production. Apart from 2008 when these populations made themselves heard through rioting, the people remain the great silent majority in this debate. They are only approached through surveys, the methodologies of which influence the results. Whether one measures the degree to which objective needs are satisfied, people's purchasing power or their opinions, the same reflection of expectations is not obtained. Populations are also studied through representatives: the state or local authorities, consumer associations or even supermarkets, which claim to speak on behalf of their clients. The legitimacy of these institutions is obviously debatable. But what is striking in all cases is the fact that the focus falls only on vulnerable or poor people. Of course, the situation for these individuals is critical and has weakened since the price increase, if we believe the institutions with responsibility for the safety nets. The attention directed to these people is of course unquestionable: they are the ones who suffer. It is the health and future of their children that are put at risk. While attention must be given to the most vulnerable, because they are often silent, more generally, the poorest populations must also not be forgotten. The deterioration in their purchasing power and resilience can tilt them towards volatile situations. It is also worth noting that no attention is given to those that have gained from the food crisis; and there are some of course! High prices, the development of agro-fuels, large-scale land acquisition and investments in the agro-food sector are all features of crises that allow some to benefit and make a profit. More generally, over-consumption in rich countries, where the demand appears infinite in a world of scarce fossil resources, is certainly a part of the problem and attention should be directed onto this matter. ■

REFERENCES

Soulé B., Blein R. and Bellora C., 2008, *Hausse des prix alimentaires en Afrique de l'Ouest. Revue et analyse des mesures engagées à court et moyen terme*. Contribution to the conference "Prix agricoles: perspectives à moyen terme et implications pour les producteurs et les politiques publiques", Paris, 16 December 2008. Lares, Bureau Issala, Fondation Farm, 65 pages.

Pretty J. (Ed.), 2008, Sustainable Agriculture and Food, Vol III, Agriculture and Food Systems. Earthcan, 465 p.

HLPE, 2011, Price volatility and food security. A report by the High Level Panel of Experts on Food Security and Nutrition of the Committee on World Food Security, Rome, 79 p.

Paillard S., Treyer S. and Dorin B. (Eds.), 2010, Agrimonde, Scénarios et défis pour nourrir le monde en 2050. Paris, Quae, 295 p.

Foresight. The Future of Food and Farming (2011) Final Project Report. The Government Office for Science, London, 208 p.

Beaujeu R., Kolie M., Sempéré J.F, Udher C., 2011, Transition démographique et emploi en Afrique subsaharienne. Paris, AFD, Coll. A savoir n°5, 217 p.

Bricas N. and Daviron B., 2008, De la hausse des prix au retour du productionnisme; les enjeux du sommet sur la sécurité alimentaire de juin 2008 à Rome. *Hérodote*, 131: 31-39

Dury S., Fouilleux E. and Bricas N., 2010, La production de statistiques pour les politiques de sécurité alimentaire: entre visions du monde et enjeux de pouvoir. Le cas du Mali. *Stateco*, 105: 7-18

Broutin C. and Bricas N., 2006, Agroalimentaire et lutte contre la pauvreté en Afrique subsaharienne; le rôle des micro et petites entreprises. Paris, Ed. du Gret, 128 p.

Headey D., 2011, Was the Global Food Crisis Really a Crisis? Simulations versus Self-Reporting. Washington, IFPRI Discussion Paper 01087.

Oxfam, 2011, http://www.oxfam.org/en/grow/pasta-pizza-paella-global-food-survey-reveals-worlds-favorite-foods

World Bank, 2007, World Development Report 2008. Agriculture for Development. Washington, World Bank, 384 p.

CHANGING FOOD SYSTEMS, CHANGING PEOPLE

Wallapa van Willenswaard,
Green Market Network, Bangkok, Thailand

Over several centuries, agriculture has slowly evolved into a branch of industry due to many factors such as enclosure (Vandana Shiva, 2005), the industrial revolution, urban growth, rural flight, the emergence of modern nation states and international economic competition. Post-WWII era changes eventually defined the present agricultural landscape: war industries were converted to agrochemical production, oil was considered inexhaustible and international development strategies replaced colonialism, addressing poverty and famines with grand strategies such as the Green Revolution.

The important question is whether the present challenges to agriculture, which include the 2008 economic crisis, peak oil, the food crisis, climate change, demographic pressure and increasing inequality, will lead to continued industrialization or to a fundamental change of course. One possible way forward is through the formation of networks of small-scale organic farms, supported by sustainable technologies, which may guarantee long-term food security more effectively than mechanistic and exploitative approaches. This article examines the potential of such an 'alternative' approach, focusing on the organic food movement in Thailand.

THAILAND'S ORGANIC ORIGINS AND CONSUMER MOTIVATION

In Asia, the alternative food movement began in 1982 with the foundation of the Pesticide Action Network (PAN). Recent estimates show that Thailand now has 18,895 hectares under organic cultivation or under conversion, and that there are around 5,000 organic farmers[1].

The organic movement is concerned about human health and the environment in equal measure. In terms of health, the organic lobby questions the nutritional value of conventionally produced food and raises fears about chemical residues. Environmental concerns include agrochemical contamination, pesticide resistance, decreasing soil fertility, landscape degradation, deforestation and biodiversity loss. Such environmental threats have led to a rise in support for the adoption of 'agroecology' (Miguel A. Altieri, 1995) in Thailand, an approach advocated by environmental groups such as BioThai.

Some consumers are motivated to buy organic foods for an additional reason: the possible health impacts on farmers that use agrochemicals. In Thailand, most farmers are caught in a debt-trap due to an inability to recoup their initial investment in chemicals, artificial fertilizers and hybrid seeds. These farmers also suffer from inequality, a lack of direct access to markets and unfair negotiation positions.

The International Federation of Organic Agriculture Movements (IFOAM)[2] has taken these issues on board and added 'Fair Trade' to its list of standards;

1. See www.organic-world.net

2. Global network of organic producers, consumer associations, traders and researchers established in 1972 in Versailles, France.

while in Thailand, the Green Net enterprise has also introduced the Fair Trade label. However, farmer groups such as Thailand's Assembly of the Poor and the global Via Campesina remain the main bodies that campaign for the rights of small-scale farmers. Smaller positive action groups are important on the regional scale, such as Bangkok's Community Supported Agriculture (CSA) consumer network, which guarantees income security to farmers through an annual advance payment.

There remains another motivation for organic consumers, which is a belief in the need for societal, economic and political change. In short, these people seek 'transformation' or systemic change, characterized by critical holism (Jan Nederveen Pieterse, 2000) and the celebration of cultural integrity.

The Bangkok-based Green Market Network, established and managed by Suan Nguen Mee Ma (Garden of Fruition), is a pioneer of the transformation movement. The Garden of Fruition also founded the School for Wellbeing Studies and Research, an independent think tank and action-research platform that is guided by activists such as Sulak Sivaraksa, Vandana Shiva and Helena Norberg Hodgea.

HOW THE GREEN MARKET NETWORK OPERATES

Professional organizations within the Thailand Organic Traders Association (TOTA) focus primarily on exports to Europe, the USA and Japan, and on high end convenience foods. Thailand's government fully recognizes the important contribution of organic agriculture to the economy and implements policies to encourage the export industry. In contrast, the Green Market Network targets the middle and lower end domestic market. It does not receive a significant amount of government support.

The Green Market Network facilitates intensive networking between producers, responsible traders and *ethical* consumers. It also supports consumer awareness with campaigns, training, publications and media relations.

The paradigm shift that pro-organic advocates seek to promote is that organic products are not in fact too expensive, but that conventional food is too cheap because of 'externalities' and hidden government subsidies that enable non-organic food to be sold at lower prices. For such a shift to be achieved, consumers must be willing to devote a bigger proportion of their household budget to direct food costs.

PIONEERING INVOLVEMENT OF HOSPITALS

The organic movement in Thailand is trialling a system to involve hospitals in the distribution of organic food (in hospital kitchens and restaurants). Direct cooperation between hospitals and groups of organic farmers can help build community networks and promote best practices to provide a foundation for a gradual, locally-induced shift towards more sustainable agricultural methods. The present pilot project involves five hospitals in Bangkok and around 100 farmers. If organic food could be promoted by public hospitals as the gold standard for health, it would send out a strong message to Thailand's consumers, many of whom cannot afford organic food in the present price regime. One way to enable hospitals to lower the cost of organic food would be for its production to be governed locally and via participatory guarantee systems, and for hospital menus to be seasonal (standard CSA practice), thus reducing transport costs, demand for non-seasonal products and certification costs.

However, this approach is problematic since Thailand's public hospitals are tied to procurement

BOX 1 THE GREEN MARKET NETWORK

- consists of a network of independent 'green' shops in Bangkok
- supports CSA projects involving an annual payment by consumers to farmers for a weekly quota of organic produce
- conducts farm visits to broaden awareness and knowledge, to build understanding and to form quality guarantee networks
- organizes weekly markets in offices and provides management support for local producers and processors
- organizes weekly markets in hospitals (currently five in Bangkok) and supports local producers in their efforts to supply hospitals
- organizes the Annual Green Fair (established in 2007) and other smaller fairs

regulations and face enormous pressures to deliver universal health care with minimal funds. To fully incorporate hospitals into local organic-based economies would require an additional paradigm shift, similar to the one that corporations must make to seriously address their social responsibilities.

NEW FOOD SYSTEMS AND THE TRANSFORMATION MOVEMENT

Researchers at the School for Wellbeing Studies and Research provide support for the hands-on approach of the Green Market Network. The School, which originated from discussions between Thailand and Bhutan on Gross National Happiness (GNH), organized a public dialogue in 2009 with Nobel laureate Joseph E. Stiglitz on the results of the Stiglitz-Sen-Fitoussi Commission that re-examined the relevance of Gross Domestic Product (GDP) for economic performance and social progress (Stiglitz-Sen-Fitoussi, 2010).

The GNH Index revealed the social divisions that exist in Bhutan, showing the discrepancy between the happiness levels in the capital Thimphu and the rural districts; while in Thailand the urban-rural divide has fuelled street protests. Unfortunately, throughout Asia, governments usually deal with such rural unrest through short-term incentives and populist policies, rather than by implementing major structural changes. Of the crucial issues that contribute towards rural dissatisfaction, two of the main ones are the laws relating to land and property, and the intellectual property rights of seeds.

The School for Wellbeing's response to the anticipated food crisis focuses on food quality and food security through the maintenance of soils and the establishment of resilient farming communities. The School aims to revitalize the "third way" debate (Anthony Giddens, 2001) by examining three basic scenarios for the future: the first is guided by socialist and communist principles, the second by neo-liberalism, while the third attempts to construct a genuine alternative, tentatively named the Well-Being Society scenario (International Exchange Platform, 2011). Organic agriculture is at the heart of the latter, which seeks to forge critical links between the best qualities of each scenario, rather than to replace the others.

The Well-Being Society scenario resonates with the 'Green Economy'[3]. It urges citizens, governments and the business sector to revitalize the core values of healthy societies: care for the individual, care for communities and care for the Earth. It advocates the implementation of food systems that maintain a high quality of life, while respecting the integrity of nature. From these fundamentals[4], other aspects of complex societies can be understood and addressed, allowing societal, economic and political transformation to gradually take place. Impulses for such transformation have emerged at various times throughout human history, often in response to warning signals. We have now reached an era when change is urgent; and changes to food production systems are where it all starts. ■

3. As promoted by UNEP towards Rio +20.

4.IFOAM Position paper *The Role of Smallholders in Organic Agriculture*: "Ninety percent of farms worldwide are less than two hectares, they provide employment to 1.3 billion people and dominate agriculture in developing countries."

REFERENCES

Altieri Miguel A., 1995, *Agroecology. The Science of Sustainable Agriculture,* Westview Press, Boulder, CO. Revised and expanded edition.

Giddens Anthony (ed), 2001 *The Global Third Way Debate,* Polity, Oxford

International Exchange Platform, 2011, *Re-thinking Property. Pathway to a Well-Being Society scenario?,* August 2011, Bangkok.

Nederveen Pieterse Jan, 2000, *Development Theory. Deconstructions and Reconstructions,* London, SAGE

Norberg Hodge Helena, 2002 *Bringing the Food Economy Home,* Kumarian Press, Sterling

Shiva Vandana, 2005, *Earth Democracy. Sustainability, Justice, Peace,* South End Press, New York

Sivaraksa Sulak, 2010, *The Wisdom of Sustainability. Buddhist Economics for the 21st Century,* Souvenir Press Ltd, London

Stiglitz, Sen, Fitoussi, 2010, *Mis-measuring Our Lives. Why the GDP Doesn't Add Up*. The New Press, New York

The question of the contribution of agriculture to economic development is hampered by the very concept of "development", which is based on a partial reading of the Western economic transition. The crucial issue for countries in sub-Saharan Africa is their ability to generate a sufficient amount of activities – including agricultural and rural ones – to absorb a growing population.

AGRICULTURE AND TRANSITION IN A GLOBALIZED WORLD

Bruno Losch,
Centre International de Recherche Agronomique pour le Développement (CIRAD)

Following a long absence from the discussions on economic development, agriculture has moved back onto the centre stage with the World Bank's 2008 World Development Report (WDR08), published in late 2007[1]. Its publication coincided with a sudden surge in agricultural prices during 2007 and 2008, which triggered multiple and disparate forms of government intervention and thus served as a reminder that agriculture – through its central function of feeding humanity – is a "state matter" (Coulomb et al., 1990). These events have come to echo the climate change debate and its global consequences (2009 Copenhagen summit), bringing back into focus perspectives that had been forgotten since the 1970s, along with the alarmist predictions of the Club of Rome.

Feeding a world of nine billion people by 2050 in a sustainable manner in a fragile natural environment is of course an agricultural issue, although other areas of activity outside agriculture in the strict sense appear essential to meet this challenge (such as consumption patterns, energy sources or the incorporation of environmental costs). However, bringing the agricultural issue back to a question of food, or even sustainable food, is a clear narrowing of the role of agriculture, when a need for multifunctionality has now been acknowledged[2]. Agriculture does not only produce

1. The arguments in this chapter are largely based on work carried out in the RuralStruc Program (see box). They are those of the author and do not reflect the views of the World Bank or any of the programme's other donors.

2. An acknowledgement that is challenged when it comes to justifying public support, as shown by the opposition at the WTO between the EU, the United States and the Cairns Group. See Losch 2004.

food or raw materials, it also has economic, social, cultural and environmental roles due to its special relationship with land area, territories and modes of life, whilst also fulfilling both commercial and non-commercial functions (Group Polanyi, 2008).

This narrow view of the role of agriculture can be a major disadvantage when discussing its full contribution to development. Indeed, while the WDR08 has reopened a path by emphasizing the role of agriculture in the various "stages" of development, it is one that remains fraught with obstacles that prevent a true rehabilitation for two reasons. The first is temporary and relates to the media coverage of issues, which is critical for attracting the attention of policy makers. The management of the financial crisis that began in late 2008 concentrated on short-term goals that had little to do with the 2050 horizon. Meanwhile, the themes of international debate are constantly changing, for example: while the WDR08 focused on agriculture, WDR09 centred on the importance of density and urbanization (World Bank, 2008) and the UNIDO 2009 report highlighted that industrialization is integral to economic development (UNIDO, 2008).

The second, more fundamental, reason for the blockages concerns the very concept of development. Forged at the end of World War II and marked by decolonization, development is rooted in the notion of progress in stages, formalized by Rostow among others, to ultimately reach the level set in Western Europe and its overseas projections (primarily the US) during the nineteenth and twentieth century. Development – which Rist (1996) calls a "Western delusion" and that, as Goody (2006) reminds us, monopolizes world history by stealing other people's histories – is, by its very foundations, of an evolutionary nature; and "emerging" states are those that follow closely the development patterns of western countries (Gabas & Losch, 2008). The fundamental flaw of such a vision is that by focusing on the actions of a territorial entity (usually one thought of as a nation state), the central role of the relationship between national and international is forgotten, along with the links between internal processes and "rest of the world" conditions, all of which are crucial to understand the dynamics of change, including the balance of power and the available manoeuvre room for local stakeholders who establish the field of possibilities[3].

BRINGING THE AGRICULTURAL ISSUE BACK TO A QUESTION OF FOOD, OR EVEN SUSTAINABLE FOOD, IS A CLEAR NARROWING OF THE ROLE OF AGRICULTURE, WHEN A NEED FOR MULTIFUNCTIONALITY HAS NOW BEEN ACKNOWLEDGED.

In this regard, what is the role of agriculture, particularly for countries characterized by a weakness in their structural change and which are most dependent on their agricultural sector? This is the angle of analysis adopted for the rest of this chapter, which first discusses the canonical model of the evolutionary scenario and

3. This view is reinforced by the widespread use of "development indicators", such as the Millennium Development Goals, helping to focus attention on internal objectives and to segment the debate between subject areas, ignoring the question of the viability of the process and their relationships.

the historical particularities of past transitions. The structural situation of "late developers", specifically sub-Saharan Africa (SSA) is then reviewed by highlighting the challenges of a double transition (economic and demographic) in the era of globalization and under the constraints of global change. Finally we discuss the particular situation of agriculture and barriers to agricultural and rural diversification before concluding on the necessary choices in terms of public policy.

THE EVOLUTIONARY SCENARIO AND ITS OMISSIONS

According to the historical sequence observed in other parts of the world and confirmed by the statistical evidence (Timmer, 2009), the conventional process of economic transition involves a gradual shift from an agriculturally based economy to one based on industry and then services, and from rural to urban. This structural transformation is made possible by the productivity gains that allow the accumulation and transfer of labour and capital from one sector to another, along with the increase of living standards and growth and diversification of demand.

Agriculture therefore plays a role as both an initiator and a driving force, which is accompanied by a massive outflow of assets from the sector. Only a minority can remain and become specialized in agricultural tasks (usually the type of agriculture with the most production factors), while the vast majority of people gain employment in other non-agricultural activities, sometimes in rural areas but more often by migrating to cities, other regions or abroad. These three options – agricultural specialization, rural diversification and migration – are those that the WDR08 presents as possible exits from rural poverty. However, if the report draws attention to the composite characteristics of the solutions, the question of the viability of these exit options today is not really debated, while previous transitions have clearly taken place at other historical moments[4].

This approach has three major omissions. First, the European transition, which serves as a reference – beyond the long-term processes related to the development of market capitalism – has been fuelled by a global geopolitical order characterized by the expansion of European domination or that of states derived from it (e.g. the US). This domination of the West has facilitated its industrial development due to – a fact that is too often forgotten – captive markets and the elimination of Asian competition (Bairoch, 1997; Grataloup, 2007).

Subsequently, labour surpluses derived from agricultural modernization, which could not find an outlet in emerging industries, were largely absorbed by large-scale international migration to the "new worlds", derived from the expansion of Europe[5]. This adjustment through European migration has been decisive and instrumental in shaping the current global state.

Finally, these western transitions have been fundamentally marked by a *national*

4. The WDR08, although given the title *Agriculture for Development*, actually focuses on exiting from rural poverty. This semantic shift complicates or prevents further attention to the transition processes.

5. This "white migration" (Rygiel, 2007) involved approximately 60 million Europeans between 1850 and 1930.

vision of development, which has experienced a widespread expansion phase from 1930 to the late 1970s – i.e. the beginning of the current period of globalization. The objectives were then to consolidate the nation state and to develop domestic markets through proactive public policies, including import substitution. This was the path followed in the interwar years by most Latin American countries – which had existed as states for over a century – and by the Asian countries from the 1950s. This national vision was reinforced by the external support of the Cold War in the form of massive aid programmes to countries that were considered strategically useful against the communist threat (South Korea, Taiwan, Singapore, India and many Latin American countries).

THE VERY SPECIFIC SITUATION OF SUB-SAHARAN AFRICA

The current period of globalization, characterized by the withdrawal of the state, trade liberalization and technological advances in communications, has of course changed the stakes profoundly for the least developed countries in their process of structural change, particularly so in SSA which is the last part of the world to engage in economic and demographic transition.

African states are among the latest arrivals on the international scene, having originated mostly in the early 1960s on the foundations of European colonial territories. They benefit from the advantages of "late developers" (technical progress, learning from the past), and from opportunities of the new system of world growth (trade openness and access to new markets). But they face the constraints of strong competition – which highlights the growing global imbalances – and the instability of the international economic environment, along with the consequences of global change, drastically reducing their manoeuvre room.

AN EMBRYONIC ECONOMIC TRANSITION Fifty years after independence, SSA economies remain characterized by the significance of the agricultural sector in the gross domestic product (GDP), in foreign trade and above all, something which is often overlooked, in the structure of employment: agriculture, which accounts for an average employment of 65% for the subcontinent (excluding South Africa) and between 75% to 85% for many countries, remains the main source of activity and household incomes.

The most striking phenomenon is the large structural inertia of SSA economies, even though their populations have shown a very high degree of mobility, expressed by the rate of urbanization: with a rate approaching 40%, the urban population has multiplied by 12 since 1960, despite the lack of an industrialization drive. This "low speed" urbanization, without industries, contrasts with other developing regions, especially Asia, where economic changes were very fast (see Figure 1)[6].

Thus, in the absence of a dynamic manufacturing sector, workforce growth in SSA

6. Latin American countries with earlier transitions are not shown here.

FIGURE 1 THE PERSISTENT SINGULARITY OF AFRICAN ECONOMIES

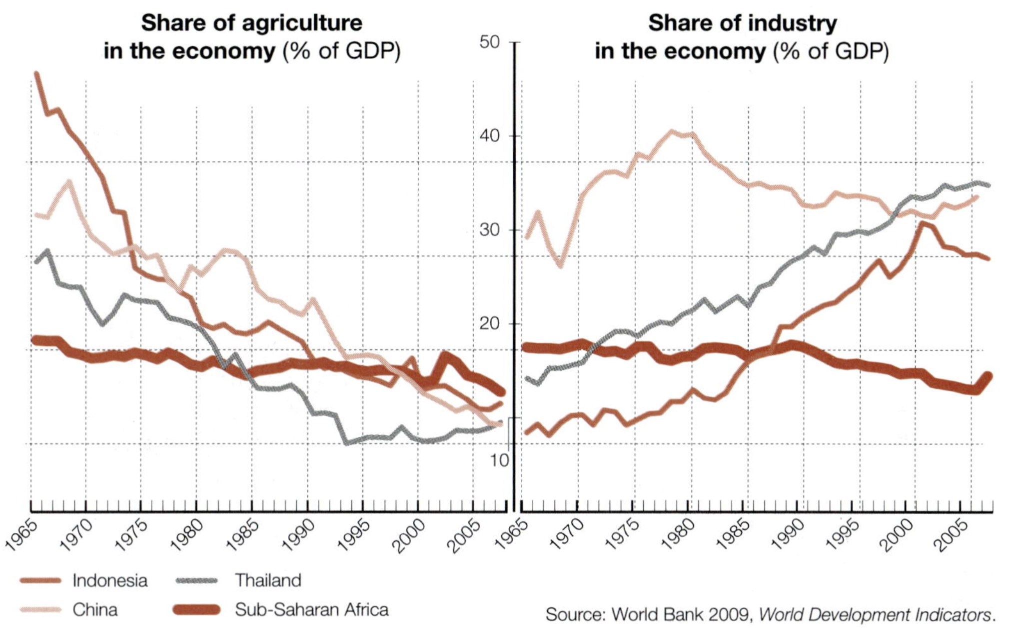

Source: World Bank 2009, *World Development Indicators*.

was primarily absorbed by agriculture and the informal urban sector[7], which has played a complementary role as a shock absorber, without enabling real dynamic growth: low productivity, underemployment, job insecurity and low income frequently serve as poverty traps, as shown by the urbanization of slums (UN-Habitat, 2003).

With domestic markets that are scattered due to fragmentation into 42 states (on the continent) and a solvent demand constrained by the extent of poverty, the growth of SSA economies remains highly dependent on the outside. These characteristics, combined with a low overall economic productivity (due to the importance of the agricultural sector and the informal economy) and frequent political instability, as well as high population growth, explain the very low growth in GDP per capita and its high volatility. The strong growth in recent years – before the financial crisis – should therefore be analysed from a long-term perspective. With few exceptions, this growth was initially "pulled" by the commodity and trade boom and has not decreased the "structural anaemia" in the subcontinent[8].

Despite rapid urbanization, the structure of economies in SSA has changed little in recent decades. This represents a very different development trajectory to that followed in other parts of the world such as East and Southeast Asia.

7. This represents between 30% and 45% of non-agricultural GDP and from 70% to 90% of the total non-agricultural employment.

8. Recent arguments in favour of "African lions" (McKinsey, 2010) should be taken with caution: they concern the entire African continent and therefore include North African countries and South Africa, which accounted for 62% of the continent's GDP in 2008.

FIGURE 2 THE CRUCIAL ECONOMIC INTEGRATION OF YOUNG PEOPLE

Sub-Saharan Africa faces major demographic and economic challenges. By 2025, 330 million young Africans will enter the labour market. Taking this demographic issue into account, and building an economy that is able to integrate these people, will be key to enabling growth and social stability in the continent.

This specific situation, given the structural changes observed in other regions of the world, can be explained by the historical conditions of the integration of the continent into the global economy (restrictive and recent colonial administration) and the young age of African states. Overtaken by globalization and structural adjustment, these states, unlike Asian and Latin American countries, have not implemented proactive policies to modernize and correct their own mismanagement.

AN UNFINISHED DEMOGRAPHIC TRANSITION This weakness of the transformation of African economies is more acute given the demographic situation in the subcontinent, since the SSA is the last region of the world to engage in a process of demographic transition. Yet this process remains largely unfinished: after four decades of strong population growth (more than 2.5% per year – with the exception of countries most affected by the AIDS pandemic) and the late manifestation of the declining birth rate, the population will double by 2050, equating to at least 900 million more inhabitants according to the UN. This evolution will result in a change to the demographic structure with a strong increase in the size of the workforce – and therefore the demand for jobs – and then in a progressive evolution of activity rate. This rate, which is an expression of the ratio between active and inactive members of the workforce, was close to 1 in the 1980s and 1990s (compared to 2 in East

Asia during the same period), thus constituting a major economic handicap at the height of the structural adjustment. Its steady increase to beyond 2050 will represent a unique opportunity in terms of growth. But for this "demographic dividend" to fully play its incentivizing role, it must be accompanied by a significant increase in productive investment, capacity for innovation and productivity.

Due to its low-diversified economic structure, marked by the weight of agriculture and the informal sector, the crucial question for SSA is its ability to generate enough activity and employment to absorb the rapid growth of its labour force. Today, in the entire SSA, a cohort of around 17 million young people enters the labour market each year. This number will increase to about 25 million in 2025 (Figure 2), which is not conjecture as these individuals have already been born. In total, over the next fifteen years, SSA economies will have to absorb 330 million new active workers (i.e. the current population of the United States). This is a major challenge that carries considerable economic, social and political pressures. The "Arab revolutions" have provided a brutal reminder, bringing youth employment back onto the list of main concerns for African leaders (African Union heads of state summit, January 2011).

A CENTRAL ROLE FOR AFRICAN AGRICULTURE

To meet the challenges of its structural change, SSA certainly has many advantages, both natural (mineral resources, land availability, young population) and comparative (low cost of labour); but these advantages are largely frustrated by the constraints linked to major delays in terms of public goods, human capital and productive infrastructure, which induce and are accompanied by numerous imperfections and the incompleteness of the market.

The recent boom of Foreign Direct Investment, caused by the attractiveness of the continent's advantages, remains largely confined to the exploitation of resources and the infrastructure associated with it. The weaknesses of local private capital and of public resources, as well as the limits of ODA, make simultaneous treatment of all other constraints difficult. However, it is clear that SSA will not, in the short and medium term, be able to rely on the security valve of international migration, such as the mass migration that accompanied the European transitions between the mid-nineteenth century and the 1920s[9]. Moreover, while there is probably a place for the industrialization of low-income countries (UNIDO, 2009), and while the information technology revolution creates new opportunities for services, these changes will be slow because of the extent of the disadvantages[10].

It thus appears that for most SSA countries, progression of their agricultural bases will remain essential for development over the next two decades, which will see an extraordinary surge in the number of young people seeking work. Of the 330

9. The situation in Mexico and Morocco, with 10% of their nationals living abroad, is not reproducible: to achieve these proportions, more than 80 million Africans south of the Sahara would have to emigrate today.

10. A major issue concerns the viability of a structural change that could "jump" the historical stage of industrialization made difficult by the asymmetries of the global economy.

million new young workers that are anticipated by 2015, 200 million will be in rural areas if current population distribution trends between urban and rural areas are maintained. This large number of workers will have to find employment in rural areas or risk increasing the pressure on urban growth, which is already difficult to manage. Also, agriculture will have to play a major role, given that it is primarily the growth of farm income – a lesson from earlier transitions – that creates a rural demand that generates other non-agricultural rural activities.

STRUCTURAL CONSTRAINTS TO AGRICULTURAL GROWTH

African agriculture remains constrained in its development by several well-known barriers: high transaction costs, lack of credit and supply, the pitfalls of commercialization, lack of information and training, etc. The stagnation of overall agricultural productivity has been increased by the fall in public investment in agriculture – states and aid agencies alike – a consequence of the adjustment options taken during the last 25 years.

However, while these findings relate to the economic and institutional environment, the situation of farm and rural households is less often considered. Household surveys conducted by the RuralStruc Program (see box) confirm the extent of rural poverty (75% of surveyed households have a total income below $2 a day per person and 40% have incomes below a $1 a day per person); but they also reveal the extent of the risk and insecurity in relation to the economy, and even to food (confirmed by converting the average income into kilocalories per adult equivalent).

More specifically, the results of the programme can highlight the importance of the relationship between household income and the diversification of activities, which determines the pace and scale of rural transformation. Indeed, the importance of economic risk for households faced with persistent poverty and low income, along with the small size of existing incomes, are a barrier to investment and innovation possibilities, which impacts structurally on the productivity and the prospects for increased agricultural performance.

BOX 1 THE RURALSTRUC PROGRAM: A COMPARATIVE APPROACH TO RURAL DIVERSIFICATION

The RuralStruc Program on the structural changes in rural economies in the globalized world arose from a joint initiative of the World Bank, French Cooperation (AFD, Ministries of Agriculture and Foreign Affairs, CIRAD) and IFAD. Launched in 2006, the programme gathers seven national teams – Mexico, Nicaragua, Morocco, Senegal, Mali, Kenya and Madagascar – in a comparative project of two phases (2006-07 and 2007-10), designed to highlight the diversity of the processes of change and the adaptation modalities of rural households, particularly through the implementation of surveys of 8,000 households in 26 regions of the countries concerned. Dissemination of results at the national level, but also to regional organizations and the community of donors, can contribute to the debate on development policies and the special place of the rural sector in the structural change of developing countries.

For more information, read: Losch B., Freguin-Gresh S., White E., 2010, *Structural dimensions of liberalization on agriculture and rural development: A cross-regional analysis on rural change*, RuralStruc Program, final report. World Bank, Washington. http://www.worldbank.org/afr/ruralstruc

BOX 1 RETHINKING INTERVENTIONS IN RURAL AREAS

Beyond its contribution to a better understanding of the rural world and the situation of rural households in the countries studied, the RuralStruc programme questions donors about their practices. Its results confirm several realities that policymakers are struggling to integrate into their strategies.

The first is the massive influx into the labour market of young people, mainly from rural areas. Yet the donor community does not consider agriculture to be an employment sector. Our modernization models, based on the trajectories of Northern countries, aim to improve the profitability of production, investment and work, without regard to employment. How can we increase global productivity without overly sacrificing employment?

Another reality is the extent and level of poverty in rural areas, particularly in countries in sub-Saharan Africa. Not only does poverty affect a large proportion of households, its severity is often high, with large proportions of the population living below the extreme poverty threshold. There is nothing new in this, and the literature on the subject is abundant. But aren't our intervention methods leading us away from these populations, although they are the majority in the poorest countries? Indeed, the financing of development is increasingly oriented towards economic investments with rapid profitability and the search for visible results. But can we ask households that are so poor, whose primary concern is to maintain a supply of food, to invest in or adopt new technologies and engage in new types of production? Support through the introduction of social nets is not sufficient. Long-term actions combining public goods, collective productive investments and support to individual economic activities are also needed.

RuralStruc also provides information that is less disseminated. Its results clearly show that food security concerns, which are reflected in auto-consumption, remain a determinant factor in farming strategies, including in regions considered to be more favoured. This observation prompts the re-prioritisation of staple crops to reduce the level of risk and unlock innovation capacity. However, aid agencies and research have often given priority to export crops which are certainly promising in terms of marketing opportunities and income, but are often less inclusive compared to food crops that involve the largest number of people while being linked to domestic and regional markets that present very high growth rates.

The Program also provides new insights into the dynamics of diversification versus specialization, which are at the heart of the transformation process for rural economies. It shows that the diversification of household income is still marked by a rationale of adaptation that does not necessarily lead to an exit from poverty. Diversification is usually a strategy of risk management, the revenues of which are constrained by the weakness of the local supply of non-agricultural employment, which is a consequence of rural demand that is insufficiently stimulated by agricultural income.

Beyond elements of understanding from studies carried out in specific national contexts, RuralStruc underlines our difficulty in envisaging agricultural transitions, economic transitions and differentiation outside of the traditional sequential evolutionary models of development. Its results lead us to accept the diversity of trajectories highlighted by the great heterogeneity of situations, and also justify a deepening of our knowledge of the current transformation process. These observations de facto advocate for the reinforcement of local statistical systems and the collection of primary data to take into account this diversity and to better understand the mechanisms of differentiation and development that are the closest to local realities. To better promote, guide and support the necessary transformations without "promoting" universal solutions, it is essential to have specific and concrete studies. To ensure that funding for the transformation of rural economies meets the goals of eradicating poverty and supporting employment and sustainable agriculture, it is better to start from a knowledge of the realities than from a trajectory of agricultural modernization copied from that which occurred in northern countries, and its social and environmental externalities that are now better evaluated.

Marie-Cécile Thirion,
Agence française de développement,
Paris, France

FIGURE 3 THE RELATIONSHIP BETWEEN SPECIALIZATION OR DIVERSIFICATION AND INCOME

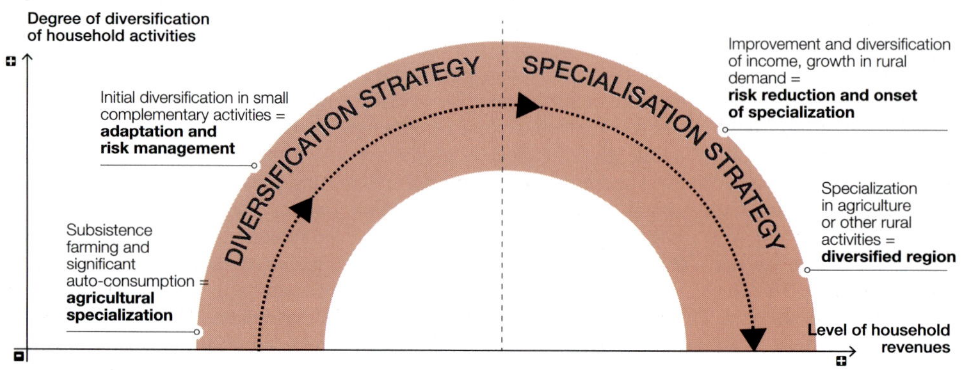

Figure 3 shows a stylized inverted U-shaped evolution where low-income households are confined to a specialization in subsistence agriculture. The growth of farm incomes enables increased access to other sources of income, derived from the emergence of rural demand which, by reducing risks, then allows household specialization in an agricultural sector that is directly market-oriented (with a decrease in auto-consumption) but also in other sectors of non-agricultural activities (production and services).

The SSA regions that have been investigated by the RuralStruc Program are almost exclusively situated at the initial stage of the process. Rural incomes are characterized by the structural weight of agricultural incomes, an important part of which corresponds to auto-consumption. And while diversification into other activities outside farming is generalised and often strong, it mainly concerns low-paid supplementary activities ("small jobs" in trades and services, casual agricultural employment), which neither enable a way out of insecurity nor lead to a true rural diversification.

GUIDANCE FOR PUBLIC POLICY

These results suggest several directions in terms of public policy. Given the magnitude of the constraints faced by African agriculture, compiling a long list of possible intervention areas does not provide the selectivity necessary to take into account the obvious resource limitations, even if governments and donors strive to put agriculture back into the foreground. Realism requires the targeting of priority areas of action that correspond to the strongest constraints.

The heterogeneity of situations – another important result of RuralStruc – implies a reinvestment in the analysis of contexts (survey and regional studies) that is necessary for the development of diagnostic assessments prior to the preparation of action strategies. Support for collaborative public policy development therefore appears necessary to define the appropriate programmes, but it is however possible to highlight three priorities:

The first corresponds to the necessary support for family farms (as opposed to agribusinesses). Beyond the false debate on small scale versus large scale[11], family farms are more inclusive because they correspond to almost all African farms. They have many assets in terms of productivity and competitiveness (World Bank, 2009) and offer considerable potential in terms of absorption of new workers (whether from the potential development of land or from labour-intensive techniques of intensification).

The second priority is to support the development of so-called substantive crops. This choice is obviously not exclusive and there are other production opportunities that offer better returns. But substantive crops are of concern to the largest number of people, and thus improvements in their yields and development are able to break the stranglehold of risks related to food and the economy, and facilitate investment and diversification into non-agricultural activities. Finally, substantive crops have a considerable market potential at the continental level due to demographic and urban growth, while providing opportunities for local processing.

DIVERSIFICATION INTO OTHER ACTIVITIES OUTSIDE FARMING IS GENERALISED AND OFTEN STRONG, BUT MAINLY CONCERNS LOW-PAID SUPPLEMENTARY ACTIVITIES.

The third priority is the need to strengthen the local level, which is the necessary condition for the consolidation of rural diversification and simultaneously to slow down the process of urban concentration, the management of which is difficult. The densification of urban-rural relationships requires better provision of public goods, infrastructure and services in rural towns and small cities, which is essential for the development of agricultural and non-agricultural activities and local networks, which provide a means for adaptation to the realities of a globalized economy. ■

11. The crisis of agricultural prices and foreign agricultural land investments have largely blurred the debate by focusing on the question of the size and organization best suited to meet the food demand (Collier, 2009), when the issue concerns the model of agricultural development most likely to facilitate the transition process and provide employment.

REFERENCES

Bairoch P., 1997, *Victoires et déboires. Histoire économique et sociale du monde du XVIè siècle à nos jours*. Gallimard, Folio, Paris, 3 volumes, 2788 p

Collier P., 2009, "Africa's Organic Peasantry. Beyond Romanticism." *Harvard International Review*, 32(2), 62-65.

Coulomb P., Delorme H., Hervieu B., Jollivet M. and Lacombe P. (Dir.), 1990, *Les agriculteurs et la politique*. Presses de la Fondation nationale des sciences politiques, Paris, 594 p.

Gabas J.-J. and Losch B., 2008, "Fabrication and Illusions of Emergence." In Jaffrelot Ch. (ed.), *Emerging States: The Wellspring of a New World Order*, C. Hurst & Co., London.

Goody J., 2006, *The Theft of History*. Cambridge, Cambridge University Press.

Grataloup C. 2007, *Géohistoire de la mondialisation. Le temps long du monde*. Paris: A. Colin, 256 p.

Groupe Polanyi, 2008, *La multifonctionnalité de l'agriculture - Une dialectique entre marché et identité*. Editions Quae, 360 p.

Losch B., 2004, "Debating the multifunctionality of agriculture: from trade negotiations to development policies by the south". *Journal of Agrarian Change*, 4 (3): 336-360.

McKinsey Global Institute, 2010, *Lions on the Move: The Progress and Potential of African Economies*. McKinsey&Co.

Rist G., 1996, *Le développement. Histoire d'une croyance occidentale*, Paris: Presses de Sciences Po

Rygiel P., 2007, *Le temps des migrations blanches. Migrer en Occident (1850-1950)*. Paris: Aux Lieux d'être, 208 p.

Timmer C. P., 2009, *A World without Agriculture: The Structural Transformation in Historical Perspective*. The American Enterprise Institute Press, Washington DC, 83p.

UN-Habitat, 2003, *The Challenge of the Slums: Global Report on Human Settlements 2003*. UN-Habitat, Nairobi, London, Sterling.

UNIDO, 2008, *Breaking In and Moving Up: New Industrial Challenges for the Bottom Billion and the Middle-Income Countries*. Industrial Development Report 2009, Vienna, UNIDO.

World Bank, 2008, *Reshaping Economic Geography*. World Development Report 2009, Washington DC, World Bank.

World Bank, 2009, *Awakening Africa's Sleeping Giant. Prospects for Commercial Agriculture in the Guinea Savannah Zone and Beyond*. FAO and the World Bank, Washington.

countries, focusing on agricultural and non-agricultural linkages to understand agriculture's role in development.

Ethiopia and Ghana were chosen to highlight the variation that exists across Africa. Ethiopia is landlocked, poor in natural resources other than agricultural land and in the early stages of economic transformation. Since large-scale, commercial agriculture is scarce, few doubt that smallholders must lead agricultural growth. In contrast, Ghana is a well-situated coastal country that benefits from trade with Europe. Due mainly to recent improvements in infrastructure, it has the potential to develop export-oriented services. Ghana thus fits well with the agro-pessimist's argument that many African countries should focus on the development of export-oriented industry rather than agriculture.

ETHIOPIA With only 15% of the population living in urban areas, Ethiopia remains one of the least urbanized and most agrarian economies in the world (Figure 1). Besides coffee, sesame seed, animal hides and a few other agricultural primary products, the domestic food economy remains insulated from global markets by high transport and marketing costs. Agriculture is dominated by locally grown non-internationally traded staples such as teff, sorghum and barley. The country has no mineral resources and exported goods and services only account for around 13% of GDP, much lower than most African countries. The main industrial imports are fuel, fertilizer and other chemical products.

Agro-optimists cite the closed nature of Ethiopia's economy as a basis for strong agricultural linkage effects. As Ethiopia is insulated from global markets and dominated by agriculture and a large rural population, when agricultural growth raises the income of farmers they spend disproportionately on domestically produced non-agricultural goods and services. Thus there are strong consumption linkages between the two sectors. A closed economy also implies that food prices fall with increased agricultural production. Therefore when growth is led by agricultural productivity, farmers who adopt high productivity farming practices benefit from the growth, while rural net buyers and urban households benefit from lower food prices, and so increase their demand for both agricultural and non-agricultural goods. Lower food prices benefit non-agricultural sectors by lowering labour costs. While this classical linkage effect has been weakened in many developing countries due to globalization, it remains dominant in Ethiopia.

Agro-optimists believe that spatial factors in Ethiopia, a large country with diverse agro-ecological conditions, are the other main cause of strong agricultural linkage effects. Domestic trade is active and markets have benefited from recent improvements in road infrastructure and market policies. When increases in local agricultural production are used to meet demand in various locations, opportunities are created for the development of agriculture-related services such as transformation, storage and trade. In this way Ethiopia has experienced rapid growth in its domestic service economy, as occurred in India during its green revolution (Hazell and Haggblade, 1991).

Since Ethiopia is labour abundant, agro-pessimists point to opportunities to develop a labour-intensive and export-oriented manufacturing sector (Ellis and Harris, 2004; Ellis, 2005). However, for a land-locked, large country like Ethiopia, high domestic and international transportation costs significantly discount the low labour cost advantage for such development opportunity. Moreover, low labour cost is only one of the many prerequisites for the development of a manufacturing sector. Agricultural growth would also be necessary to support an increased food demand, since urban populations in small towns consume mainly locally produced agricultural products as they are cheaper than imports. Although Addis Ababa's urban economy appears able to develop independently of agricultural development, it is already one of Africa's largest cities and any benefits from manufacturing development will be diminished by the increased costs of living in such a city. Urban costs of living would be further increased if more food is imported, and thus the traditional link between urban wages and agricultural productivity remains (World Bank, 2007).

GOVERNMENTS OF MANY AFRICAN COUNTRIES ARE DISAGREEING WITH THE AGRO-PESSIMISTS AND ARE PUTTING THE FOCUS BACK ONTO AGRICULTURE FOR DEVELOPMENT.

Agriculture's linkage effects are an important guide for Ethiopia's Agricultural Development Led Industrialization (ADLI) strategy, which aims "to bring about a structural transformation in the productivity of the peasant agriculture and to streamline and reconstruct the manufacturing sector, so that it makes extensive use of the country's natural resources and manpower" (GOE, 1993). The strategy has been central to the government's development programme since 1993, undergoing fine-tuning as understanding has deepened on the role of agriculture in growth, as reflected in Ethiopia's 2005 Plan for Accelerated and Sustained Development to End Poverty (MoFED, 2005).

Despite the challenges of a large and rapid population growth and limited and deteriorating land resources, recent investments and policy decisions have had a strong impact on agricultural development. Since 2004 Ethiopian agriculture, as well as its whole economy, has witnessed the most rapid and sustained growth in the country's history, with an average annual growth rate of around 10% from 2005 to 2009.

Agriculture accounts for 40% of Ethiopia's economy and so its direct contribution to economic growth is large. However, agriculture also indirectly contributes to non-agricultural sectors. This can be demonstrated through the application of the Computable General Equilibrium (CGE) model for Ethiopia (Dorosh and Thurlow, 2010). Our simulation suggested that a slowdown in agricultural growth would negatively affect growth in non-agricultural sectors, 70% of the observed decline in overall growth being directly due to agriculture, while the rest is through agriculture's linkages that negatively affect other sectors. In contrast, when non-agricultural growth slows, only 20% of the decline in overall growth is due to linkage effects between non-agricultural and agricultural sectors. It therefore seems reasonable to conclude that agricultural growth has directly contributed to Ethiopia's recent period of strong economic growth, while indirectly stimulating growth in the non-agricultural sector through its linkage effects.

The poverty reduction effect of agricultural growth on the role of agriculture is less controversial. Our CGE analysis suggested that a slowdown in agricultural growth results in a much higher poverty rate than a similar slow down in non-agricultural growth (Figure 2).

GHANA Ghana has long been regarded as one of sub-Saharan Africa's "star performers" (Coulombe and Wodon, 2007). In the past three decades, Ghana has had positive per capita GDP growth every year, a highly unusual phenomenon globally (Breisinger and Diao, 2008). Sustained economic growth has not only raised per capita income to more than US$1,000 in recent years, but also helped the country reduce poverty rates from 51.7% in 1991/92 to 28.5% in 2005/6 (GSS, 2007). Ghana's growth rate further accelerated after 2006, so expectations are for Ghana to achieve the Millennium Development Goal 1 by halving the 1990s poverty rate before 2015 (Kolavalli et al, 2011).

Sustained growth has been accompanied by rapid urbanization. In 2009, around half of the population lived in urban areas (Figure 3). However, progress in urban industrial development has been slow and government efforts to promote it have largely failed. Following independence, a state-led development strategy created several large industrial enterprises, but few have survived in today's liberalization and globalization era. Agriculture's share in the economy has been replaced by informal services, while manufacturing growth has stagnated. Therefore, the key issue for Ghana is not whether it should develop its industrial sector, but that it must determine the role of agriculture in such development. This potential role is examined below through an analysis of agro-processing.

The pink shading represents urban peripheries. Here road conditions are similar to urban areas but population density is lower. The situation is similar for medium size cities and their peripheries, represented by the green colours.

Ghana's agro-processing industries account for 60% to 70% of its manufacturing, which as a whole accounts for less than 10% of GDP (Breisinger et al., 2009). They are concentrated in basic foods, textiles, furniture and other wood products. In general, the domestic agricultural sector is able to provide the necessary input materials.

While traditional food processing – often small-scale and informal – has long existed in Ghana, the consumption of foods processed by formal manufacturing industries has rapidly increased, spurred by rising incomes, urbanization and concomitant increase in the opportunity cost of women's time (Reardon, 2009). This increases market opportunities to develop agro-processing to meet local demand without necessitating the level of competitiveness required for export. However, the increased domestic demand for processed foods is often met by imported goods, a consequence in part of trade liberalization and globalization (Robinson and Kolavalli, 2010). This trend has not only challenged the development of Ghana's agro-processing industries but also threatens its agricultural sector, which loses out to imported processed substitutes.

Ghana's agro-processing situation was analysed by Robinson and Kolavalli (2010). They conducted a case study on tomato processing which suggested that production capacity is not a constraint for development. Indeed, Ghana has sufficient tomato processing capacity to meet domestic demand; in fact it is significantly underutilized. However, consumers are substituting imported paste for domestically produced fresh tomato, and smallholder tomato production has stalled. The industry fails because the productivity and quality of fresh tomato is too low to ensure a regular supply of tomatoes at competitive prices. Thus, whether African countries should develop agro-processing as the first step of industrialization or not, actually depends on agricultural productivity.

The tomato example is not unique, with increased demand for processed food, the failure of large-scale food processing is common in Africa. For example, except for cocoa, Ghana has recently become a net agricultural importer of primary foods and agro-processing products. About one-third of foreign exchange earned by exporting cocoa is needed to pay for imported foods and other agricultural products, many of which could be produced locally. Rice and chicken imports have surged after Ghana liberalized its domestic market by lowering its import barriers in the early 1990s; and processed vegetable and oilseed product imports have increased on average by 20% per year in the past decade. Ghana currently imports around 60% to 70% of rice and 80% to 90% of poultry meat consumed domestically. Local producers cannot compete with imports and either reduce production or switch to other products (MoFA, 2009). Considering demand for most imported foods is highly income-elastic, such imports are expected to continue to rise in the absence of competitive domestic production.

AGRICULTURAL TRANSFORMATION IS DIFFICULT AND THERE IS NO MAGIC BULLET. PERHAPS THE MOST IMPORTANT LESSON FROM ASIA'S GREEN REVOLUTION IS THAT SERIOUS AND LONG-TERM POLITICAL COMMITMENT IS ESSENTIAL.

This suggests that competitive agriculture can make inroads into manufacturing through food processing to meet domestic demand. However, when agricultural productivity stalls, development of some manufacturing such as agro-processing is constrained because cheaper imported processed foods can substitute primary non-traded agricultural products. Indeed, with globalization, agriculture and manufacturing have become increasingly integrated and interdependent, suggesting that African manufacturing should be a mechanism to transform rather than replace agriculture.

OPPORTUNITIES AND CHALLENGES FOR AFRICAN AGRICULTURAL TRANSFORMATION

Governments of many African countries are disagreeing with the agro-pessimists and are putting the focus back onto agriculture for development. For example, in reply to Collier's agro-pessimism, an African Minister commented, "This is not the

Africa we live in" (Brooks, 2010). Moreover, more than thirty Sub-Saharan African countries had signed Comprehensive African Agriculture Development Program compacts by 2010 and agreed to increase agricultural public spending to 10% of total government budget. Ten countries have already reached the target. Donors and development partners also believe in the important role of agriculture in Africa for economic growth, food security and poverty reduction, suggesting we seize the moment and make an irreversible commitment to agriculture as a driver of growth (Brooks, 2010).

Encouragingly, Africa's agricultural performance, aggregate and per capita, has improved recently. But most output growth has come from area expansion, rather than growth in total factor productivity or increased use of modern inputs. So, although growth has recently improved, productivity and international competitiveness lag.

Africa's challenges are considerable and in some respects unique in world history, with simultaneous economic and demographic transitions occurring in the context of globalization and climate change (Naylor 2011). The most taxing issue is how to develop agriculture and what role governments should take to support private sector-led agricultural transformation. As de Janvry (2010) stated, it is a serious mistake to believe that we know what should be done. While African countries have much to learn from Asia's example, Africa's green revolution will be different given the initial conditions and the current global environment.

Agricultural transformation is difficult and there is no magic bullet. Perhaps the most important lesson from Asia's green revolution is that serious and long-term political commitment is essential. Benefits will only follow if the effort is sufficiently large, concerted and sustained (de Janvry, 2010). Asian experiences emphasize the active and consistent engagement of government in technology adoption and in the transformation process (for example, Asian countries often have a strong and effective public extension system). With the expectation that the private sector will lead African agribusiness development, the way a country's public sector engages in improving agricultural productivity will be different. What the government should do, how it should do it and the requirements for institutional revolution in smallholder farming, are challenging issues.

There is more than one path to effective transformation. Mistakes are inevitable and lessons must be learned. Success in export-oriented agriculture, private sector-led in particular, has been encouraging. However, it remains to be seen whether the lessons from export agriculture can be transferred to traditional agricultural components that primarily cater to the domestic market. More effort is needed to critically examine experiences at sectoral and country levels and to build appropriate frameworks for analysis. ■

REFERENCES

Barrett, C. B., Carter M. R., and Timmer C. P. 2010. A Century-Long Perspective on Agricultural Development. American Journal of Agricultural Economics 92 (2): 447-468.

Breisinger, C., Diao X., and Thurlow J. 2009. Modeling Growth Options and Structural Change to Reach Middle Income Country Status: The Case of Ghana. Economic Modeling 26: 514-525.

Breisinger C. and Diao X. 2008. Economic Transformation in Theory and Practice: What are the Messages for Africa? IFPRI Discussion Paper 797. Washington, DC: International Food Policy Research Institute.

Brooks, K. 2010. African agriculture: What do we not know? What do we need to know? Dinner address delivered at the International Speaker Series and Banquet, AAEA Annual Meeting, Denver, Colorado, July 26, 2010.

Coulombe, H. and Wodon Q. 2007. Poverty, livelihoods, and access to basic services in Ghana (draft). Paper presented at the Ghana CEM technical review growth workshop, Accra, Ghana, May 2-3, 2007. Available at http://siteresources.worldbank.org/INTGHANA/Resources/CEM_poverty.pdf

De Janvry, A. 2010. Agriculture for development: new paradigm and options for success. Agricultural Economics 41: 17–36.

Dercon, S. 2009. "Rural poverty: Old challenges in new contexts." The World Bank Research Observer 24(1): 1-28.

Dorosh, P. and Thurlow J. 2009. Implications of accelerated agricultural growth on household incomes and poverty in Ethiopia: A general equilibrium analysis. ESSP Discussion Paper No. 002. Washington, DC: International Food Policy Research Institute.

Ellis, F., and Harris N. 2004. New thinking about urban and rural development. Keynote paper for Department for International Development sustainable development retreat, University of Surrey, Guildford, July 13, 2004.

Ellis, F. 2005. Small farm, livelihood diversification, and rural-urban transitions: Strategic issues in Sub-Saharan Africa. In The future of small farms, Research Workshop Proceedings, 135–149, Wye, U.K., June 26–29, 2005. Jointly organized by the International Food Policy Research.

GSS (Ghana Statistical Services). 2007. Ghana Living Standard Survey Round 5 (GLSS5). Accra, Ghana: Ghana Statistical Office.

GOE (Government of Ethiopia). 1993. An economic development strategy for Ethiopia; A comprehensive guidance and a development strategy for the future. Ministry of Planning and Economic Development, Addis Ababa, September 1993.

Gollin, D., Parente S., and Rogerson R. 2002. The role of agriculture in development. American Economic Review 92 (2): 160–164.

Hayami, Y., and Ruttan V. 1985. Agricultural Development: An International Perspective. Baltimore: Johns Hopkins University Press.

Hazell, P. and Haggblade S. 1991. Rural-urban growth linkages in India. Indian Journal of Agricultural Economics 46(4): 512-529.

Hirschman, A. 1958. The Strategy of Economic Development. USA: Yale University Press.

Johnston, D. G., and Mellor J. W. 1961. The role of agriculture in economic development. American Economic Review 51(4): 566–593.

Kolavalli, S., Robinson E., Diao X., Alpuerto V., Folledo R., Slavova M., Ngeleza G., and Asante F. 2011. Economic transformation in Ghana. Paper presented in International conference on understanding economic transformation in Sub-Saharan Africa, Accra, Ghana, May 10-11,.

MoFA (Ministry of Food and Agriculture). 2009. Ghana National Rice Development Strategy (NRDS) (draft). Available at http://www.jica.go.jp/activities/issues/agricultul/pdf/NRDS_gha_E.pdf. Last accessed May 27, 2011.

MoFED (Ministry of Finance and Economic Development). 2005. Ethiopia: The Millennium Development Goals (MDGs) Needs Assessment Synthesis Report. Addis Ababa: MoFED.

Naylor, R. 2011. Expanding the boundaries of agricultural development. Food Security 3:233-251.

Reardon, T., Barrett C.B., Berdegue J.A., and Swinnen J.F.M. 2009. Agrifood industry transformation and small farmers in developing countries. World Development 37(11): 1717-1727

Robinson, E.J.Z., and Kolavalli S. 2010. The Case of Tomato in Ghana: Processing. GSSP Working Paper No. 21. Washington, DC: IFPRI.

World Bank. 2007. World Development Report 2008: Agriculture for Development. Washington, DC: The World Bank.

MASSIVE INVESTMENT IN AGRICULTURE IN AFRICA: A DRIVER OF DEVELOPMENT?

Vatché Papazian,
Agence française de développement, Paris, France

Reinvesting in agriculture, doubling its production, and making the production accessible are all facets of the current consensus to ensure food security for nine billion people in 2050. This consensus, however, does not provide answers to the following questions: Which investments should be prioritised? Which farm types should be encouraged (small, large, family or managerial)? How should agriculture be developed to achieve food security and to respond to the current environmental, political and social challenges?

Since a 2008 report from GRAIN (an NGO supporting small farmers), and the announcement from the Korean company Daewoo Logistics of their appropriation of 1.7 million hectares in Madagascar, land grabbing by foreign investors in Southern countries has been a hot topic in the media. The issue has motivated various initiatives that have produced numerous publications[1] to record this trend, and has led to the emergence of consensual principles to regulate these practices: the acknowledgement of rights, the safeguarding of family farming and the encouragement of investor responsibility. Some advocacy NGOs consider that these statements of principles are insufficient and impractical, with Oxfam even calling for a moratorium. In fact, several Southern countries make major agricultural land resources available to investors, with low or even no costs, in the hope of boosting production, modernizing agriculture and ensuring food security (Figure 1). Smallholder and civil societies are rightly concerned by these initiatives, which seem to privilege "big investors" (private, national and/or foreign), but not family farms.

How do these asymmetrical positions and interests manifest themselves, in the particular case of the Office du Niger (ON)? Due to the desirability of the region of irrigable land in the interior delta of the Niger River, Mali is often cited in regards to the subject of land grabbing, as is also the case for Sudan and Madagascar. Located in the heart of Mali, in the dead delta of the Niger River in the south Sahelian zone, the ON zone extends over more than two million hectares, and is one of the oldest hydro agricultural development projects in West Africa, initiated in 1932 by the French engineer Emile Bélime. It today consists of aging infrastructure that is spread across an area of 100,000 ha[2] of irrigated land, which is allocated to 35,000 family farms[3]. According to the 2008 Master Plan for the development of the ON zone (SDDZON), an additional 200,000 ha could be irrigated by means of costly investments (€6,000 to €10,000 per hectare) for smallholder rice farming, but there are insufficient public resources to realize

1. World Bank, FAO, IFAD, EU, Bilateral cooperation, NGOs, think tanks

2. 55,000 ha developed in 1950

3. 420,000 people live here, on farms of 2 to 3 ha that utilize few tools. They pay a €100/ha/year fee to the ON.

this plan. Beyond the potential offered by this land, both the developed and undeveloped areas, the rest of the ON zone is owned by the nation and governed by customary rules, despite the decree that entrusts its management to the ON. These areas are predominantly used for pastoral farming (nomadic herders), along with small-scale agriculture (millet/sorghum in rainfed culture) and fisheries.

Mali justifies the use of new sources and modes of financing (private investors, sovereign wealth funds) to enable the production of cereals, oilseeds and biofuels for the domestic market. In recent years, different types of private initiatives and sovereign funds have placed demands on an area of nearly 800,000 ha in the ON zone (i.e. eight to ten times greater than the area developed since the 1930s), and been granted provisional allocations.

Who are the beneficiaries? According to the ON, the 800,000 ha that have been subject to allocation requests are divided between private national interests and foreign investors (400,000 ha each). The demands made by national private interests (90%) are for small areas (38% of applicants request between one to five ha), while around ten of them demand areas larger than 500 ha. Foreign investors (who request between 500 to 100,000 ha) include sovereign wealth funds, parapublic companies (MALiBYA[4]), Chinese companies, bilateral and regional institutions, sugar agro-industries such as SOSUMAR, etc. The Malian government is responsible for and/or participates in Malibya and Sosumar projects, theoretically keeping a certain amount of control over the situation. The lack of transparency in these major land allocations is a cause of aggravation in the national debate on land tenure. Farmer organizations and the civil society, that have become very involved in the subject, organized an international conference ("Stop the land grab") in Nyeleni (Sélingué) in November 2011, with the help of, among others, Via Campesina.

But how advanced is the actual development? To move from a provisional allocation to a lease (an ordinary one of 30 years or an emphyteutic lease of 50 years), private developers are required to submit studies of feasibility and environmental and social impacts, and a financing plan. At the present time, almost 250,000 ha have been leased to major national and foreign private companies that were able to conduct the required studies and provide operational financing plans.

National investors have been less successful, finding it difficult to gain access to water and finance. Thus, a significant proportion of potential beneficiaries are unable to obtain a lease: over the past year, the ON has cancelled 200,000 ha of provisional allocations, particularly those to small national private companies. In each instance, there are disputes that occur at the start of the process, which can escalate, between investors and the current land occupants or users who recognize a different legitimacy regarding their custodianship of the land.

FAMILY FARMS AND PRIVATE INVESTORS: DIVERGENT PERSPECTIVES

The land tenure situation thus created is characterized by conflictual bipolarization. At one end of the scale are family farms situated on small state-developed irrigated sites, to whom have been granted either annual contracts (revocable on non-payment of the €100/ha/year fee) or exploitation permits without legal validity; the production costs for these farms curtail their ability to develop or invest. On the other side are new investors who have access to land and water, through state contracts or through leases with a legal validity granted by the ON, who must ensure the delivery of final developments and pay a €3/ha contribution to the maintenance of the primary water network. This system seems to function as if the status of potential investors opens up a whole range of possibilities to those that can mobilize the necessary finance for development (€8000/ha on average), possibilities that are closed to family farms who are confined to a poor/"recipient" status, which disqualifies them in advance from being recognized as potential "private" investors.

In this region of Mali, land value is measured in terms of the water that it can expect to receive. In the absence of adequate data, land allocated by the ON or the state is not made according to water availability. In this context, it is likely that the dry season will see the emergence of a new dividing line in relation to

4. A Libyan company that was allocated 100,000 ha. The content of the agreement between Mali and Malibya is not known.

FIGURE1 AFRICA, BETWEEN INVESTMENT AND LAND GRABBING

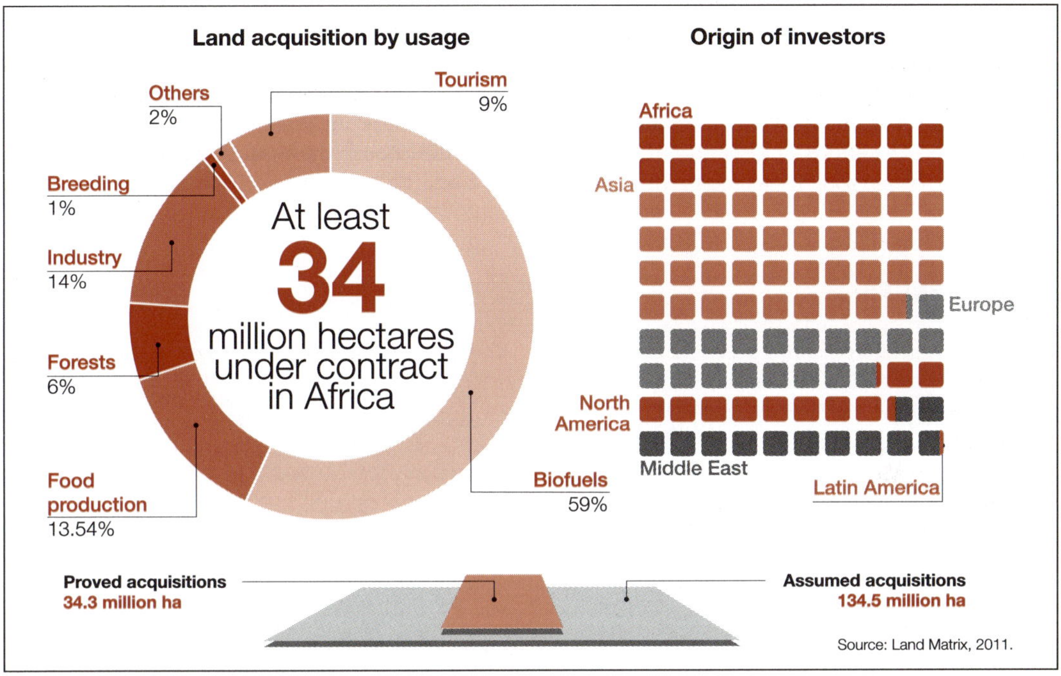

Data relating to land transactions remain difficult to access. The latest estimates, however, suggest that there is more than 34 million hectares of farmland under contract in Africa, but the actual total could be four times higher. These lands are dedicated to a variety of usages - biofuels, forest plantations, etc. - but only 13% provide for human consumption. Investments are therefore significant but are only marginally beneficial to the needs of African smallholder agriculture.

access to water resources, there will be those that are "first served" (situated at the head of the network, such as the sugar industries and major water users), those with contracts stipulating a preferential water supply, and then the family farms that are located at the end of the network. For 70 years, the ON's public regulation capacity and its specific skills as a manager of water and land on behalf of the state, has remained the system in place that determines the fate of the area, both in terms of development and for the prevention of potential social conflicts, which are caused by an asymmetrical access to land and water.

WHAT INCLUSIVE DECISIONS ON LAND AND WATER COULD DEVELOP INVESTMENT IN AGRICULTURE?

Mali has taken action on the ON area in awareness of these issues, commencing a rereading of the Decree of land management[5], developing a hydraulic master plan and through the adjustment of developmental models of "modern" and economically viable family farms. Any effort to bring together these two "engines of development" (family farms and private investment) is beneficial for easing potential conflicts.

In this way, it would be possible to overcome the controversy regarding massive land allocations or the facilities granted to large private investors, while raising and addressing fundamental questions: What are the potential models for the inclusive development of the ON area, and by extension for agriculture

5. Supported by technical and financial partners of Mali (World Bank, France, Netherlands, European Union) and in consultation with farmers.

in Mali? How can the area's development objectives be aligned with social peace?

Mali, like other countries in the sub-region, is caught up in a land dilemma that raises issues of resource governance and on the multiplicity of rights operating in the area. Appropriate land use choices reflect the efforts of a country to reduce its asymmetries, in line with the Voluntary Guidelines on Responsible Governance of Tenure of Land and Other Natural Resources, proposed by the Committee for Food Security (CFS) in October 2011. Other stakeholders concerned (investors, civil society, technical and financial partners) will have no other option but to "choose sides" and to adjust their actions or support in the direction of these efforts, or to look elsewhere for land... ■

REFERENCES

FAO, 2011, High Level Panel Experts, Study on land tenure and international investments in agriculture.

GRAIN NGO, 2008, "Seized: The 2008 landgrab for food and financial security", available on: http://www.grain.org/, last viewed on 29 November 2011.

IIED, FAO and FIDA, 2009, "Land Grab or Opportunity? Agricultural investment and international land deals in Africa", Lorenzo Cotula, Sonja Vermeulen, Rebeca Leonard and James Keeley.

OXFAM, 2001, "Terres et Pouvoirs: le scandale grandissant qui entoure la nouvelle vague d'investissements fonciers", September 2011, available on: http://www.oxfamfrance.org/, last viewed on 29 November 2011.

Office du Niger, "*Situation des attributions de terres en bail dans la zone Office du Niger*", May 2010, internal report.

CHAPTER 5

Rural reform has shaped the Chinese development model by contributing to the low cost of unskilled labour, and thus to the competitiveness of the export industry. There are still many challenges to overcome in the modernization of the countryside, while at the same time agriculture must feed a growing and increasingly wealthy population.

LAND, PEASANTS AND MIGRANTS: THE HEART OF CHINA'S DEVELOPMENT

Maëlys de la Rupelle,
Centre de Recherche en Economie du Développement, FUNDP, Namur, Belgique

Li Shi,
Beijing Normal University, Beijing, Chine

Thomas Vendryes,
ADIS, Université Paris XI, Paris, France

During the last three decades, China has continued to develop at a fast pace. Between 1981 and 2001, the number of individuals living on less than one dollar per day decreased by 400 million (Chen and Ravallion, 2004). A major part of this success is due to rural reform. While the country was experiencing rapid growth, agricultural production and land rights underwent deep changes. At the same time, rural outmigration was strictly regulated, a policy which confers on rural-urban migrants a very peculiar economic and social role, and shapes the pattern of Chinese development, contributing to the low labour cost of unskilled workers, and therefore to the attractiveness of export oriented industry. However, many rural challenges are left unsolved due to such constraints on rural-urban migration, which can reduce remittance flows and limit the modernization of rural areas; while at the same time agriculture is required to feed a population that is becoming larger and richer.

In this chapter we describe agriculture and rural land in China, and investigate their links with China's development path. In the first section we highlight the main characteristics of agriculture and rural areas throughout China's history. The second section examines rural outmigration, which is of crucial importance for urbanization, industrialization and agricultural development, and the relevant regulatory institutions. The third section addresses the rising land tensions that are caused by the conflict between productivity requirements and ecological concerns. The fourth section focuses on the challenges faced by agricultural policies, such as dramatic pollution levels, water shortage and food safety.

CHINA'S AGRICULTURE: GEOGRAPHY, HISTORY AND INSTITUTIONS

After hundreds of years of progressive evolution, Chinese agriculture was subjected to a radical experiment in collectivism from 1949. Partial decollectivization began in 1978, which created a unique institutional system.

THE TRADITIONAL CHINESE COUNTRYSIDE Ever since the Han dynasty (206 BC to AD 220), imperial China seemed to fit quite well with Wittfogel's concept of "Oriental despotism" (Wittfogel, 1957): an agricultural society organized by a bureaucratic state.

This could appear paradoxical, as China's geographical and climatic features are not especially favourable to agriculture. Only 15% of the total area is arable (Naughton, 2007), most of which is in the East where the land is relatively low and exposed to the moist "monsoon" air from Asian seas. Despite these difficult conditions, China has remained one of the most densely populated regions in the world. In the East, today's population density reaches 260 people per square metre, six times the world average (Naughton, 2007). An agricultural worker farms only five *mu* on average, equivalent to one third of a hectare (Maddison, 2007). This astonishingly intensive exploitation of a scarce amount of farmland necessitates massive investment in infrastructure and the use of highly labour-intensive techniques (Naughton, 2007).

Chinese agriculture is generally considered to have benefited from the early emergence of a centralized bureaucracy (Maddison, 2007). Indeed, as agriculture constituted the main basis for taxation, it was strongly supported by officials through, for example, hydraulic projects and the diffusion of new seeds and new techniques. China's overall trade integration also led to a highly commercialized form of agriculture (Maddison, 2007).

However, Chinese farmers were not significantly richer than their counterparts in other regions of the world (Maddison, 2007) and they achieved subsistence incomes at the cost of strenuous labour. In a situation of widespread misery, peasants were often indebted to local landlords and, as a result, land ownership tended to be concentrated (Naughton, 2007).

THE MAOIST EXPERIENCE It is in this context, and after decades of foreign and civil wars, that the Chinese Communist Party (CCP) founded the People's Republic in 1949, with rural rather than urban support strategies.

The first policies were quite favourable to rural citizens. An extensive land reform redistributed half of the total land area (Lardy, 1987) from landlords to 300 million poor peasants, and farmers were given extended private rights over their plots. Political stabilization also allowed agricultural production to rise significantly (Lardy, 1987).

But this increase was short-lived and the CCP, under the impulse of Mao Zedong, then decided to radically accelerate the collectivization of agriculture and the establishment of a soviet-style economic plan, to transfer a maximum amount of resources from agriculture into industrial sectors (Lardy, 1987). The apex of this policy was reached during the Great Leap Forward (1958-1961). China's entire society and economy were organized into People's Communes, gathering on average 5,500 households. In rural areas, commune regulations covered the minutiae of the daily lives of citizens, to the point where even the ownership of private kitchen utensils was forbidden. The results were catastrophic, and the consequent famine killed between 25 and 30 million people (Naughton, 2007).

Major adjustments to the system were then decided. Thus, the "60 articles on Agriculture", approved by the CCP in 1961 and which prevailed up to 1978, established small-scale collectivization centred around production brigades of 45-50 households, with additional private plots and private markets tolerated (Lardy, 1987).

These reforms stabilized the situation in rural areas, but maintained the general bias in favour of industry and urban areas. In 1978, heavy industry, with very high capital investment but relatively low employment, was then significantly more developed than other sectors including light industry, services and, above all, agriculture. Workers in state-owned plants and urban companies had a disproportionately favourable status. At the beginning of the reform era, the dual structure of China's economy and society was quite pronounced.

REFORMS AND THE "HOUSEHOLD RESPONSIBILITY SYSTEM" In 1978, when Deng Xiaoping became the head of state, China was still an overwhelmingly rural (82% of the population) and agricultural (70% of the workforce) country (Naughton, 2007). But rural real incomes were very low, especially compared to urban ones.

Policies to increase rural incomes were therefore adopted (Vendryes, 2010): state prices for agricultural products and the supply of agricultural inputs were raised, while the scope of the tolerated private economy was extended.

The main evolution, however, came from local experimentation: some localities granted land use rights to rural households for specific plots, thus conceiving the "Household responsibility system" (Vendryes, 2010) which rapidly spread across China and continues to prevail today. This system was officially acknowledged in the 1982 Constitution, in the 1986 Land Management Law and in a series of No.1 Documents from 1982 to 1986[1]. A second wave of laws and regulations on this issue began with the revision of the Land Management Law in 1998, followed by the 2002 Law on Land Contract in Rural Areas and the 2007 Property Law, and paralleled by a series of No.1 Documents from 2004 to 2011.

The system's principle is simple: agricultural land remains the property of local collective authorities and rural households are granted land use rights. Officially, these individual land use rights are extremely long-term and secure. Their duration,

1. The SCNPC and CCPCC Documents are issued on a yearly basis by central authorities in order to give political and administrative directions. The "No.1 Document", the first published in a given year, indicates the political priority for the year to come.

which was initially set at 15 years, was extended to 30 years in 1997 and, since the 2007 Property Law, has now become perpetual. The only official limitation is that farmers cannot use their plots for non-agricultural purposes.

Two related institutional reforms also took place in the late 1990s and early 2000s (Vendryes, 2010). First, the 1998 revised Organic Law on Villagers' Committees organized village-level voting, with elected authorities put in charge of local affairs, especially local land management. Second, between 1998 and 2002 agricultural taxation was put under strict control through the "Tax-for-fee reform", before being completely suppressed in 2005 to 2006. These reforms were meant to reduce the power of local cadres and their incentive to manipulate land allocation.

However, despite these improvements, administrative plot seizures and the collective reallocation of land take place frequently, while land markets have remained very limited. Thus, individual land use rights are much less secure and long-term than the central government intended.

At the same time, since the beginning of the reforms in 1978, agro-industry policies were modified significantly to support agricultural development, giving a renewed and dramatic drive to the "green revolution" that began in the mid-1970s (Naughton, 2007): irrigation works were significantly extended, the use of chemicals and fertilizers considerably developed, while mechanization and the use of new hybrid high-yielding seeds spread throughout China. By the end of the 1990s, the foundations for the modernization of Chinese agriculture, i.e. its insertion into a comprehensive agro-industrial sector, had been deeply laid.

"LEAVING AGRICULTURE BUT NOT THE COUNTRYSIDE": THE CONSTRAINTS ON CHINESE RURAL CITIZENS

From 1978, China promoted rural industrialization, encouraging peasants to "leave the soil, but not the countryside" and applying strict limitations on rural outmigration to large cities. However, the development of township and village enterprises in rural areas slackened sharply after the mid 1990s, putting migration policies into question.

CONSTRAINTS ON RURAL MIGRATION The dismantling of the rural People's Communes lifted the constraints on the labour allocation of rural citizens, allowing the emergence of rural outmigration. In 1989 the number of migrant workers stood at 30 million, by 2009 this figure had risen to 145 million (Chan, forthcoming) – a quarter of the rural labour force[2]. Predictably, most of these migration flows were directed towards urban areas in China's more industrialized and developed Eastern regions (Chan, forthcoming).

2. Migrants mainly travel towards the Eastern region, Guangdong being the province that receives the largest number of migrants. As migration flows are triggered by unbalanced development, most originate from poor Western areas and head for the urban and developed East. However, significant rural-urban migration also occurs within regions.

However, most migrations have not implied permanent settlement. First, within a year, many migrants change location or return home to participate in household business. Second, they usually return to their village of origin after a few years of migration. Indeed, a set of policies was put in place to prevent rural people from definitively settling in urban areas, constraining urbanization in a unique way. While the Chinese urbanization rate has risen from 29% in 1978 to 49.7% in 2010 (National Bureau of Statistics of China, 2010), it remains far lower than other developing countries at the same development level (Chang and Brada, 2006). This "paradox of China's growing under-urbanization" (Chang and Brada, 2006), owes much to the Chinese household registration system, or *hukou*. Since the 1950s, the *hukou* has classified the Chinese population according to two dimensions: activity and residence (Naughton, 2007). Regarding activity, households can have either agricultural or non-agricultural status. Agricultural households are entitled to usage rights on plots of land while non-agricultural households have subsidized access to urban benefits regarding education, healthcare, employment and housing. The place of residence defines where these benefits can be obtained. In the 1960s and 1970s, the *hukou* was a cornerstone of collectivization and economic planning, and it made individual migration impossible. After 1978, the dismantling of rural People's Communes (1984), along with the implementation of a temporary residence permit and an identity card, conferred a legal status on rural workers in cities (1984-1985), while the end of grain rationing in 1992 progressively alleviated constraints on migration. Yet, the remaining rights and duties associated with the *hukou* render urban life extremely difficult and costly for rural workers. In addition, it is still extremely difficult for an agricultural worker to obtain an urban *hukou* and applications are subject to quotas at the municipal level.

CHINA'S REGULATORY INSTITUTIONS FOR LAND AND MIGRATION SHAPE ITS DEVELOPMENT: RURAL PEOPLE TEND TO MIGRATE LESS AND FOR SHORTER PERIODS, HOUSEHOLDS ARE DETERRED FROM LEAVING AGRICULTURE.

Besides *hukou*, the land rights further limit rural outmigration. Since farmers cannot sell their land or use it as collateral, it is hard for those without capital to leave agriculture and start an off-farm activity. Furthermore, outmigrants are at risk of losing some of their plots because collective authorities have the right to redistribute the land allocations of absent farmers to the villagers currently present, or even to seize it for their own benefit. This insecurity over land rights hinders migration, as migrant workers fear the loss of their land and thus tend to shorten their time away (de la Rupelle, Deng, Li and Vendryes, 2009).

DEVELOPMENT CONSEQUENCES China's regulatory institutions for land and migration shape its development: rural people tend to migrate less and for shorter periods, households are deterred from leaving agriculture, and the pace of urbanization is relatively slow. The transfer of the labour force from the agricultural sector to modern urban activities is allowed to take place without uncontrolled urbanization or a sudden drop in agricultural production. The *hukou* also segments the labour market, as agricultural *hukou* holders working in cities do not have the same rights as their urban counterparts. This has contributed to limiting wage growth and to maintaining the competitiveness of export-orientated sectors (Chan, forthcoming).

Regarding rural issues, the system of collective land ownership, with periodic redistributions, plays an insurance role for migrant workers who only have limited access to formal social insurance or credit markets (Murphy, 2002). Although agriculture is not very profitable, access to land guarantees at least a subsistence income for rural households. For example, in the aftermath of the financial crisis, many redundant migrant workers simply went back to their rural homes (Kong, Meng and Zhang, 2009).

These institutions, through restricting outmigration and the rights of rural migrants, also have negative consequences on rural development. First, as rural industrialization has slowed since the mid-1990s and also because rural credit markets remain underdeveloped, migration constitutes a crucial route to enable access to off-farm jobs and capital (Murphy, 2002). Second, returning migrants become key agents for rural development, bringing not only money but also skills and networks (Murphy, 2002). Third, because rural outmigrants are limited to "3-D" jobs (dirty, dangerous and difficult) and cannot access urban public schooling, existing and potential migrants tend to reduce their investment in children's education (Chan, forthcoming). Fourth, insecurity over land rights hinders productive investment in agriculture (Vendryes, 2010). Finally, and perhaps most importantly, conflicts over agricultural land and the "second-class" status of rural populations have aroused deep feelings of injustice and social unrest (Ding, 2007; Chan, forthcoming) Such tensions relate to a much larger problem faced by agriculture: the multifaceted pressure on land.

TENSIONS OVER AGRICULTURAL LAND

China's very rapid development has dramatically raised the stakes regarding land usage. Industrial and commercial pressures, as well as ecological concerns, are challenging agricultural development, even though agricultural production is a key political objective.

PRESSURE FROM BOTH SIDES: INDUSTRIAL DEVELOPMENT AND ECOLOGICAL CONCERNS

It is well known that for the last 30 years China has experienced extremely rapid economic development and that a huge proportion of the workforce has moved out of agriculture (70% of workers were engaged in the primary sector in 1978, compared to significantly less than 50% since 2004 (Naughton, 2007).

This industrial and urban development has generated a huge demand for land, constituting an important pressure on agricultural land, especially on the fringes of the most developed urban areas. The development of satellite imagery in the 1990s has allowed land use changes to be closely monitored, unsurprisingly revealing that the built-up area has significantly increased. Estimates are that China's total area of built-up land has grown by a quarter between 1986 and 2000 (Deng, Huang, Rozelle and Uchida, 2006).

This trend has tended to reduce the arable land area and has created intense social tensions because farmers do not have the right to change the land use of their allocated agricultural plots themselves. Only the state has the formal right to own non-agricultural land, and so a change in land use implies a change in land ownership (Ding, 2007). At the root of the issue is the fact that the compensation offered to farmers whose plots are seized is often far below the benefits associated with the land use change (Ding, 2007), a phenomenon that is responsible for the recent multiplication of conflicts in rural areas.

The legal response to the conflicts over land seizures and fair compensation has so far been quite insufficient. The 1998 Land Management Law requires such seizures to be in the "public interest" (article 2), and the 2002 Law on Land Contract in Rural Areas compels authorities to give a "fair compensation" to expropriated farmers (article 16). Yet, both laws are not sufficiently specific, leaving the door wide open for many types of local misuse. This is today probably the single most important problem for the social and economic sustainability of China's agricultural development.

On the opposite side, growing ecological concerns have encouraged authorities to take measures to protect soils, especially to mitigate agricultural land loss due to erosion and desertification. Data show that such concerns have a very real basis, for example, over the period 1991-2002, it is estimated that land resources degradation led to a net loss in agricultural land of 19.66 million hectares (Qu, Kuyvenhoven, Shi, Heerink, forthcoming).

From the beginning of the 1990s onwards, important policies have been implemented to address this phenomenon. The Water and Soil Conservation Law of 1991 prohibited the exploitation of land with a slope above 25 degrees. This was followed in 1998 and 1999 by two extremely ambitious programmes: the Natural Forest Protection Programme (NFPP) and the Sloped Land Conversion Programme (SLCP) (Qu *et al.*, forthcoming). The NFPP has led to an increase of 1.72 million hectares of forest, while the SLCP aims to reconvert all sloping lands[3] and help farmers move to more sustainable production processes (Qu *et al.*, forthcoming). Contrary to the NFPP, household participation in the SLCP is voluntary, and farmers who convert plots to forest or grassland receive significant compensation. State compensation offered for the conversion of agricultural land to forest or farmland, mainly through the SLCP, is also significant, high enough to attract the voluntary consent of over 32 million households between the years 1999 to 2006 (Qu *et al.*, forthcoming). Results have been satisfying: nine million hectares had been converted under the SLCP by the end of 2006. The ecological results have also been encouraging, erosion has been reduced, soil quality has increased and water resources have been protected.

However, this success comes at the cost of an increasing pressure on agricultural soils, as plots are reconverted to forest and grassland.

3. With a slope higher than 15° in Northwestern China and 25° elsewhere.

DECREASING FARMLAND AND STATE REACTION Industrial and urban development, along with programmes motivated by ecological concerns, are together significantly increasing the pressure on land availability. As land availability decreases, the value of land for non-agricultural, commercial or industrial use rises.

Studies based on the aforementioned satellite imagery concur with the notion that agricultural land is facing strong pressures. Even though the overall area of farmland may not have decreased, this is only because agricultural areas have been shifted away from urban activities and moved onto less productive soils. In fact, studies (Deng *et al.*, 2006) estimate that losses in the quality of agricultural land have almost been offset by increases in quantity during the period between 1986 and 2000. However, many of the more recent studies (Vendryes, 2010) show that the decrease in agricultural land area has accelerated since the beginning of the 21st century, despite the exploitation of new resources in marginal agricultural regions. It is therefore likely that future losses in land quality will no longer be compensated by increases in quantity.

Due to concerns regarding this issue, in the mid-1990s Chinese authorities initiated a policy of farmland protection to limit the decrease in exploited arable land acreage (Vendryes, 2010). In 1994, the Regulations on the Protection of Basic Farmland aimed to identify "basic farmland", i.e. land that was of above average quality, and to protect it by making land-use changes subject to the approval of provincial or national authorities. In 1998, the scope of these regulations was extended to include all farmland under the revised Land Management Law, which set a compulsory objective of no net decrease in farmland. Finally, in 2008 the Ministry of Land and Resources set a minimum limit for arable land in China of 1.8 billion *mu* (120 million hectares)[4]. However, in terms of actual land resources, China is already edging dangerously close to this "red line".

CURRENT AGRICULTURAL PRODUCTION AND THE CHALLENGES AHEAD

AGRICULTURAL PRODUCTION TODAY During the nineteenth century, China experienced famines and food crises at such a regular frequency that geographers in the 1920s named it the "land of famine" (Mallory, 1926). One century later, China seems far removed from such a description. Nevertheless, food provision and grain self-sufficiency in particular remain national priorities. The Chinese government has constantly stressed that the country should be independent of world markets as regards feeding its population, an intention that was underlined once again in March 2011, when China's Vice Premier Hui Liangyu stated that China was to uphold its policy of food self-sufficiency[5]. As a result, there is still much emphasis on grain production, as illustrated in the following graph.

4. Ministry of Land and Resources of the PRC, *Communiqués on Land and Resources of China 2007*, available online at: http://www.mlr.gov.cn/mlrenglish/communique/2007/

5. Xinhua, "China upholds policy of food self-sufficiency", *China Daily*, 26 March 2011

With this political willpower, China has just about been able to achieve a limited degree of self-sufficiency (United States International Trade Commission, 2011). However, following WTO accession in 2001, China was obliged to open up its agricultural market to a certain extent (United States International Trade Commission, 2011), and since then the level of liberalization has increased significantly. Trading has, unsurprisingly, followed the law of comparative advantage: the two following graphs show that China, which has plenty of labour but limited land resources, has imported land-intensive products (e.g. soybean) and exported labour-intensive ones (e.g. prepared food, fruits and vegetables). The only exception to this general pattern is grain, the trade of which is strictly controlled due to the political pressure for self-sufficiency. Grain imports therefore remain highly dependent on government policies. For instance, in 2010 China imported an unprecedented two million tons of US maize and wheat and then a further 116,000 metric tons of maize in March 2011, causing an 11% increase in world prices[6]. This rise in imports may be partially explained by low international prices, but other factors such as China's growing hog industry are also responsible.

FIGURE 1 EXPORT OF AGRICULTURAL LABOUR

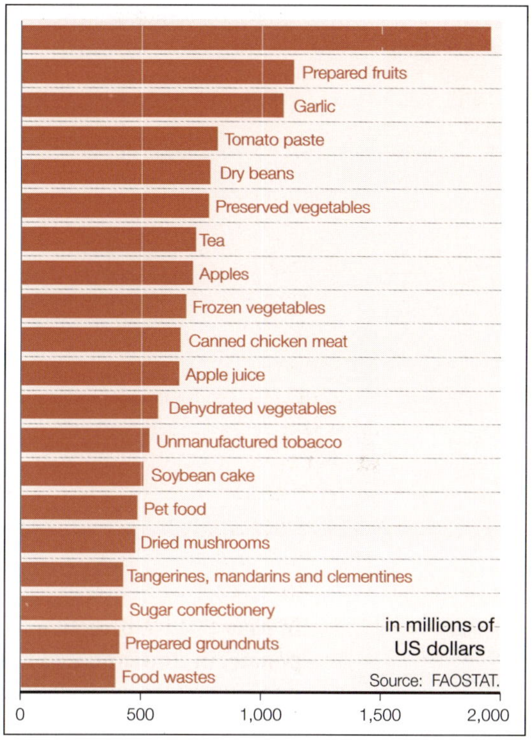

Chinese agricultural exports consist primarily of labour-intensive products, such as: preserves, prepared food and processed products.

CHALLENGES AHEAD Agriculture has to address the conflicting objectives of productivity increase and environmental protection. As China's income per capita continues to grow, food consumption patterns are changing and the demand for meat and fish is rising. In 1980, average meat consumption per capita was 20 kg per year (Fuller, Tuan and Wailes, 2002), by the early 2000s this figure had reached 52.4 kg per capita[7].

This shift in demand also increases the need for grain and water for livestock, challenging the goal of food self-sufficiency and increasing the already intense pressure on water resources (Naughton, 2007).

Raising agricultural productivity is also problematic since a heavy reliance on fertilizers already exists: China uses 35% of the world's nitrogen fertilizers (Greenpeace, 2010), contributing towards worrying levels of pollution. This was illustrated by the first national pollution census in 2010, which took agricultural effluents and landfill discharges into account for the first time, thus revealing that rural

6. "China buys corn after long break", *The Wall Street Journal*, 19 March 2011.

7. In 2002, the meat consumption per capita was 13 kg in Sub-Saharan Africa, 27.8 kg in Asia (excluding Middle East), 82 kg in Brazil, and 124.8 kg in the US (Source: FAO).

FIGURE 2 **REDUCING LAND USE THROUGH IMPORTS**

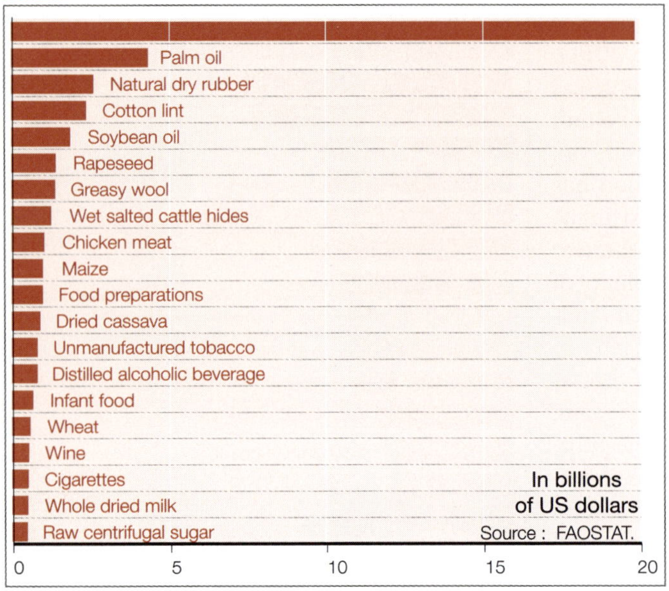

The main agricultural products imported by China are those that require the mobilization of extensive farmland, including: soybean and its products, palm oil, cotton, rapeseed and livestock products. Indeed, nationally owned land is reserved for cereal production.

contamination[8] was much higher than previously estimated. For example, the census showed that water pollution had more than doubled and that agriculture was responsible for 67% of phosphorus and 57% of nitrogen discharges nationally[9]. Water shortages and pollution incidents raise questions about the enforcement of environmental regulations, the tensions between different government levels and, more generally, the fight against corruption.

Similar issues arise in the food industry, which in recent years has seen a multiplication of contamination scandals (Ni and Zeng, 2009). The most dramatic example occurred in 2008 when melamine added to milk caused 300,000 children to be taken ill, resulting in six deaths. Stricter policies were adopted in the aftermath: in May 2011 a directive from the Supreme Court sanctioned the use of the death penalty in contaminated food cases involving fatalities[10]. Nevertheless, more efficient quality checks are called for, especially as official restrictions on media coverage further complicate the disclosure of health scandals[11].

China's increased demand for land and water is certain to raise many geopolitical questions, with worldwide implications for the future. There are at least three main issues that are likely to feature on the global agenda: China's extension of its land area by buying agricultural land abroad, particularly in Africa; the potential impact on world water resources due to China's increasing meat consumption (in 2008 China was already the number one importer of virtual water (Chenoweth, 2008), i.e. water used in the production of imported food and goods); and finally, China's growing demand for grain, if translated into imports, may cause world prices to increase, with knock-on effects on the socio-economic equilibrium worldwide. ■

8. "China says water pollution doubles official figures", *China Daily*, 10 February 2010

9. Jonathan Watts, "Chinese farms cause more pollution than factories, says official survey", *The Guardian*, 9 February 2010

10. Barbara Demick, "China wrestles with food safety problems", *Los Angeles Times*, 26 June 2011

11. Bei Fangshuo, "Food won't be safe unless journalists are", *The Economic Observer*, 22 June 2011, reports about plans of the Ministry of Health to "blacklist journalists who mislead the public about health concerns", as the Department of Food Safety and the Ministry of Health were organizing a conference on food additives.

REFERENCES

CHAN Kam Wing, "China, Internal Migration", *in* Immanuel Ness and Peter Bellwood (eds.), *The Encyclopedia of Global Migration*, Blackwell Publishing, forthcoming.

CHANG Gene Hsin and Brada Josef C., 2006, "The Paradox of China's Growing Under-Urbanization", in *Economic Systems* 30.1, p. 24–40.

CHEN Shaohua and RAVALLION Martin, 2004, "How Have the World's Poorest Fared since the Early 1980s?", in *World Bank Research Observer* 19.2, p. 141–169

CHENOWETH Jonathan, "Looming water crisis simply a management problem", *New Scientist,* 28 August, 2008, pp. 28–32.

DE LA RUPELLE Maëlys, DENG Quheng, LI Shi and VENDRYES Thomas, 2009, *Land Rights Insecurity and Temporary Migration in Rural China*, IZA Discussion Papers 4668, Institute for the Study of Labor (IZA).

DENG Xiangzheng JIKUN Huang, ROZELLE Scott and UCHIDA Emi, 2006, "Cultivated land conversion and potential agricultural productivity in China", *Land Use Policy*, 23, p. 372-384.

DING Chengri, 2007, "Policy and Praxis of Land Acquisition in China", *Land Use Policy* 24.1, p. 1 –13.

FULLER Frank, TUAN Francis and WAILES Eric, 2002, "Rising Demand for Meat: Who Will feed China's Hogs?", China's Food and Agriculture: Issues for the 21st Century/AIB 775. Economic Research Service/USDA.

GREENPEACE, August 2010, "The Real Cost of Nitrogen Fertilizer in China", *Greenpeace China Report*.

KONG Sherry Tao, MENG Xin and ZHANG Dandan, 2009, "Impact of economic slowdown on migrant "workers, Chapter 12 in R. Garnaut, Lina Song & Wing Thye Woo (eds.), *China's New Place in a World in Crisis Economic, Geopolitical and Environmental Dimensions.* Acton, A.C.T.: ANU E Press.

LARDY Nicholas, 1987, "Economic Recovery and the 1st Five-Year Plan," *in* Roderick MacFarquhar and John Fairbank (eds.), *The Cambridge History of China, Volume 14, The People's Republic, Part 1: The Emergence of Revolutionary China, 1949-1965*, Cambridge University Press.

MADDISON Angus, 2007, *Chinese Economic Performance in the Long Run*, OECD.

MALLORY Walter H., 1926, *China: Land of Famine*. New York: American Geographical Society.

MURPHY Rachel, 2002, *How Migrant Labor is Changing Rural China*, Cambridge University Press.

NATIONAL BUREAU OF STATISTICS OF CHINA, 2010, *China Population Census*. Beijing: National Bureau of Statistics of China.

NAUGHTON Barry, 2007, *The Chinese Economy: Transitions and Growth*, MIT Press.

NI Hong-Gang, ZENG Hui, 2009, "Law enforcement is key to China's food safety", *Environmental Pollution*, vol. 157, n° 7, p. 1990-1992.

QU Futian, KUYVENHOVEN Arie, SHI Xiaoping, HEERINK Nico, "Sustainable natural resource use in rural China: Recent trends and policies", *China Economic Review*, forthcoming.

UNITED STATES INTERNATIONAL TRADE COMMISSION, March 2011, *China's Agricultural Trade: Competitive Conditions and Effects on U.S. Exports*, Investigation No. 332-518 USITC Publication 4219.

VENDRYES Thomas, 2010, "Land Rights in Rural China: Reforms, successes and shortcomings", *China Perspectives* 2010/4.

WITTFOGEL Karl, 1957, *Oriental Despotism; a Comparative Study of Total Power*, Yale University Press.

CHANGING PATTERNS OF FOOD CONSUMPTION IN INDIA AND CHINA

Zhang-Yue Zhou,
School of Business, James Cook University, Townsville, Australia
Hong-Bo Liu,
School of Business, James Cook University, Townsville, Australia
Vasant P. Gandhi,
Indian Institute of Management, Ahmedabad, Gujarat, India

Feeding the populations of India and China is a challenge that has long captivated the interest of thinkers and researchers worldwide. In 2010, India and China's populations reached 1.19 and 1.34 billion respectively, accounting for around two-fifths of the global population. Expectations are that these figures will continue to rise, exceeding a combined total of 3 billion by 2050. The fact that both countries have largely managed to feed their populations with domestically produced food represents a remarkable achievement. However, recent economic development has generated rapid transformations in food consumption patterns that have raised many questions, including: In what way are consumption patterns changing? Will such changes result in a greater demand for food and for higher quality food? Can India and China continue to produce enough food? What impact will changing consumption patterns have on global markets and food security?

CHANGING DIETARY PATTERNS

Transformation of the proportion of income spent on food

Food expenditure has increased substantially in both countries since the 1970s. However, the proportion of income spent on food, known as the Engel coefficient, is declining, as shown in Figure 1 China's Engel coefficients are lower than India's, while in both countries the coefficients are higher in urban areas compared to rural ones. To look at this in financial terms, in 2005 rural residents in China and India spent US$125 and US$84 per capita per year respectively on food, while the corresponding expenditures of urban residents were US$356 and US$122. This shows that Chinese rural and urban residents have higher food expenditures compared to their Indian counterparts, but the consumption gap between urban and rural dwellers is wider in China.

Changing food baskets

Changes in food basket composition are a significant component of food pattern trends in China and India. The share of cereals, for example, has shown a marked decrease: in rural India, 54.4% of food expenditure went on cereals in 1970/71, reducing to 32.7% in 2004/05 (the corresponding figures for urban India are 35.6% and 23.7% respectively). While the data do not allow an exact comparison, a similar trend nevertheless emerges in China: in 1978, the share of

FIGURE 1 THE BENEFITS OF URBAN LIVING

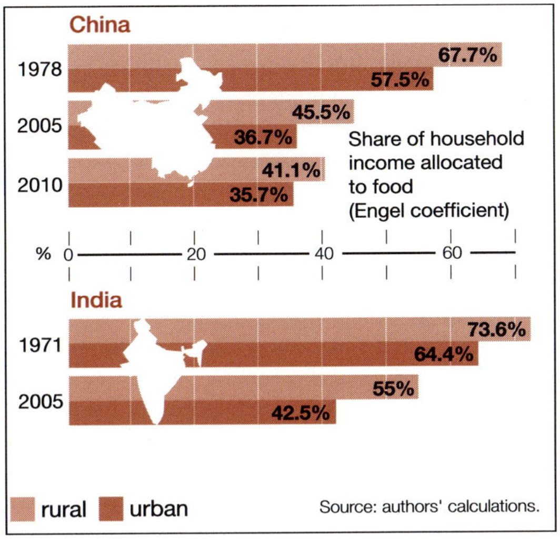

In India, as in China, the share of income allocated to food expenditure has steadily declined, reflecting the overall increase in the average incomes of the two populations. However, the proportion of food expenditure remains higher for rural citizens, who have lower incomes, than for urban dwellers. In India, this gap between the two populations is increasing.

FIGURE 2 LESS GRAIN, MORE PROTEIN

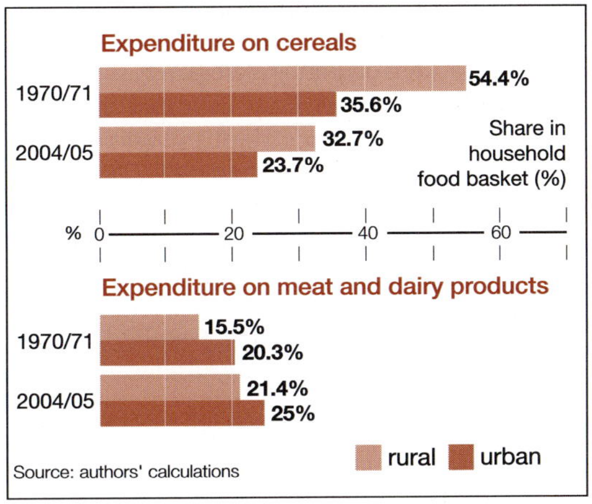

The evolution of the Indian food basket is representative of many emerging countries: an increase in living standards results in lower cereal consumption, even in rural areas, and an increased consumption of animal protein.

cereals in food expenditure in China was 65% for rural and 22% for urban consumers, by 2009 the share for urban consumers had declined to 8% (rural data not available).

In contrast, both countries have experienced a rise in the consumption of animal products, which can once again be seen by analysing trends in food basket composition: rural India's expenditure on animal products has risen from 15.5% in 1970/71 to 21.4% in 2004/5; while this figure has risen from 20.3% to 25% for urban residents, now exceeding the share of cereals in their food baskets (see Figure 2). Similar data are unavailable for China but it is highly likely that a parallel change has occurred.

Due to a unique prevalence of lacto vegetarianism, India's consumption of animal products is dominated by milk, while meat consumption is low. The reverse is true in China, with pork consumption being particularly significant. However, these historical preferences are starting to change as diets diversify. For example, chicken, goat and – to a lesser extent – beef consumption have risen considerably in India, and in urban China the consumption of fish and dairy products is rapidly increasing.

There is much regional variation in animal product consumption within India and China, caused by many factors including differences in income, food habits and product availability. In China for example, the uneven distribution of certain ethnic minority groups across the country is a major reason for regional variation in the consumption of products such as dairy foods, beef and mutton.

THE DRIVERS OF DIETARY CHANGE

Many factors are driving dietary change in India and China; of these, rising income is the most important. As observed throughout the world, as incomes rise, consumers reduce their consumption of foods of plant origin, switching to more expensive foods, particularly animal products.

Urbanisation is another important influence since urban residents tend to consume less cereals and more animal products than rural populations. Furthermore, this factor is becoming increasingly significant: China's urbanisation level has increased from 20% in 1980 to 50% in 2010 and during the same period India's urbanisation has increased from 23% to 30%.

The combination of increased income and urbanisation leads to major lifestyle changes, which are themselves another cause of dietary change. For example, eating out, taking holidays and convenience foods are becoming increasingly popular in China; while in India a young and more affluent workforce is going out to restaurants and opting for convenience foods more often.

Greater opportunities for cultural exchange are also a factor for change, with Indian and Chinese citizens becoming increasingly influenced by worldwide cultures, stimulating interests in untraditional foods – beef for example in India, or dairy products in China, the consumption of which is increasing in both cases.

A final influence to mention is improvements in food production and marketing, such as the introduction of modern cold chain transportation and storage facilities. Such developments are improving food availability throughout these countries. In addition, supermarkets, which are emerging particularly rapidly in China, are also having a key effect.

FUTURE OUTLOOK

The drivers of dietary change will continue to exert their influence for the foreseeable future. While rural consumption of animal products in both countries will lag far behind their urban counterparts for some time, it is likely that rural trends will emulate urban ones to some extent. It can also be anticipated that marketing improvements will make different food products more widely available throughout China and India, although regional differences will remain. In addition to dietary changes, consumers in both countries will increasingly demand safer and higher quality products.

Research conducted by the authors, along with the findings of other studies (e.g. Bhalla, Hazell and Kerr, 1999; Zhou and Tian, 2005) suggest that India and China should be able to produce sufficient amounts of cereals to meet domestic demand in the near future; however, their ability to produce enough animal products is questionable. Both governments have emphasised the importance of self-sufficiency in cereal supply. The main objective of India's 2007 National Food Security Mission (NSFM) is "to increase production and productivity of wheat, rice and pulses on a sustainable basis so as to ensure food security of the country" (Government of India, 2007). In 2008, the Chinese government published its 'Outlines of medium- and long-term national grain security plan (2008-2020)', in which the government set a target to achieve a self-sufficiency rate of no lower than 95% for its major staple cereal supply (self-sufficient for rice and wheat, largely self-sufficient for maize) (Government of China, 2008).

Both governments have implemented policies to boost cereal production. In India, the NFSM has pledged to enhance the production of rice, wheat and pulses by a total of 20 million tonnes by the end of the Eleventh Five-Year Plan, i.e. 2012. In China, due to strengthened government support, grain production has increased successively over the past seven years, reaching 546 million tonnes in 2010.

It thus appears that, although facing major challenges due to urbanisation, industrialisation, resource depletion and environmental degradation, India and China both intend to meet their rising demands for cereals largely through domestic production for the near future. Long-term cereal self-sufficiency for India however looks more problematic due to declining resources (limited land and increasing water scarcity) together with an increasing population (Gandhi, Zhou and Mullen, 2004). China, in contrast, is in a stronger position to remain self-sufficient for cereals in the long term, particularly because its population will start to decline by around 2035. Therefore, China is unlikely to have a major direct effect on global markets, particularly for rice and wheat.

In contrast, both countries will have major difficulties in producing enough animal products to meet growing domestic needs. In India, the animal industry would need an unprecedented boost to avoid large gaps between supply and demand that are likely to emerge for many types of meat (except chicken) as well as eggs and dairy products (Gandhi and

Zhou, 2010). For China, according to Wang (2010), a shortage of meat and eggs may not emerge until 2025 or 2030, triggering the need to import. For dairy products, however, China has already experienced shortages and resorted to importation; a phenomenon that is likely to continue.

Wang's projection is based on the assumption that China has sufficient feedstuffs for domestic animal production. However, this may not be the case. In 2010, China imported 55 million tonnes of soybean (85% of which was for animal feed) and 1.5 million tonnes of maize. Three million tonnes of corn meal were also imported in 2010. In future, it remains to be seen whether China will need to import animal feeds, animal products or livestock and to what extent. However, whatever action China and India take, they are sure to have a major impact on the world food situation. ■

REFERENCES

BHALLA, G.S., HAZELL, P. and KERR, J., 1999, 'Prospects for India's cereal supply and demand to 2020', *Food, Agriculture and Environment: Discussion Paper No. 29*, IFPRI, Washington, DC.

GANDHI, V.P. and ZHOU, Z.Y., 2010, 'Rising demand for livestock products in India: nature, patterns and implications', *Australasian Agribusiness Review*, Vol. 18, pp. 103 - 135.

GANDHI, V.P., ZHOU, Z.Y. and MULLEN, J., 2004, 'Wheat economy of India: Will demand be a constraint or supply?' *Economic and Political Weekly*, 39(43) October 23.

GOVERNMENT OF CHINA, 2008, 'Outlines of medium- and long-term national grain security plan (2008-2020)', available on www.gov.cn, 13 November 2008, accessed 15 December 2008.

GOVERNMENT OF INDIA, 2007, 'National food security mission: operational guidelines', Ministry of Agriculture, New Delhi.

WANG, J.M., 2010, 'Project report on sustainability of China's livestock industries', Institute of Agricultural Economics, Chinese Academy of Agricultural Sciences, Beijing.

ZHOU, Z.Y. and TIAN, W.M., 2005, *Grains in China: Foodgrains, Feedgrains and World Trade*, Ashgate, Aldershot.

CHAPTER 6

As we enter the Anthropocene, the question of the impact of human activities, particularly agriculture, on the planetary system becomes more acute. What are the "planetary boundaries" and how can we ensure "a safe space for humanity"?

GLOBAL CONDITIONS FOR THE FUTURE OF AGRICULTURE IN THE "ANTHROPOCENE"

Uno Svedin,
Stockholm Resilience Centre, Stockholm University, Stockholm Sweden

Current global agriculture and food systems are characterized by a failure to deliver food security for all, as illustrated for example by the effects of the present price volatility, which is severely restricting access to food for the poor, while simultaneously placing an increased strain on the environment. Addressing these challenges may seem daunting for a sector that has long been considered as conventional and has only recently made a return to the centre stage. However, it is essential that such challenges are vigorously tackled given the tasks ahead for the future agricultural system. There is thus a wide and growing consensus that a major challenge in the 21st century will be to produce sufficient food of good quality, as well as sufficient amounts of feed, fibre and fuel – all in a sustainable manner. As discussed in this paper, modern agricultural practices have contributed to the dawning of the "Anthropocene", a new historical era characterized by the fact that human intervention, and not the Earth's natural processes, is now the main driving force behind systemic changes (such as climate change, biodiversity loss, etc.). Accepting that agriculture is at the centre of human interference in natural cycles (phosphorous, nitrogen, water, etc.), this paper points to the need for deep reforms in agriculture and food systems over the coming years if we are ever to live within planetary boundaries (Rockström et al., 2009), by acting inside a "safe operating space for humanity".

LIVING IN A NEW GEOLOGICAL ERA: THE ANTHROPOCENE

It is important to consider a relatively long timeline when discussing the possible future conditions for agriculture – and thus humankind – i.e. from the perspective of the next 50 to 100 years. At the present moment it seems as if we are positioned in a new and rapidly changing situation for humanity at a planetary level. Within this new period, human actions appear to run the risk of taking us outside of the frames of familiarity that we have learnt to live within during the last ten thousand years (often referred to as the Holocene). During this recent era, very favourable and unusually stable conditions have prevailed, allowing humans to develop and thrive. One expression of this stability is, for example, that it has been possible to develop agriculture as a major civilization innovation.

However, strong signs now suggest that we are heading into a situation where we are about to exit this established frame of stable conditions. Nobel Prize winner Paul Crutzen and others such as Will Steffen (Crutzen and Stoermer, 2000; Crutzen, 2002; Steffen et al., 2004, 2007, 2011; Richardson et al., 2009) have drawn attention to a simultaneous set of accelerating and interconnected phenomena in terms of global environmental change during the last half century, such as the degree of growth in the world's population, the rate of climate change, etc. The origins of these processes seem to correspond with the beginning of the industrial revolution, around the middle of the 18th century, but show a clear and remarkable escalation after World War II. This suggests that we are facing major and ongoing – but sometimes insufficiently acknowledged – structural change. In qualitative terms, especially as this period has already been in existence for at least 50 years, it seems that we may now have sufficient understanding to decide that this new era warrants a new name, for which some scholars have already suggested the "Anthropocene". As noted above, the key characteristic of this era is that humankind is providing the dominant influence on the planet's mechanisms. This has never before been the case in the Earth's history.

ANTHROPOGENIC INFLUENCE – OR INTERFERENCE – HAS NOW REACHED SUCH A LEVEL THAT NONE OF THE NATURAL CYCLES – CARBON, NITROGEN, WATER, ETC. – ARE UNTOUCHED BY HUMANS.

Anthropogenic influence – or interference – has now reached such a level that none of the natural cycles – carbon, nitrogen, water, etc. – are untouched by humans. In most cases human behaviour interconnects with and impacts upon these cycles to such a degree that they cannot be regarded as independent of the human sphere. Thus humans act at the planetary level as a collective causal force, strongly setting the stage for most of the Earth's processes. However, this increased impact that humans are having on their environment has not yet been met by a corresponding level of responsibility (nor the institutional, technical or social capacity to face the task). The best example to illustrate this emerging situation is of course the climate change challenge. The Intergovernmental Panel on Climate Change (IPCC), which started its work at the end of the 1980s, has given a step-by-step assessment of the degree to which the Panel has considered human interference to be of distinct and visible character. This has provided very important input into the political discussions on if, and how, society should intervene through major actions in terms of

Figure 1 CLIMATE CHANGE HAS ALREADY STARTED

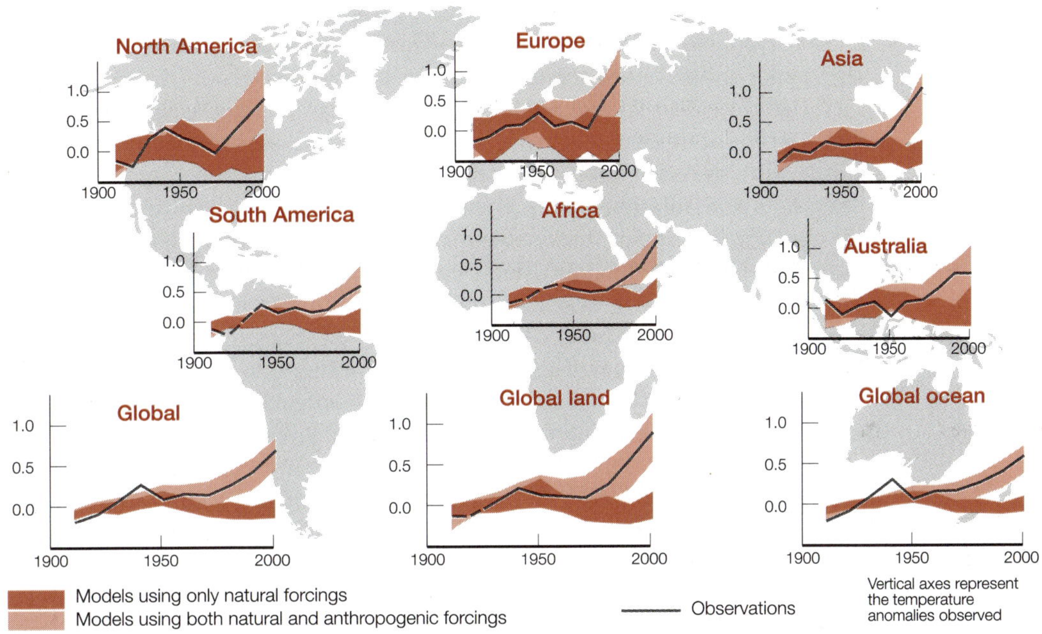

Source: *Climate Change 2007: The Physical Science Basis.* Contribution of Working group I to the Fourth Assessment Report of the Intergovernmental Panel on Climate Change, FAQ 9.2, Figure 1. Cambridge University Press

mitigation and adaptation measures. The decision facing society on whether or not to take action has been dependent on the level of consensus on the dominance of human influence as a cause of the observed climate change indications. If research shows that the human element accounts for a sufficiently strong impact, amongst the consolidated explanations, then this is strong motivation to strengthen counter-measures, especially considering the risks involved with continued emissions of greenhouse gases at the current level. The outcome is then in principle in the hands of humanity[1].

The latest (fourth) IPCC assessment in 2007 clearly showed that the influence of anthropogenic factors has now become clearly visible in the empirical data (see Figure 1). This IPCC diagram presents data on temperature change for the major regions of the world, while three graphs in the lower part of Figure 1 show different forms of global aggregates of the temperature anomaly from 1900 to 2000. The lower left figure provides the global average situation. The core interest is to reflect on how the results from two different types of climate simulation models compare

The work of the IPCC has found that the average global temperature has increased by 0.7°C between 1990 and 2000. Modelling has also revealed the impact of human influence on these changes - without human activity, temperature changes during this period would have been much less marked.

1.Recent developments in the international negotiations on climate change show that action is beginning, at many levels and through a wide array of means, although not at the scale that the IPCC indicates is necessary.

with the actual observations. The blue band shows models that only take natural forcing into account, while the pink curve also includes anthropogenic forcing. It is clearly visible that the results from the "combined" simulation models provide the most reasonable fit to the actual data. This relationship holds true for the aggregated data and also for each of the continental graphs.

It is therefore evident that not only does the anthropogenic aspect have to be taken into account, but also that from around 1950, the anthropogenic component of climate forcing has been significantly "boosted" (without allowing for this increase, the results of simulation models are unable to fit well with the observed temperature changes). It thus seems reasonable to conclude that we are indeed emerging into a new historical era, which justifiably earns the title "Anthropocene".

The need for decisive action by the global community to address this situation should thus no longer be in doubt. It is essential that any scenarios for 2025 and beyond should take human-induced climate forcing into account.

However, it is insufficient to focus only on climate change, since the dawn of the Anthropocene is not only characterized by the degree of human influence in one natural cycle – carbon – but in many others as well. It is also particularly important to understand how the influence differs between the various cycles. The planet is a connected and combined natural system, within which the different natural cycles interact. This interconnection can be illustrated by the transfer of material (dust, surface water runoff, etc.) from land to the oceans. These transfers are feeding biological ocean life, which in turn provides a feedback loop to climate since the biology of the sea influences its absorptive capacity of carbon dioxide. This in turn changes the climatic conditions, which influence precipitation patterns over land and sea. These interactions, at regional scales, drive feed-forward and feedback loops between different natural cycles that are complex in character and are only just beginning to be explored by researchers, leaving the possibility open for future surprises. What is sure, however, is that the human influence on these interactions is growing, including the impact from agricultural practices. Nitrogen leaching from Scandinavian agriculture into the Baltic Sea, causing eutrophication, provides one such example. Another is the role of agriculture and forestry in influencing the carbon cycle in terms of carbon dioxide capture, storage and release under the different conditions found throughout the world.

MECHANISMS AND RESOURCES HAVE BEEN ADDRESSED IN TERMS OF FINDING THE DEGREE TO WHICH "A SAFE SPACE FOR HUMANITY" COULD BE OUTLINED.

TOWARDS A SAFE SPACE FOR HUMAN OPERATION

We have looked at the major influence of humans at planetary and regional scales with regard to natural cycles. However, these impacts – and potential options for deliberate interference in the natural systems – raise questions about the potential restrictions on how far humankind can, and ought to, interfere with these natural cycles. Such considerations for a number of interrelated processes, mechanisms and resources have been addressed in terms of finding the degree to which "a safe

FIGURE 2. THE PLANETARY BOUNDARIES

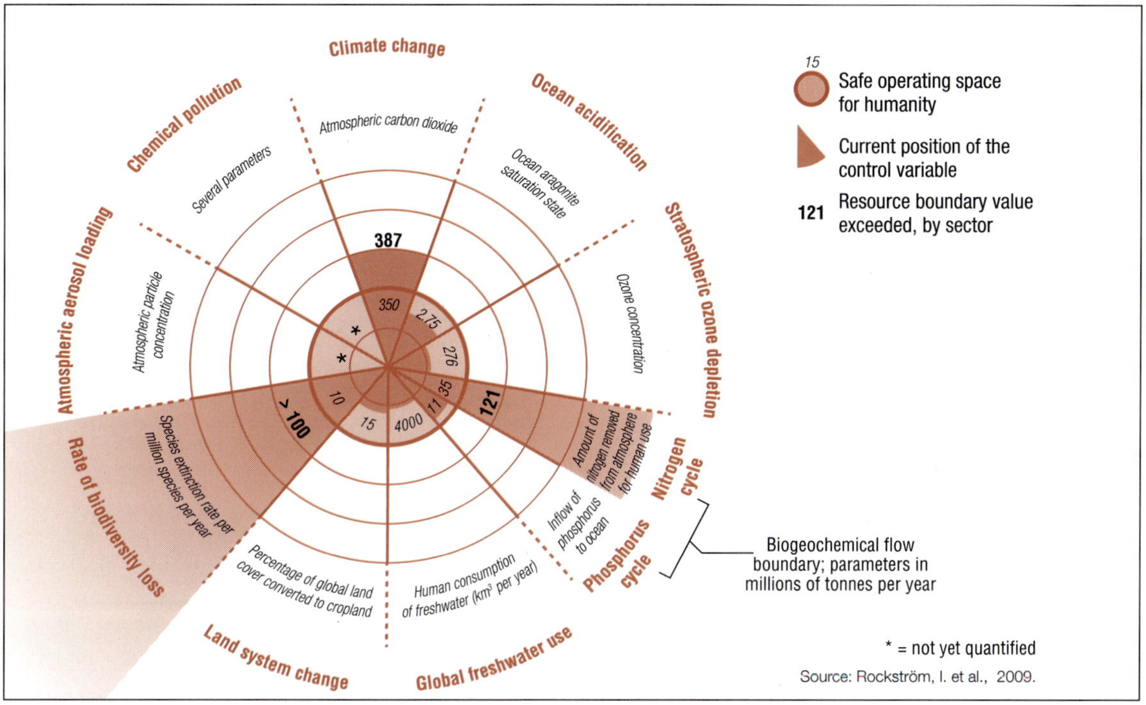

Source: Rockström, I. et al., 2009.

space for humanity" could be outlined; or in other words: to investigate what the safe boundaries of interference may be, within which the risk of the occurrence of unexpected and excessive systemic shifts remains limited. The reason for focusing attention on this issue is the potentially grave consequences that may emerge if such conditions are not respected. The causal nature of many of these processes and their impacts are not fully clear at the moment. However, these phenomena have an inherent potential to drastically change key conditions for humankind, e.g. to exert a major influence on vulnerability related to reduced food security and other safety hazards.

Like any ecosystem, the Earth has a significant, but not unlimited, capacity for resilience against natural or anthropogenic changes. Resource overexploitation and changes in global natural cycles, beyond certain thresholds, could lead to disastrous effects for humanity.

In the abovementioned 2009 Nature article (Rockström et al.), the authors (which included the author of this chapter) addressed the question of if, and how, it would be possible to identify "a safe operating space for humanity" considering the Earth's interconnected features and non-linear behaviour patterns. In this study nine connected issues were investigated – including climate change. Each investigation showed a detailed risk panorama (Figure 2). In three of the nine cases, humankind now already faces serious threats (i.e. approaching a global planetary boundary), which is shown in Figure 2 as red-coloured segments. The three problematic areas (i.e. that involve a high level of threat) are climate change, the nitrogen cycle and biodiversity loss.

FIGURE 3 AGRICULTURE IS AMONG THE MOST VULNERABLE SECTORS

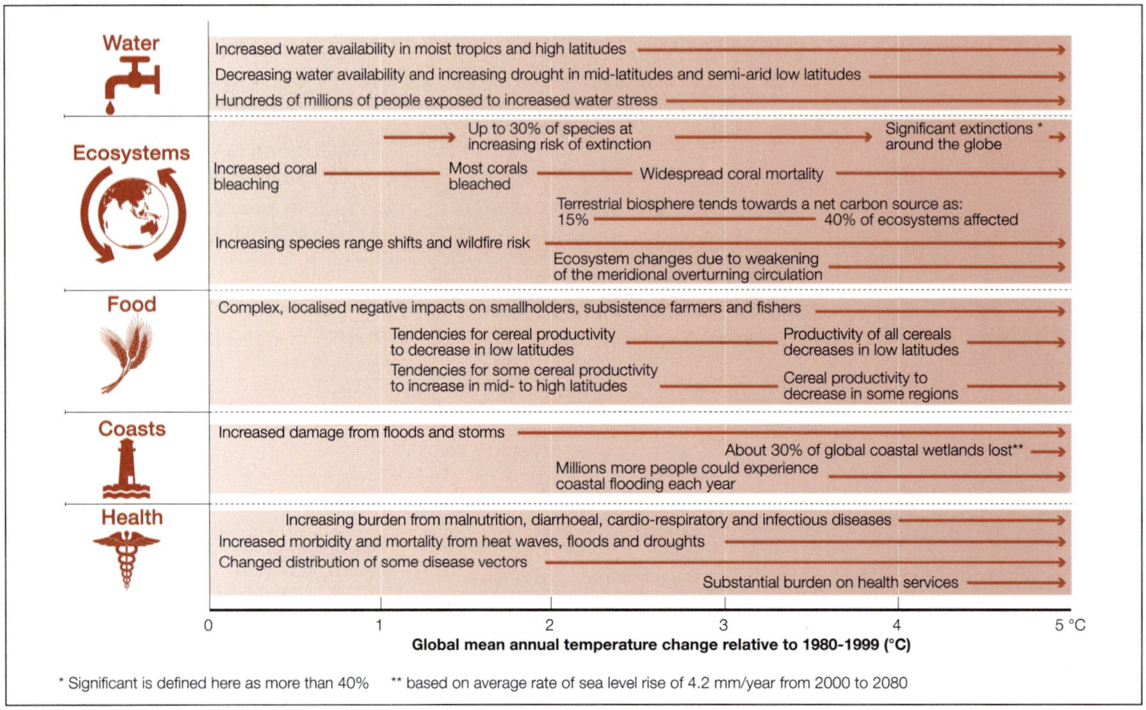

Several sectors, including agriculture and food production, will be profoundly affected by the impacts of climate change as soon as the temperature increase exceeds 1°C. Beyond 4°C, disruptions are likely to be profound and adaptation is difficult to imagine.

AGRICULTURE AND THE PLANETARY BOUNDARIES

The three highly problematic threat sectors all have direct linkages with agriculture. In addition, agriculture is involved in, and dependent upon, several of the other factors e.g. the phosphorus cycle, global freshwater use and land use – either directly or indirectly. Thus the world's capacity for food generation and the possibility for long term resilience in environmental services (related to long term food production and production of renewable resources) are under scrutiny with regard to the safe operating space. The connections between climate change and agriculture were addressed by the 2007 IPCC report, not least in the discussions around the risk panorama demonstrated in Figure 3 below. This was done for different sectors under various climate change assumptions (especially with regard to temperature increase).

We see that impacts worsen rapidly for temperature increases over 2 degrees (some impacts, such as coral bleaching for example, would be dire even before a 2 degree rise), while at even higher temperature increases there is a multiplication and intensification of the negative impacts. This observation holds true for the food sector, which would face issues such as a change in the productivity of agricultural soils even at relatively "low" temperature changes. The graph thus demonstrates that the "food sector" will need to develop a distinct adaptive capacity to cope with even "small" temperature changes. In addition, regional variations must be considered as they may provide even stronger shifts than global mean temperatures currently indicate.

Since discussions on the potential outlook for 2050 indicate that there is a considerable risk of experiencing a global temperature rise of at least 3 degrees in half a century, it is therefore clear that agriculture (including forestry) is severely threatened by climate change within the current century. Thus, this issue clearly constitutes one of the world's major challenges because it pertains to global food security.

Regional climate shifts will also cause shifts in ecological zones with related movements in patterns of crop preferences, but also changes in the spectrum of biological hazards that affect crop and animal species. The erratic nature of these phenomena – and the diminished stability of conditions, including eruptions of unforeseen natural hazards – seems likely to increase. The social capacity to withstand these new conditions has not yet been developed sufficiently. Thus, the combined socio-ecological system shows a strong indication that its "resilience" will be eroded.

HUMANKIND ALREADY FACES SERIOUS THREATS: CLIMATE CHANGE, DISRUPTION OF THE NITROGEN CYCLE.

The issues are not only related to climate change, e.g. in terms of the overall change to the carbon cycle (with its close links to whatever happens in agriculture and forestry), but also to other cycles. In addition to water conditions (in either "green", "blue" or other forms), the nitrogen and phosphorous cycles have an integral relationship to agriculture – as does biodiversity (see Chapin et al., 2000 for example).

The connection between agriculture and the water cycle is obvious – too little or too much water is very harmful and both these aspects are connected to climate change. In the words of Wolfgang Ritter referring to the European Commission SCAR foresight reports (European Commission, 2011) in connection to the 2011 formulation of the Budapest declaration on food security: *"Water is being withdrawn at increasing rates, and the ratio of irrigated areas to population is decreasing. This means that food production in the world is increasingly dependent on rain, and climate change is reducing rainfall rates in many parts of the world."*

The nitrogen (N) and phosphorus (P) cycles are also essential for agriculture. On the human modification of the N cycle the "Planetary Boundaries" article (Rockström et al., 2009) states, and references that: "Human activities now convert more N_2 from the atmosphere into reactive forms than all of the Earth's terrestrial processes combined". This reactive N is used as fertilizer to enhance food production, but a significant portion of it runs off and pollutes waterways and coastal areas. This problem has not been addressed at levels above the local and regional.

Phosphorus is a finite fossil mineral, mined for human use, and its use leads to increased amounts of P leaking into soils and waterways causing pollution. A major concern – highlighted in the "Planetary Boundaries" article – is the growing interaction, through human interference, of the N and P cycles, which causes eutrophication (notably algae blooms) in rivers and coastal areas, as well as anoxia (severe drops in water oxygen levels that kill aquatic life) (Zillén et al., 2008).

The rate of biodiversity loss is another associated factor strongly connected to agriculture. Loss of biodiversity *"can increase the vulnerability of terrestrial and aquatic ecosystems to changes in climate and ocean acidity, thus reducing the safe boundary levels for these processes. The current and projected rates of biodiversity loss constitute the sixth major extinction event in the history of life on Earth – the first to be driven specifically by the impacts of human activities on the planet"* (Chapin et al., 2000).

LIVING WITHIN PLANETARY BOUNDARIES

In a presentation by the Swedish International Agricultural Network Initiative (SIANI) it was noted that agriculture covers 40% of the Earth's total ice-free land area. It accounts for 70% of global fresh water use and emits more greenhouse gases than any other human activity. It employs three quarters of the world's poorest people and feeds us all. It is the key foundation of human civilization and it is also where most of our present day development problems such as poverty, hunger, environmental degradation and climate change converge. Therefore, it is immediately evident that if we are to meet these global challenges, we must certainly address agriculture. There is thus a wide and growing consensus that probably the biggest challenge in the 21st century will be to produce enough food, feed, fibre and fuel, while aiming to stay within the safe operating space defined by the interconnected environmental dimensions presented above. Innovative research will be essential to make progress on the path of sustainable development to meet the five-fold challenge of growing more food, fibre and fuel on limited land, for a global population that continues to grow and exhibits changing consumption patterns. All this has to be achieved with less energy and less usage of other scarce inputs, while at the same time improving the resilience capacity of ecosystems and exploring all possibilities for adapting to and/or controlling climate change.

It is thus apparent that we need to learn to live within the planetary boundaries, and in order to do so we will need new forms of governance, a new economic development model, a strong knowledge and education pathway and creative changes in behaviour patterns – i.e. a celebration of creativity in a diverse set of forms. All these elements, and maybe others, are needed to transform the present agricultural system and to create a new role for the sector that is more strongly based on biological resources. In other words, to create a new global economy with more reliance on biological resources and ecosystem services.

The much-needed agricultural reform must consider the combination of agriculture, agribusiness and the new bioeconomy. It must implement an approach that is based on biological, physical, chemical, geological and geographical components, but also simultaneously drawing on knowledge mobilized by expertise from social science and the humanities that could illuminate various concerns about the development aspects, and the "human dimensions" of global change. This includes the cultural context that sets an important part of the frame for these endeavors. There is also a critical need for innovation – technical, but also societal and institutional.

Underlying all this has to be an ethos of concern for the service of humankind.

At a more "operational level" the following issues can be emphasised: the connections between different policy domains must be strengthened e.g. between the food and non-food sectors; there must be greater cooperation between the agricultural, energy, industry, and health sectors; and there must be better linkages between microeconomics and macroeconomics.

With regard to research and innovation, the following aspects are very important: the institutional organisation of research must be addressed, to take into account the various balance issues in the knowledge production system in terms of, for example, timing, financing and the development of human capital and the promotion of solutions that will be of service to the next generation. The importance of trust between the various actors involved connects to the need to earn public understanding and respect for the endeavour in the midst of normative conflicts and tensions between various stakeholders. The aspects of research, innovation, transfer, education and development need to be better linked and more carefully connected to policy.

THE NEXT STEPS, WHAT TO BE ALERT FOR IN A CHANGING WORLD – AND WHAT TO DO?

That the next half-century will provide exceptional challenges for humanity is a highly probable prospect supported by several foresight and other study types, such as the IAASTD report "Agriculture at a crossroads" (2009), and the Swedish report "Future Agriculture – Livestock, Crops and Land Use. A strategic programme for research" (Swedish University of Agricultural Sciences, 2010) or more recently from the UK "The future of Food and Farming" (UK Government Office for Science, London, 2011). There is less certainty, however, on how events will proceed in terms of the capacity to develop relevant and timely solutions, and in terms of the innovative mobilization of entire societies in order to provide the necessary support for the much-needed transformations. Perhaps humanity has the ability to mobilize itself – when on the brink of very serious consequences – to provide the required capacities for innovation and social transformation in an equitable, fair and sharing manner. But there is also the potential to find ourselves – or our children – in a situation of serious conflict, where individuals withdraw to their own "backyards" and show a drastically diminished consideration for whatever happens in the rest of the world – if in the long term such a posture is indeed possible. Perhaps such a strategy is feasible for the wealthy for a limited period, but natural or social events on a scale that is currently difficult for us to perceive, are likely to remove this option eventually.

It appears that agriculture will play an increasingly important role in future, and it faces a need to adapt to a society that must rely more on biological resources. This society must establish appropriate links between urban and rural environments, or the total global development pattern will not be functional in a world where a majority of people live in urban settlements. In this regard, there may also be new challenges for agriculture, such as for example, a need for food production to expand geographically into areas that we would now consider to be non-rural areas, i.e. into current and future types of urban settlements. Thus agriculture may emerge as a varied activity, with technological, social and management aspects, that will diversify into a pattern of symbiotic functions that perform the necessary tasks for the survival of the global population – both rural and urban; a population that will at the same time expand towards 9 or 10 billion people by the end of this century. The changes that will take place during that time are likely to not only be in terms of the size of the world population, but will also be in the form of increased needs and demands. It remains to be seen whether by the end of the century it will be possible to meet all of these emerging demands. Recent studies indicate that it might be possible to expand the use of as yet non-mobilized capacities in, for example, land use, direction of consumption patterns, resource efficiency, reduction of waste, etc. However, it is not known whether it will be possible to realize these potential responses due to social, economic, political and cultural challenges. The final verdict has not yet been written – and it will not be possible to consolidate for a long time. Much depends on the capacity to mobilize a spirit of knowledge development and innovation, but also to create the right conditions to enable the necessary transformation of a society facing the challenges of the Anthropocene. ■

REFERENCES

Chapin F.S. et al., 2000, Consequences of changing biotic diversity. Nature 405: 234-242.

Crutzen. P.J. and Stoermer, E.F., 2000, The "Anthropocene". *Global Change Newsletter* 41: 17-18.

Crutzen. P.J., 2002, Geology of mankind: the Anthropocene. *Nature* 415: 23.

European Commission, Standing Committee on Agricultural Research (SCAR), 2011, The 3rd SCAR Foresight Exercise (FEG3) (expert report), "Sustainable food consumption and production in in a resource-constrained world".

Richardson K. et al., 2009, Synthesis Report. Climate Change: global risks, challenges and decisions, University of Copenhagen, Denmark.

Rockström, J., Steffen, W., Noone, K., Persson, Å., Chapin, F.S., Lambin, E.F., Lenton, T.M., Scheffer, M., Folke, C., Schellnhuber, H.J., Nykvist, B., de Wit, C.A., Hughes, T., van der Leeuw, S., Rodhe, H., Sörlin, S., Snyder, P.K., Costanza, R., Svedin, U., Falkenmark, M., Karlberg, L., Corell, R.W., Fabry, V.J., Hansen, J., Walker, B., Liverman, D., Richardson, K., Crutzen, P., and Foley, J., 2009, A safe operating space for humanity. *Nature* 461: 472-475.

Steffen, W., Sanderson, A., Tyson, P.D., Jäger, J., Matson, P., Moore III, B., Oldfield, F., Richardson, K., Schellnhuber, H.J., Turner, B.L., and Wasson, R.J., 2004, *Global Change and the Earth System: A Planet Under Pressure*. The IGBP Global Change Series. Berlin Heidelberg New York: Springer-Verlag.

Steffen, W., Crutzen, P.J., and Mc Neill, J.R., 2007, vol 36, no 8, p. 614, *Ambio*.

Steffen, W., Persson, Å., Deutsch, L., Zalasiewicz, J., Williams, M., Richardson, K., Crumley, C., Crutzen, P., Folke, C., Gordon, L., Molina, M., Ramanathan, V., Rockström, J., Scheffer, M., Schellnhuber, J., and Svedin, U., 2011, The anthropocene: from global change to planetary stewardship. Contribution (background report Nr. 2) to the 3rd Nobel laureate Symposium on global Sustainability, Stockholm, May 2011.

Swedish University of Agricultural Sciences (SLU), (Bengtsson et al.), 2010, Future Agriculture – Livestock, Crops and Land use. A strategic programme for research.

UK Government Office for Science, 2011, *The Future of Food and Farming*, Final report, London.

Zillén, L., Conley, D.J., Andrén, T., Andren, E., and Björk, S., 2008, Past occurrences of hypoxia in the Baltic Sea and the role of climate variability, environmental change, and human impact. *Earth Science Reviews* 91: 77-92.

THE RESISTIBLE RISE OF THE WATER SHORTAGE OR THE SOCIAL CONSTRUCTION OF A "NATURAL FACT"

Sara Fernandez,
Centre Alexandre Koyré, Ecole des Hautes Etudes en Sciences Sociales, Paris, France

While to some extent water scarcity is defined through scientific observation, it is also largely a social and political construction (Trottier, 2008). This conclusion is supported by the joint analysis of management and water science in the Garonne basin in France, the history of which shows that water scarcity is not based on inevitable biophysical determinisms.

As long ago as 1752, Toulouse merchants complained about the difficulties their boats encountered during the Garonne's low-flow periods. Over a century later, in 1879, the Academy of Sciences of Toulouse registered concern about a lack of water in the Garonne. Then, in 1996, according to most water management stakeholders, "*the Garonne is deficient, or even very deficient*". While on 27th September 2009, *Le Monde* published an article on the impacts of global warming which, with regards to water, stated that: "(...) *The most affected areas would be the areas already affected today by structural deficit, such as the south-west [of France]*" (Fernandez, 2009). On this evidence it would appear that the Garonne has lacked water for over two and a half centuries, and it seems that this trend is not likely to change soon!

The Garonne basin is today characterized by large hydraulic infrastructures. Since the early nineteenth century, decentralized state agents in particular have sought control of the river and of the use of its waters, primarily for navigation but more recently for irrigation and power generation. The legitimacy of such projects was based on the need for development in an area long regarded as economically backward (Fernandez, 2009). In the mid-nineteenth century, the problem of the Garonne's low water level was reformulated to support the construction of navigation canals, which were soon to become obsolete once the railway had been completed. Irrigation became the next argument used to justify public expenditure, with claims that it would enable the development of land that was "*uncultivated and burned by the sun*". However, it wasn't until the 1970s that irrigation became a significant practice within the Garonne basin, earlier projects being mostly targeted at grassland and limited to the Pyrenean valleys and areas upstream of St. Martory, where the collection and transport of water using gravity was straightforward and required little investment.

Since the 1950s, agricultural and hydroelectricity policies were reinforced nationally, with the creation of the Electricité de France (EDF) and of regional development companies, including the Compagnie d'Aménagement des Coteaux de Gascogne (CACG).

The 1960s saw the beginning of the deployment of significant resources at local, national and European levels, enabling an agricultural socio-technical

FIGURE 1 STAKEHOLDERS AND DISCOURSES RELATED TO HYDRAULIC DEVELOPMENTS IN SOUTH-WEST FRANCE, FROM THE EARLY 19TH CENTURY"

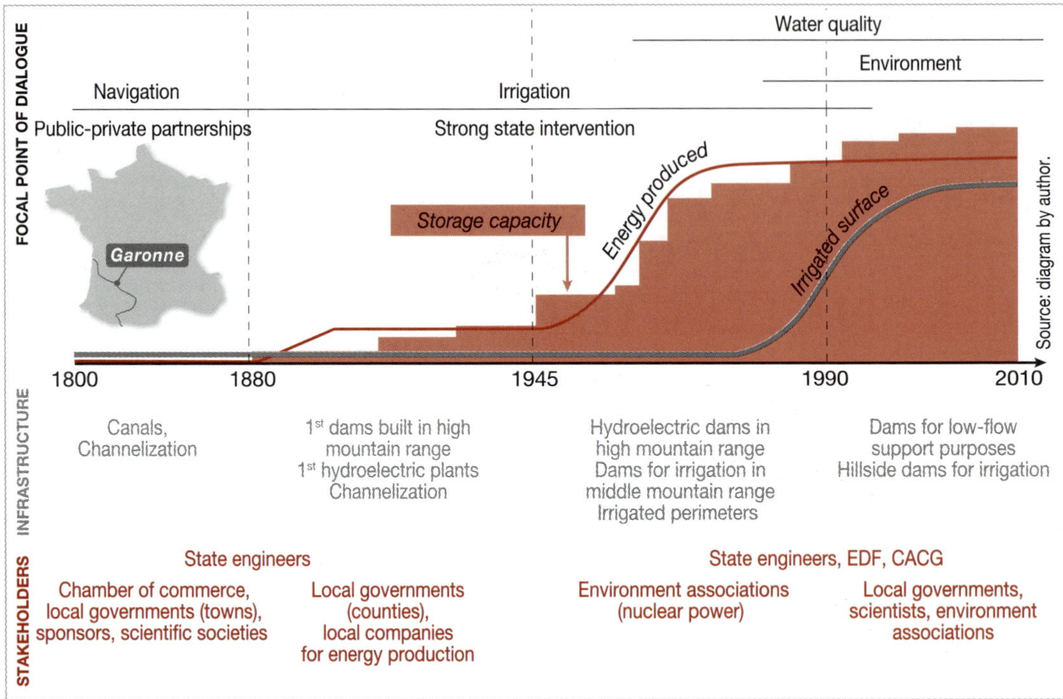

Since the early nineteenth century, the governance modes for the water in the Garonne basin have involved both management systems (for water infrastructure, indicators, contractual arrangements, management agencies...) and dialogue that seek to legitimize these systems, the ultimate goal being to control the water.

network to form in the Garonne basin, involving businesses, cooperatives, inter-professional associations, standards organizations, consultants, state engineers, researchers and others, who shaped the territory's "maize-growing suitability". This obscured the river's links with industry, particularly power generation, yet hydropower and then nuclear power developed widely between the 1950s and 1990s (Figure 1).

The French Water Agencies were created in 1964 to levy charges related to water pollution across entire river basins and to fund the restoration of water quality, but only had incentivizing powers. Agency funding was therefore directed towards environmental remediation or to increasing river flows, rather than to reducing emissions or water demand. The agencies also favoured commensuration processes (Levin & Espeland, 2002), involving the establishment of equivalents between quantitative and qualitative issues. Thus, the Adour-Garonne Water Agency (AEAG) addressed pollution in the Garonne by identifying a need to dilute discharges arising from the fertilizer production industry which, similarly to agricultural sectors and the electrical industry, seemed powerful enough to assert non-negotiable positions regarding their activities. Using water abstraction rates or pollutant emissions as given, the AEAG defined a minimum flow for rivers based on the maximum allowable concentration of pollutants that would not generate fish mortality.

In the late 1980s, however, while river flows were largely modified by the electrical and agricultural sectors, the increase in frequency and intensity of droughts was reflected in a significant rise in the tensions between the different water users. At this time, environmental spokespersons were also beginning to gain power at the national scale.

In the early 1990s, AEAG managed to enlist EDF along with the representatives of irrigated agriculture to incorporate them into its system of rules by enforcing minimum flows or "low-water target flows" (known as DOE, Débit Objectif d'Etiage), which contributed implicitly towards allocating more water to these stakeholders, by justifying the construction of new dams by the need to maintain the dilution capacity of water courses. This was particularly true for the Charlas dam project. Since 1996, the DOE has been defined as "*the rate above which the normal coexistence of all usages and the proper functioning of the aquatic environment are guaranteed and that thus must be respected each year during the low-flow period with predefined tolerances*" (Adour-Garonne basin Committee, 1996).

Initially promoted by the CACG in the early 1980s as of benefit for agricultural land irrigation on the Garonne's left bank, the Charlas dam project acquired a double purpose in the 1990s since it was alleged that the infrastructure would also help limit the severe low flows that occurred upstream of Toulouse in summer. The wider interest of the project then became the subject of an intense controversy that the Ministry of Environment decided to solve with a study aiming to determine whether compensating the Garonne's "structural" deficit was necessary. This study, which was conducted over a four-year period, made an active contribution to changing the very definition of low-water target flows. Originally calculated in terms of objectives for pollution dilution, low-water target flows were from then on regarded solely as a hydrological concept, based on an estimate of "reconstituted natural flows".

From a hydrological perspective, it proved difficult to distinguish the impacts on water flow from different users in summer, given the quantity of dams and water abstractions that are spread widely across the region and function at varying intervals. Thus, estimations of average low-water natural flow throughout the summer were based on series of statistical calculations that eliminated the distinction between periods influenced by irrigation (until late August), and those that were not (September-October). By disregarding time periods and the accountable water users, the study convinced state officials, EDF and CACG that it was in their interest to meet the DOE from late June to late October. Thus, in the 1990s the historical link between flow and dilution was broken: DOEs would become the new targets per se, and their calculation would no longer be discussed.

Today, the issue of whether to build the Charlas dam has still not been decided. Meanwhile, the use of DOE has become hegemonic in the Garonne basin. It is only by searching the archives that we can examine the assumptions that were used to justify the need to maintain such flows. In 2001, when the major AZF fertilizer plant exploded in Toulouse and was not subsequently rebuilt, the DOE in Portet-sur-Garonne, which had been set to dilute AZF's contribution of nitrogen into the river in the event of accidental pollution, was not however redefined. Indeed, by this time, water managers were already using DOEs as a given for their management plans (Fernandez, 2009).

DOE is not the only possible management indicator, it is simply the one that arose from negotiations between AEAG and powerful stakeholders in the Garonne basin, which formed the basis for further negotiations to share the economic rents derived from territorial policies related to water. While initially aimed mainly at limiting water abstraction, DOE was then used to justify the building of new dams, before being black boxed, i.e. becoming an end in itself.

Like any observation and any representation of nature, the shortage of water in the Garonne necessarily takes place in a specific political, economic and social context that influences which elements are considered as important or not, and what is deemed necessary to measure and how this should be done (Forsyth, 2003). Indicators such as DOEs are "agreements" between heterogeneous stakeholders, which are used as tools in negotiations that go beyond the scope of these indicators. The social construction of the Garonne's water shortage is thus based on very specific causal relations between hydraulic construction, agricultural and broader economic developments, between water management and science and technology developments. ■

REFERENCES

COMITÉ DE BASSIN ADOUR GARONNE, 1996, *Schéma directeur d'aménagement et de gestion des eaux du basin Adour-Garonne.*

FERNANDEZ S., 2009, *Si la Garonne avait voulu... Etude de l'étiologie déployée dans la gestion de l'eau de la Garonne, en explorant l'herméneutique sociale qui a déterminé sa construction.* Thesis for a doctorate in water sciences, UMR G-Eau, AgroParisTech, Paris, 653 p.

FORSYTH T., 2003, *Critical Political Ecology. The Politics of Environmental Science*. London and New York, Taylor and Francis Group, 272 p.

LEVIN P. and ESPELAND W. N., 2002, Pollution futures: commensuration, commodification and the market for air. *In*: A. J. Hoffman & M. J. Ventresca (éd.), *Organizations, policy and the natural environment: institutional and strategic perspectives*. Stanford, California, USA, Stanford University Press, p. 472.

TROTTIER J., 2008, Water crises: political construction or physical reality? *Contemporary Politics*, 14 (2), 197-214.

CHAPTER 7

Is agricultural intensification sufficient to stall deforestation? This idea does not stand up well to economic analysis. Indonesia was unable to limit widespread deforestation despite increased productivity of food and cash crops. This confirms the need for the support of public policies to combine agricultural production with the control of cultivated land expansion.

CAN AGRICULTURAL INTENSIFICATION SAVE FORESTED LAND? A REVIEW AND CASE STUDY IN INDONESIA

Anne Booth,
School of Oriental and Africain Studies, University of London, London, United Kingdom

Romain Pirard,
Institute for Sustainable Development and International Relations, Paris, France

Ahmad Dermawan,
Center for International Forestry Research, Bogor, Indonesia

Heru Komarudin,
Center for International Forestry Research, Bogor, Indonesia

Traditionally, agriculture and forests represent competing land uses, with land as a fundamental and scarce production factor that can be utilized in several ways, including the preservation of forest ecosystems, the establishment of agricultural fields and urban expansion. In the tropics this competition is fiercer than ever as, while tropical deforestation is topping the international agenda, between 1980 and 2000, 83% of new cropland in the tropical zone came at the expense of forests (Gibbs et al., 2010).

The rapid disappearance of tropical forests has attracted a great deal of attention in the context of the negotiations under the Convention on Climate Change (UNFCCC), leading to an agreement in Cancun in 2010 to establish a funding mechanism to fight deforestation and degradation in developing countries (REDD+[1]).

Stakeholders have become progressively aware that the fate of tropical forests depends significantly on agricultural policies. Therefore agricultural development is analysed not only for its positive contribution to food security, but also for its negative impacts on forest ecosystems, particularly in developing countries. The compatibility between both objectives depends considerably on the strategies to be prioritised for the future of agriculture worldwide.

1. REDD+ stands for Reducing Emissions from Deforestation and Forest Degradation. The "+" refers to the inclusion of the increase in carbon stocks, for example through appropriate forestry practices or plantations. The principle is to pay developing and emerging countries by means of contributions from industrialised countries, either through a market or a fund.

Agricultural intensification is the most commonly proposed solution to combine both objectives, and is usually understood as a green revolution based on new crop varieties, more capital-intensive technologies and the use of chemical inputs. The Borlaug hypothesis (*land sparing*) provides the underlying theoretical framework for this solution, according to which increased productivity per hectare leads to smaller areas under cultivation as less land is required to produce the same amount of agricultural commodities. Yet other development pathways exist for agriculture, notably organic agriculture for the supply of local markets, or the support of agroforestry to provide ecosystem services.

In this chapter, we examine the theory of land sparing in light of results from its actual application (section 1). We present an Indonesian case study and analyse the green revolution that took place here to determine the nature of the agricultural strategy, the types of agricultural technologies promoted and whether there has been any evidence of land sparing (section 2). Indonesia provides a relevant example as it faces extremely rapid deforestation, is highly populated and largely dependent on agriculture for its development and poverty alleviation. There also exists, both presently and historically, a clear competition between standing forests and agriculture. Finally, the Indonesian REDD+ oil palm strategy is addressed to assess the expected contribution of agricultural intensification in policies and measures to ease the pressure on remaining forests (section 3).

AMBIVALENT LINKS BETWEEN AGRICULTURAL TECHNOLOGIES AND DEFORESTATION

The outlook for tropical forests depends on public policies that are directed towards sectors of the economy indirectly linked to forestry (e.g. taxation that encourages extensive land use such as cattle ranching in Brazil). Agricultural land policies are one such example: they aim to modify agricultural practices, ownership and usage regimes. They entail changes to agricultural technologies; and agricultural intensification, typically implying green revolution, represents just one, rather restricted, interpretation of the range of possible technological changes.

THEORETICAL FOUNDATIONS: THE BORLAUG HYPOTHESIS The principle of minimising the extension of land area under cultivation by increasing yields from existing cropland is known as the Borlaug hypothesis, taking its name from one of the fathers of the green revolution that took place in Mexico from the late 1960s and then later in Asia.

Although the hypothesis seems initially rather evident, it is actually based on the simplistic idea that food production will increase until the initial demand is met. This is not the case, however, as economic forces are in fact the key determinant: investment decisions are made by farmers – both small farmers typically working less than one hectare per household as well as entrepreneurs that provide capital and modern production techniques to cultivate large areas – in reaction to price signals, while taking into account their own needs, outlets and property rights.

From an economic perspective, the Borlaug hypothesis is thus reformulated as follows: an increase in yield leads to a fall in agricultural commodity prices due to excess supply over demand, and therefore a decrease in profit per hectare, ultimately generating a supply adjustment through smaller growth in cultivated areas.

There is an inherent complexity in the analysis of the decision-making process of farmers since numerous variables are involved (including food prices that depend on public intervention). This problem is best addressed by taking a simplified look at the two analysis levels for agricultural expansion according to yields: the micro- and the macroeconomic levels.

FIGURE 1 THE EXTENSION OF AGRICULTURAL LAND IN INDONESIA

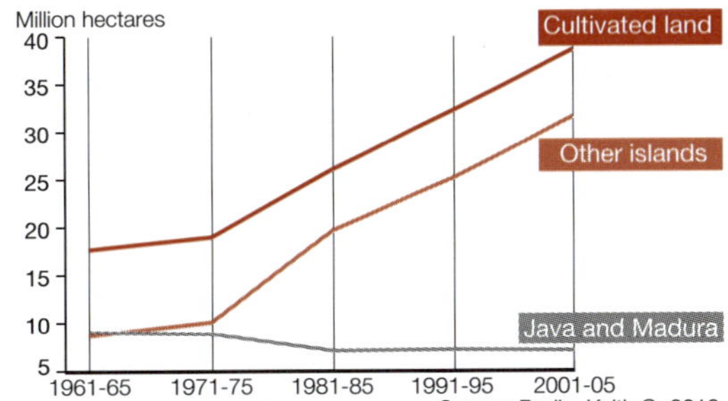

Source: Fuglie, Keith O. 2010.

Cultivated land grew steadily in Indonesia, more than doubling in forty years. The overcrowded island(s) Java (and Madura), at the socio-economic centre of the country, reached the limit in the 1960s, and saw a decline in cultivated areas during the same period.

The microeconomic level refers to decisions on new agricultural land, which are made by agents who are supposedly rational and well-informed. When farmers have the possibility of selling surplus production, technological progress generally leads to more land being converted to cropland. This has always been the case during "commodity booms" for export markets, such as bananas in Ecuador (Wunder, 2001). Such a trend apparently runs counter to the Borlaug hypothesis. Arguably, however, this tendency towards expansion may be moderated by the imperfect nature of markets: shortage of production factors locally (labour or capital available), high transaction costs to adopt new technologies or the presence of risks that influence investment decisions.

The macroeconomic level involves the functioning of the overall system, which can be visualised at the country level and beyond. The significance here relates to the repercussions of the adoption of new agricultural technologies at the macro level, and not in the policies and measures that could either facilitate their adoption or influence their impacts (e.g. a specific fiscal regime, human migrations, land tenure changes, etc). These economic repercussions occur when new technologies and associated higher yields spread to the extent that greater supplies lead to lower prices and reduced profitability. Human migrations are also a key feature, as the Indonesian case study suggests.

At the macroeconomic level, the elasticity of demand can be critical. Indeed, the simple interpretation of the Borlaug hypothesis is based on a demand that is *fixed* at the outset. Accordingly, meeting this demand should in theory prevent new conversions to cropland, since surplus production has no outlets. In reality, the way in which this demand is met will determine its evolution. But other factors will also influence this evolution, such as the existence of a food deficit and the elasticity of demand (see figure 1). For example, if the initial level is reached through productivity gains and lower production costs per unit, the expected drop in prices may generate an automatic increase in demand. This is known as the "rebound effect" and it occurs

FIGURE 2 INDONESIAN RICE, SUCCESSFUL INTENSIFICATION

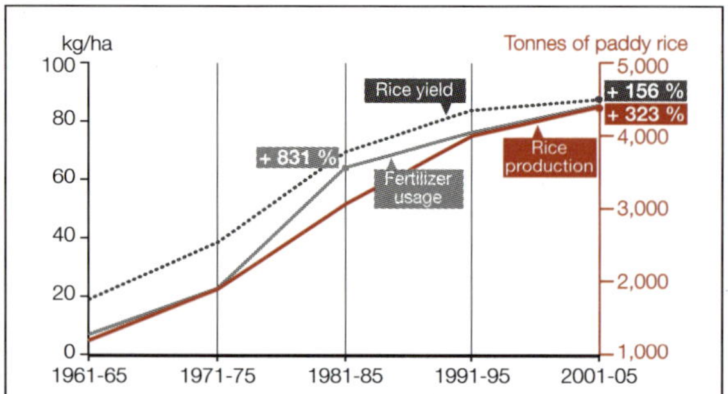

Source: Fuglie, Keith O. 2010.

The increase in rice production in Indonesia is largely correlated with higher yields, demonstrating the relevance of the intensification policy conducted. Fertilizers have played a significant role in this; their use has increased more than ten fold since the 1960s.

when demand is elastic (e.g. fluctuates in response to prices), which is likely to be the case for non basic needs.

Another important factor is the diversification of production. Aside from the fact that agriculture also produces non-food goods (oil for cosmetics, cotton for textiles, etc.), food production may diversify considerably to the point of creating new demand. Not everything boils down to a question of protein or staple goods: "luxuries" – i.e. non-essential foods that are consumed in addition to daily requirements – may also significantly influence production. Figure 1 provides a schematic summary of the interactions between levels and factors.

EMPIRICAL CONFIRMATION OF THEORETICAL HYPOTHESES UNDER VERY SPECIFIC CIRCUMSTANCES Two papers on land sparing were published recently that the authors believe to be the most comprehensive to date. Of these, one is presented below in detail, while reference is made to the conclusions of the other.

Rudel et al. (2009) analysed the Borlaug hypothesis using historical data from 161 countries and 10 crops for the period between 1970 and 2005. They cite the existence of two contradictory forces resulting from intensification – higher profits per hectare and a fall in the sales price due to excess supply – and the need to test which was stronger historically. They remind us that "*analysts working at the global scale have modeled the land-sparing effect rather than examining historical instances*".

Their econometric study initially focused on the global situation. The trends for the 1970-2005 period indicate that cultivated areas grow less rapidly than the population and the per capita income. An absolute reduction is also noted from 1980-85. However, the study does not reveal a significant correlation between crop productivity and the global evolution of cultivated areas. Nevertheless, a negative correlation (which therefore agrees with the Borlaug hypothesis) exists in 34 countries considered individually from 1900-2005. It is interesting to note that correlations do emerge when crops are analysed separately, rather than by country or groups of countries. For example, the increase in yields for wheat and coffee seem to have resulted in a reduction in the area allocated to these crops.

The value of this study is that tests were conducted at different levels to enable the authors to identify areas that potentially confirm the Borlaug hypothesis, so that these areas could then be studied in detail to include factors not present at the econometric stage. It shows that substitute imports and national land conservation programmes have an important role. For example, conservation has taken root in China (the Grain for Green programme) and in the United States (the Conservation Reserve Program) since the 1990s, with strengthened impacts for wheat due to the possibility of imports. For other crops, aspects of international economics are critical, for example when Cuba lost its Soviet sugar market there was a consequent reduction in the area dedicated to this crop; or the implementation of the North American Free Trade Agreement, which affected Mexican soybean, maize and wheat crops. All cases illustrate the crucial role of public policies, whether for the environment or trade, to limit agricultural expansion and hence conform with the Borlaug hypothesis. The Indonesian case is interesting in this respect as the authorities did not seriously consider such interventions and forest cover has declined despite a successful green revolution.

In the second of these major studies, Ewers et al. (2009) also conducted econometric studies for many countries and crops, concluding that the Borlaug hypothesis is confirmed to a certain extent. This study took agricultural diversification into account, with regards to the development of non-staple crops and biofuels, and also considered dietary changes, such as an increase in meat consumption, which threaten the positive effects of agricultural intensification.

The next section presents an Indonesian case study. This country provides a valuable example because it faces fierce deforestation pressures and, despite a green revolution several decades ago, food production remains a great challenge. It thus provides important insights into the land sparing debate.

GREEN REVOLUTION IN INDONESIA

POLICY DEBATES 1960 TO 2000: LIMITS OF INTENSIFICATION AND THE NEED FOR LAND OUTSIDE JAVA Since the 1950s, policy debates in Indonesia have revolved around

BOX 1 INDONESIAN AGRICULTURAL DEVELOPMENT

During the first two decades of Indonesian independence (1949 onwards), agriculture dominated the economy (half of GDP, two-thirds of the labour force). In spite of a growth in mineral exports, including petroleum products, agricultural exports still accounted for around two-thirds of total export revenues, with rubber being the most important export crop. While some exports were grown on large estates, by the early 1960s smallholders were also producing large quantities of rubber and dominated the production of secondary export crops such as coffee, copra and pepper. The most important food crop was rice, but maize and cassava were widely consumed, especially in poor rural areas. The most densely settled island was Java which accounted for only 7% of the land area but 56% of the land under rice cultivation, and around three quarters of the land under maize and cassava cultivation. By 2008, agriculture only accounted for around 14% of GDP, and employed 40% of the labour force, although many of these workers were also engaged in other activities, including agricultural processing and trade. Exports of agricultural products, including processed rubber and palm oil, had fallen to less than 20% of total exports and palm oil had become the most important export, increasing from over one billion dollars in 2000 to 12.4 billion dollars in 2008. Rice production increased from an estimated 12.4 million tons of paddy rice in 1961-65 to 62.6 million tons in 2008. Much of this increase derived from higher yields, while in Java the area under rice probably fell, although data on this decline are contested. The area planted by large estates has increased to around five million hectares, which is entirely due to a very rapid growth in palm oil cultivation (4.1 million hectares in 2008). A further 526,000 hectares were under rubber. Most other cash crops are now grown largely by smallholders. In 2008, 14.45 million hectares are under smallholder cultivation for tree crops and seasonal crops such as tobacco.

FIGURE 3 QUESTIONABLE LINKS BETWEEN AGRICULTURE AND FOREST

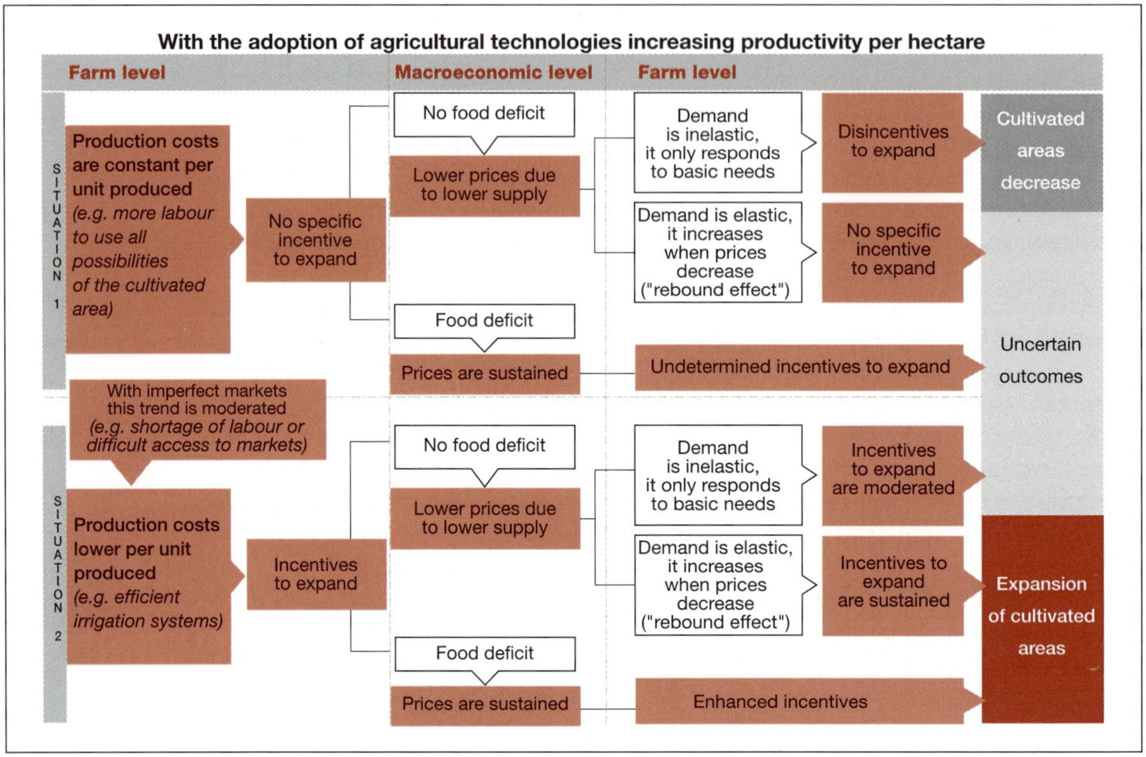

The work of Norman Borlaug, among others, argues that an increase in productivity per hectare automatically reduces the area under cultivation and thus protects forests. This overlooks a set of economic factors that determines both the choices made by farmers and the possibility of their realization. Ultimately, this relationship is rarely observed at the global level, and locally it makes virtually no sense.

three main crucial issues. The first was the perceived problem of rural overpopulation in the inner islands of Java and Bali, and the limits to agricultural intensification usually associated with the absorption of more labour, a process which had been ongoing in these islands since the late 19th century at least. More labour was used because of the introduction of double and triple cropping of irrigated rice land, more intensive use of dry land, and the cultivation of more labour intensive crops. But by the 1960s many agricultural experts thought that diminishing returns had set in and more labour could not be productively used in agriculture without new technologies which would raise yields per hectare. It was widely known that yields of rice in Indonesia were low compared with Taiwan and Japan, and agricultural economists argued for the development of higher yielding varieties and the greater application of chemical fertilisers.

A second focus for policy debate was the potential for large-scale land settlement outside the densely-populated regions of Java, Bali and Nusantara. One solution, started by the Dutch colonial government in the early 20th century, was transmigration, i.e. the movement of rural households to areas with lower population densities and more land of arable potential. This policy was continued, albeit spasmodically, after 1950.

The third area for policy debate concerned the potential to increase production of smallholder cash crops which had been grown in Sumatra, Kalimantan, Sulawesi and Maluku for decades. Comparisons with neighbouring Malaysia showed that smallholder rubber yields were much lower; the question was how to extend knowledge about new varieties and better cultivation practices to millions of farmers scattered over vast areas.

When Suharto took power in 1966, he recruited a team of mostly Berkeley-educated technocrats, known as the "Berkeley Mafia", to advise him on economic policy. They had a different approach to the Chicago-oriented economists that were influential in Latin America in the 1970s and 1980s, advocating cautious pragmatism and a strong role for government in guiding policy. They argued that agricultural development should be the top priority for government expenditure, and the first five-year plan (1969-74) allocated considerable funds to the rehabilitation and development of rural infrastructure, including roads and irrigation systems (22% of the plan's expenditure went on agriculture and irrigation projects (Department of Information, 1974)). Although subsequent five year plans during the Suharto era gave less prominence to agriculture, rural infrastructure remained a priority. Furthermore, the transmigration programme received additional funding especially in Suharto's third plan (1979-84): over this period, around 366,000 families were moved to Sumatra, Kalimantan, Sulawesi and Irian Jaya as sponsored migrants, and a further 170,000 as spontaneous migrants (World Bank, 1988).

The transmigration programme involved bringing new land under cultivation. Inevitably there was some encroachment on forest areas, although the Directorate-General of Forestry often prevented families settling on land under its jurisdiction, whether forested or not. In addition, local populations grew outside Java and took more land for cultivation. Large estates, especially in the palm oil sector, also expanded at the expense of forests. However, it must be acknowledged that these degradations were concomitant with the acceleration of food crop yields, especially rice, in most parts of the archipelago.

GROWTH IN AGRICULTURAL OUTPUT AND EXTENSION OF CROPPED AREA Between 1960 and 2000, Indonesia's growth in agricultural output was 3.5% per annum; between 1968 and 1992 it was 4.8% (Fuglie, 2004: Table 5). These growth rates were rapid by international standards, and Fuglie's estimates suggest that at least half was due to total factor productivity growth, i.e. output growth not caused by increased inputs. Both labour and land productivity grew at over 2% per annum between 1968 and 1992. After 1992, agricultural growth slowed and productivity growth was negative. Fuglie suggested several reasons for this decline including the El Nino-induced drought of the late 1990s, concomitant to the severe macroeconomic collapse in 1997/98. Furthermore, by the early 1990s, rice productivity gains due to the successful dissemination of International Rice Research Institute varieties had reached a maximum, and a general decline in government assistance to agriculture forestalled new R&D investment.

The rapid growth in food crop agriculture over the 1970s and 1980s, and especially the growth in rice production, can be seen as a partial vindication of the intensification argument. A considerable proportion of the growth was due to higher yields and to more intensive cropping of existing agricultural land, resulting from improved irrigation. As figure 2 shows, rice yields more than trebled from the 1960s to the 1990s, accompanied with a considerable increase in fertiliser use.

TABLE 1. TRENDS IN INDONESIAN AGRICULTURE (ADAPTED FROM FUGLIE AND PIGGOT, 2006: TABLE 4.2 P 89)

	1961-65	1971-1975	1981-85	1991-95	2001-03
Total cropland (million hectares)	17.6	18.9	26.0	32.2	39.3
Cropland Java and Madura	9.0	8.8	7.0	7.1	7.1
Cropland other islands	8.6	10.0	19.6	25.1	32.2
Rice yields (kg/ha)	1,761	2,542	3,786	4,352	4,465
Rice outputs (million tons of paddy rice)	12.4	21.2	35.8	47.5	51.3
Fertiliser use (kg/ha)	6.9	22.7	63.3	73.9	74.3

Total land under all forms of agriculture increased rapidly from 1961-65 to 1991-5. Most of this growth was outside Java and included growth in land under staple food crop cultivation (rice, corn, cassava, sweet potato, soybeans, sago, etc.). But there was also rapid growth in area under smallholder cultivation of cash crops such as cloves, coffee, palm oil, rubber and cocoa. Area under palm oil cultivation by large estates also increased (see box). All land area estimates in Indonesia must be treated with a certain scepticism and Fuglie's estimates, based on national and FAO data, may be overstated. Meanwhile, cropland in Java has declined, especially since the 1980s.

The reason for this decline is the rapid growth in urban and peri-urban settlements which has meant that much cropland has been used for housing, industry and commerce. Outside Java, the expansion of cultivated land is due to population growth, leading to the use of more land for smallholder food crop and tree crop agriculture, land settlement schemes and the expansion of large estates. While the official transmigration programme has attracted considerable criticism from environmental organizations, it only accounted for a small part of the growth in area under food crops and tree crops. Most of the increase in food crop area has been due either to local populations cultivating more land, or to unofficial migrants settling on land without permission. The growth in tree crop hectarage is due to smallholders and commercial estates.

IMPACT OF AGRICULTURAL POLICIES ON FOREST COVER For most of the past century, Indonesian farmers have responded vigorously to national and international price incentives, as well as to new technologies, and have increased the cultivation of many crops for different markets. Until the 1980s, the environmental costs of increased agricultural growth attracted little attention. Regarding food crops, the official view, supported by evidence, was that most production growth came from increased yields and the more intensive cultivation of existing land.

What then was driving deforestation, which accelerated during the 1980s and 1990s? For many years it was thought that forest loss was due mainly to shifting cultivation, but this view was challenged as more sophisticated satellite imaging technology became available. It emerged that large plantation companies and forest conglomerates were the main actors driving deforestation. By the end of the century, there was much pressure on the Indonesian government to conserve pristine forests and to consider the consequences of non-food agricultural growth. Oil palm expansion by smallholders and large estates has driven such growth, but it is the latter that are considered responsible for the majority of forest loss. A World Bank report (2001) placed the blame for deforestation squarely on commercial developers, especially those involved in palm oil. Given that world demand for palm oil and other vegetable oils will continue to grow, is more forest loss inevitable? What about the 15 million hectares officially classified as temporarily unused arable land (Statistics Indonesia 2009: 92)? What must be done to turn this land to profitable agricultural use? These are the issues that will challenge policy makers over the next decade and beyond.

The Indonesian case study is not a straightforward validation of the Borlaug hypothesis because it exhibits many factors that blur the links between intensification and agricultural expansion. Certainly, yields have rapidly increased during the green revolution, however there has also been diversification with oil palm. The initial food deficit was filled (self-sufficiency in rice was achieved in 1984), while the population and economy have grown spectacularly, clearly impacting on agriculture. In this context it is difficult to ascertain whether higher yields have limited expansion. The case study illustrates perfectly the contradictory nature of certain policies from an environmental perspective: while intensification was supported, populations were simultaneously moved to forested areas where licenses for oil palm plantations were freely issued. Taking these complexities into account, the next section attempts to assess whether intensification can be used as a tool to combine higher productivity with forest conservation, addressing the burning question of oil palm expansion.

BOX 3 REDD+ LETTER OF INTENT (LOI) BETWEEN NORWAY AND INDONESIA

Signed in May 2010, the LoI established a partnership between Norway and Indonesia to contribute to the reduction of emissions from deforestation, forest degradation and peatland conversion. There are three phases: preparation (2010), transformation (2011- 2013) and implementation (from 2014).

At the time of writing, the following policies are in development: establishment of degraded land databases for economic development purposes, enforcement of existing laws against illegal logging and trade in timber and related forest crimes, elaboration and implementation of appropriate measures to address land tenure conflicts and compensation claims. The flagship of these measures is the two-year "moratorium" suspending all new licences for the conversion of peat and natural forest.

IMPLEMENTING REDD+ AGAINST DEFORESTATION: THE OIL PALM CASE

While several initiatives are currently underway in Indonesia in the framework of REDD+, we focus here on the deal with Norway that may have the greatest impact.

THE OIL PALM CHALLENGE Indonesia is the world's largest palm oil producer, a position it aims to maintain by doubling its production level to 40 million tonnes by expanding its plantations to 16 million ha by 2020 and increasing productivity to approach Malaysian levels. Between 1990 and 2010, Indonesian production increased eight fold; while over the same period the area of oil palm cultivation grew from 1.1 million ha to 7.8 million ha. About 40% of oil palm plantations are managed by smallholders.

PLANNED ACTIVITIES Indonesian oil palm targets are, however, constrained by its commitments with Norway. Indeed, in May 2011 a presidential instruction, known as the "moratorium", postponed the issue of new licenses for forestry and agriculture, notably oil palm plantations, on primary forests and peatlands for at least two years.

Some NGOs have criticised the moratorium for failing to adequately address the threats to forests. Indeed, it has been shown that the presidential instruction can only protect 19% of the 45.3 million ha of total primary forests from further deforestation.

The business sector has also criticised the moratorium. The Indonesian Oil Palm Producer Association estimates that the annual land expansion for oil palm will be reduced from 350,000 ha to less than 200,000 ha for the next two years.

Great confusion remains regarding the impact of the moratorium, especially when considering the conflicts with regional development initiatives. For example, in 2009 the Merauke Integrated Food and Energy Estate initiative was promoted in the Indonesian province of New Guinea island to secure food and energy resources. Initially, expectations were for the project to cover 2 million ha, supported by significant foreign and domestic investment. However, following public pressure over possible environmental and social damage, a reduced 1.2 million hectare plan was suggested, which is now obtaining a consensus. This draft regulation proposes the development of 420,000 ha of plantations during 2010-2014, a further 630,000 ha during 2015-2019 and finally an additional 230,000 ha of plantation during 2020-2030. These figures do not take commitments at the national level into account.

OIL PALM INTENSIFICATION ISSUES REGARDING PRODUCTION TARGETS The 2008 Indonesian Forest Climate Alliance report recommends a strategy which involves: consolidating policy and approval criteria for the release of convertible forests for oil palm development, reviewing of plans to optimize degraded land, intensification of oil palm estates, and zero burning for land preparation. None of these aims are easily implemented and, compared to other components, intensification has not been adequately addressed so far.

Two of the main challenges facing the oil palm sector are low productivity and limited access to capital. Large companies generally have their own R&D units which produce high quality seedlings, whereas smallholders often have difficulty obtaining good seedlings and lack essential support (e.g. infrastructure and fertilizers). Collaboration between farmers and companies is often ineffective and there is much room for improvement.

To tackle these issues, the government has initiated a revitalization programme that is projected to help establish 730,000 ha of oil palm plantations by 2014. The programme enables plantation developers to access credit at preferential rates, as well as providing high-quality seeds and extension services on production methods and the optimisation of land resources.

CONCLUSION

The complexity involved in the transformation of agricultural production systems and the links to land availability highlight the necessity for public support policies to obtain the potential land sparing benefits. Indeed, an analysis of the deal between Indonesia and Norway suggests that the former does not prioritize intensification as a land sparing strategy or as a means to achieve REDD+ objectives.

Evidence shows that we cannot rely on spontaneous changes in technology to contribute to forest conservation. For example, Boserup (1965) showed that spontaneous innovation generally brings about an intensification of scarce resources. This means that farmers show a strong tendency to adopt extensive systems when land is abundant, to compensate for the scarcity of other production factors such as labour and capital. This holds true for both slash and burn farmers in Borneo and cattle farmers in Latin America. Accordingly, regarding deforestation in Indonesia, a reliance on spontaneous innovation and the adoption of new agricultural technologies to solve the problem would imply accepting an increase in forest clearing until the resource has almost disappeared, and then depending on farmers to implement conservation measures. A certain number of macro tools are therefore required, for example appropriate taxation on the agricultural sector or the creation of a network for technology dissemination.

FOR MANY YEARS IT WAS THOUGHT THAT FOREST LOSS WAS DUE MAINLY TO SHIFTING CULTIVATION, BUT THIS VIEW WAS CHALLENGED AS MORE SOPHISTICATED SATELLITE IMAGING TECHNOLOGY BECAME AVAILABLE.

One tool with great potential is the Payments for Environmental Services (PES) scheme, where beneficiaries (or intermediaries) of an environmental service establish voluntary contracts with the providers of this service and condition their rewards on the maintenance of the service. For Indonesia, a PES could consist of measures to support the adoption of sound agricultural technologies on the condition that forests on nearby land were left intact. Farmers and landowners would benefit from technologies capable of increasing their production and income, while the adverse consequences of forest clearance could be minimised. Ideally, a PES would secure the benefits (an increase in food production) while mitigating the adverse effects (expansion to the detriment of tropical forests).

Finally, it is critical that the "rebound effect" is tackled to avoid an increase in consumption when food prices decrease. One suggestion is for action to be taken on diets throughout the world to bring demand per capita into line globally, for example by reducing the consumption of dairy products and meat. However, this is an extremely ambitious aim and the effectiveness of associated educational programmes is uncertain. Nevertheless, food transition studies enable different scenarios to be envisaged, for instance the implementation of alternative models in the agro-food industry (Paillard et al., 2010), or the establishment of systems to tax agricultural products based on their carbon content, similar to commitments made under the Climate Convention (Zaks et al., 2009). ■

REFERENCES

Boserup, E., 1965, *The Conditions of Agricultural Growth: The Economics of Agrarian Change under Population Pressure*, Aldine, New York.

Paillard, S., Treyer, S. et Dorin B. (Coord.), 2010, Agrimonde: Scénarios et défis pour nourrir le monde en 2050, Editions Quae, Versailles, France.

Department of Information, 1974, *Pidato Kenegaraan Presiden R.I. Didepan Sidang Pleno DPR Tgl 15 Agustus 1974 (Speech of the President to the DPR)* Jakarta: Department of Information.

Fuglie, Keith O., 2004, "Productivity Growth in Indonesian Agriculture, 1961-2000", *Bulletin of Indonesian Economic Studies*, Vol 40 (2), 209-25.

Fuglie, Keith O. and Piggott Roley, 2006, "Indonesia: Coping with Economic and Political Instability" in P.G. Pardey, J.M. Alston and R.R. Piggott (Editors), *Agricultural R&D in the Developing World: Too Little, Too Late?* Washington: International Food Policy Research Institute.

Gibbs, H.K., Ruesch, A.S., Achard, F., Clayton, M.K., Holmgren, P., Ramankutty, N. and Foley J.A., 2010, Tropical forests were the primary sources of new agricultural land in the 1980s and 1990s, *PNAS*, In Press.

Ewers, R.M., Scharlemann, J.P.W., Balmford, A. and Green R.E., 2009, Do increases in agricultural yield spare land for nature? *Global Change Biology*, 15, pp. 1716-26.

Rudel, T.K., Schneider, L., Uriarte, M., Turner, B.L., DeFries, R., Lawrence, D., Geoghegan, J., Hecht, S., Ickowitz, A., Lambin, E.F., Birkenholtz, T., Baptista, S. and Grau R., 2009, Agricultural intensification and changes in cultivated areas, 1970-2005, *PNAS*, 106 (49), pp. 20675-80.

Statistics Indonesia, 2009, *Luas Lahan menurut Penggunaannya (Land Utilization in Indonesia)*, Jakarta: Central Board of Statistics.

World Bank, 1988, *Indonesia: The Transmigration program in Perspective*, Washington: World Bank.

World Bank, 2001, *Indonesia: Environment and Natural Resource Management in a Time of Transition*, Washington: World Bank.

Wunder, S., 2001, "Ecuador Goes Bananas: Incremental Technological Change and Forest Loss", *in*: Angelsen, A. and D. Kaimowitz (eds.), *Agricultural Technologies and Tropical Deforestation*, CABI Publishing, New York.

Zaks, D.P.M, Barford, C.C., Ramankutty, N. and Foley J.A., 2009, Producer and consumer responsibility for greenhouse gas emissions from agricultural production – a perspective from the Brazilian Amazon, *Environmental Research Letters*, 4.

IS CULTIVABLE LAND A SCARCE RESOURCE?

Laurence Roudart,
Université Libre de Bruxelles, Bruxelles, Belgium

Many publications warn about the scarcity of cultivable land and how it is being degraded and diminished due to various development types (including urbanization), in the context of an increasing human population with expanding needs (Millennium Ecosystem Assessment, 2005; Oldeman et al., 1991). Research aimed at assessing global and regional areas of cultivable land, whether in use or not, is even scarcer still (Fischer et al., 2002; Ramankutty et al., 2005).

There are four reference databases available on this subject: the FAO's FAOSTAT, which provides information on the area under cultivation; the fieldwork element of the *Global Agro-Ecological Zones* (GAEZ) study by the *International Institute for Applied Systems Analysis* (IIASA) and the FAO, which provides information on cultivable land; the database of the *Center for Sustainability And the Global Environment* (SAGE) of the University of Wisconsin and also that of the University McGill which contains data on cultivable and cultivated areas.

This article focuses on a comparison of the GAEZ and FAOSTAT databases.

CULTIVABLE AND CULTIVATED LAND: DEFINITIONS AND UNCERTAINTIES

The GAEZ study evaluates global land suitability for the cultivation of 154 plant varieties and the potential yields in function of three main crop management methods – "advanced", "improved" and "traditional" – and depending on whether crop production is rain-fed or irrigated. The study classifies land into five categories according to its capacity for cultivation, by comparing an area's yield potential with the maximum yield attained in the same climatic zone. On this basis, land is classified as: *very suitable* if the yield potential exceeds 80% of the maximum yield attained, *suitable* if this ratio is between 60% and 80%, *moderately suitable* between 40% and 60%, *marginally suitable* between 20% and 40%, and *not suitable* below 20% (Fischer et al., 2002). The GAEZ database shows corresponding areas of land, as well as those under forest cover, and indicates the infrastructure range.

The FAOSTAT database provides estimates of areas of "arable land[1]" and land under "permanent crops[2]", the sum of which we call "cultivated land".

There are of course many uncertainties in the FAOSTAT and GAEZ databases. A. Young, soil scientist, claims that cultivable land might be overestimated by 10% to 15% and that cultivated land might be underestimated by 10% to 20% (Young, 1999).

From these two databases, land currently under cultivation represents nearly 40% of useable land in *rain-fed agriculture* (irrigation not necessary)

1. "Arable land" is defined as "land under temporary agricultural crops, temporary meadows for mowing or pasture, land under market and kitchen gardens and land temporarily fallow (less than five years)" (FAO).

2. "Permanent crops" are those that are "sown or planted once and then occupy the land for some years and need not be replanted after each annual harvest, such as cocoa, coffee and rubber. This category includes flowering shrubs, fruit trees, nut trees and vines, but excludes trees grown for wood or timber." (FAO)

worldwide, i.e. about 1,560 million hectares out of 4,150 million. By taking into account the maximum error suggested by A. Young, this proportion would increase to 53%, or 1,874 million hectares out of 3,529 million.

THREE HYPOTHESES FOR THE EXPANSION OF CULTIVATED LAND, BASED ON IIASA AND FAO DATA

Three hypotheses

With knowledge of the various land category areas that can be used in rain-fed agriculture (GAEZ study), and the areas of cultivated land in 2005 (FAOSTAT), one can calculate the theoretical possibilities for cultivated land expansion. With this intention, we considered three hypotheses: 1) a very restrictive scenario, which assumes that land classified as very suitable, suitable and moderately suitable could be cultivated, except for forested land and land required for urban and other infrastructure[3]; 2) a less restrictive scenario, which includes marginally suitable land as well as the three categories of the first hypothesis, except for forested land; and 3) a least restrictive hypothesis which also integrates all arable land under forest, i.e. one third of the world's forests.

Results and interpretation

Some aspects of this type of analysis tend to lead to an overestimation of the expansion possibilities for cultivated areas. This would occur mainly in areas currently classified as "grassland", "shrubland" or "permanent meadows and pastures", some of which are possibly located in protected areas, although it is not possible to distinguish this from the data. In addition, to be classified as suitable, an area of land only has to be suitable for growing one of the 154 species considered. However, other aspects of the analysis tend to lead to an underestimation of the expansion possibilities: the GAEZ study assumes low-yielding land to be unsuitable for cultivation and does not give due consideration for possible improvement by human intervention that could render some land cultivable.

Thus, the expansion of cultivated area on a global scale was calculated according to the three hypotheses; the results were, respectively: 1) approximately 1000 million hectares, which leads to 1.7 times more than the present area; 2) 1450 million hectares, 2 times more, and 3) 2350 million hectares, 2.5 times more.

The potential for expansion, however, varies considerably from one region to another (see figure), for example in hypothesis 1 South America and sub-Saharan Africa have a particularly high capacity for expansion, while in Asia there is almost none due to the fact that the currently cultivated area already exceeds that which could be cultivated according the three hypotheses analyzed. This means that for Asia at least, the IIASA methods clearly tend to underestimate the potential of land for cultivation, and by implication the ingenuity of people to develop land to make it cultivable.

The GAEZ study also examines the consequences of seven global warming scenarios on the areas cultivable with wheat, maize or rice: all of which predict a limited extension (1% to 6%) of cultivable land globally for these crops, though a decrease in developing countries (1.3% to 11%) and an increase in developed countries (11% to 25%).

AGRICULTURAL DEVELOPMENT AND PUBLIC POLICY

Thus, according to IIASA and FAO data, land that can be used in rain-fed agriculture and is presently uncultivated is not, and will not be soon, a globally scarce resource. This is true even if all forests and currently protected areas remain uncultivated and when the possible effects of global warming are taken into account. However, uncultivated cultivable land is indeed a limited resource, or even almost non-existent in the Middle East and Asia. In addition, only *local* ecological, technical, economic and social studies can provide a full understanding of whether uncultivated cultivable lands are actually available for cultivation, whether they are accessible and by whom they could be cultivated.

The fact remains that those responsible for agricultural public policy, whether national or stemming from international cooperation, have flexibility in the type of international agricultural development they will prioritize. A first option, to which most

3. These areas are calculated by assuming a 50% increase on the area already designated for this purpose in 2000, which is in proportion to the expected population growth up to 2050. However, this method tends to overestimate the increase because, in general, such areas grow at a slower rate than population.

FIGURE 1 CAN CULTIVATED LAND BE EXTENDED?

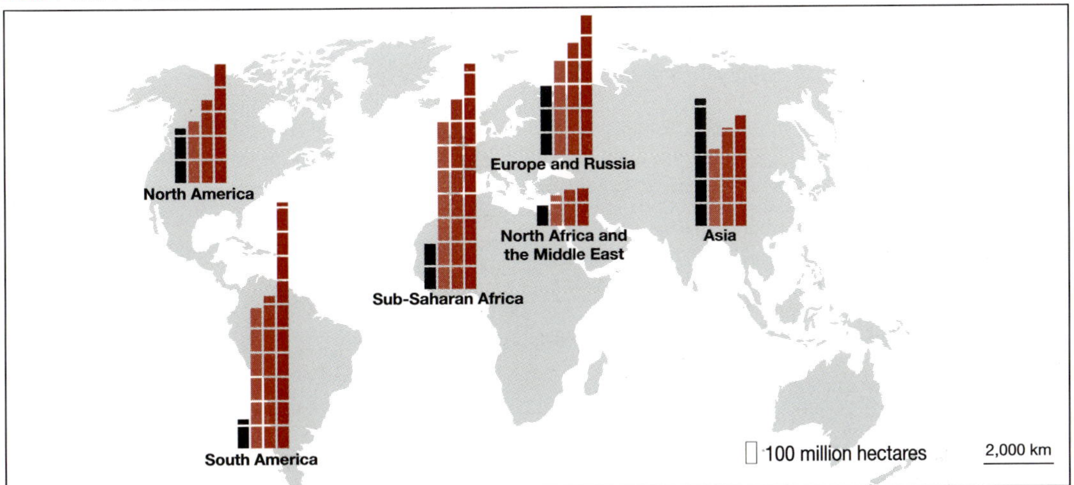

Cultivated land in 2005 and areas that could be cultivated according to the three hypotheses analyzed

- Area cultivated in 2005 (FAO)
- Cultivation of all land classified as very suitable to moderately suitable, except for forests or land required for infrastructure.
- Cultivation of all land classified as very suitable to marginally suitable, except for forests or land required for infrastructure.
- Cultivation of all land classified as very suitable to moderately suitable, including forest land, except land required for infrastructure.

Source: FAO

The imperative to increase global agricultural production is often opposed by the limits of arable land. The three hypotheses presented here indicate different possibilities for the extension of cultivated land, showing that the potential offered by these possibilities is very unevenly distributed around the world.

institutions currently adhere, is to pursue policies and practices that are conducive to a competitive mode of agricultural development that is uneven and indeed contradictory. Such an approach has entailed a sharp increase in labour productivity and crop yields for some of the world's production units, while hundreds of millions of other farmers have fallen into poverty, malnutrition and possibly exodus and emigration. In some areas, added to these serious social issues are environmental problems, such as salinity, pollution, biodiversity loss and more (IAASTD, 2008; Mazoyer, Roudart, 2006).

Given that in many parts of the world it is possible to cultivate larger areas than at present, an alternative solution would be to promote diversified agriculture, with relatively low yields, low inputs, that use little or no fossil fuels, have few negative effects on the environment or perhaps even positive effects, and that could ensure a decent livelihood for nearly three billion people – the world's farming population – in the true spirit of the "Doubly Green Revolution".

For this approach to succeed, the right problem must be addressed: the main challenge is political, economic and social, it concerns the development of new types of agricultural policies; it is not about the scarcity of cultivable land. ■

REREFENCES

Fischer, G., van Velthuizen H.T., Shah M., Nachtergaele F.O., 2002, Global Agro-ecological Assessment for Agriculture in the 21st Century: Methodology and Results. IIASA, Laxenburg.

IAASTD, 2008, Agriculture at a Crossroads. Washington DC, Island Press

Mazoyer, M., Roudart L., 2006, A History of World Agriculture. From the Neolithic Age to the Current Crisis. Columbia University Press, New York; Earthscan, London.

Millennium Ecosystem Assessment, 2005, Ecosystems and Human Well-Being: Current State and Trends. Island Press, Washington DC

OECD, FAO, 2009, OECD-FAO Agricultural Outlook 2009-2018. OECD, Paris.

Oldeman, L.R., Hakkeling R.T.A., Sombroek W.G., 1991, World Map of the Status of Human-Induced Soil Degradation: An Explanatory Note. International Soil Reference and Information Centre, Wageningen. United Nations Environment Program, Nairobi.

Ramankutty, N., Hertel T.W., Lee H.-L., 2005, Global Land Use and Land Cover Data for Integrated Assessment Modelling, GTAP Technical Paper.

Roudart L., 2010, Terres cultivables et terres cultivées: apports de l'analyse croisée de trois bases de données à l'échelle mondiale, Notes et études socio-économiques, n° 34, p. 57-95.

Young, A., 1999, Is There Really Spare Land? A Critique of Estimates of Available Cultivable Land in Developing Countries. Environment, Development and Sustainability 1, 3-18.

Webography

http://faostat.fao.org/site/377/default.aspx#ancor visited on 22/09/11

http://www.iiasa.ac.at/Research/LUC/GAEZ/home.htm visited on 22/09/11

https://www.gtap.agecon.purdue.edu/resources/res_display.asp?RecordID=1900 visited on 22/09/2011

http://www.geog.mcgill.ca/~nramankutty/index.html visited on 22/09/2011

Today's agriculture is at a crossroads. Climate change is already impacting negatively on food production while there are expectations for the sector to meet a rise in demand of 70% to 100% within only 40 years. Failure to reach this target would further reduce food security, while success may commit the world to further warming by accelerating greenhouse gas (GHG) emissions. Can we avoid a spiral of poverty, hunger, environmental degradation and conflicts by developing climate smart agriculture?

CLIMATE CHANGE AND FOOD SECURITY. A CRUCIAL TEST FOR HUMANITY?

Jean-François Soussana,
Institut National de la Recherche Agronomique, Paris, France

Nothing is more important to humanity and the stability of societies than a reliable and affordable supply of food. One of the great challenges of the 21st century will be to increase the global food supply to accommodate a world growing to 10 billion or more people while preserving a safe operating space for humanity by avoiding dangerous environmental change (Rockström et al., 2009).

Of the 14 billion hectares of ice-free land on Earth, 10% are used for crop cultivation, while an additional 25% of land is used for pasture. Over 2 billion tonnes of grains are produced yearly for food and feed, providing roughly two-thirds of total direct and indirect protein intake; a mere 10% of this total, or 200 million tonnes, is traded internationally. Resource management is key to achieving current production levels; for instance, although irrigated land occupies only 17% of total arable land, irrigated crops supply a much higher proportion of total production (ca. 40% for cereals) and consume 2,500 billion m^3 of water, or 75% of the total fresh water resources consumed annually. Finally, agriculture is a significant contributor to land degradation and anthropogenic global GHG emissions, being responsible for 25% of carbon (largely from deforestation), 50% of methane, and more than 75% of N_2O emitted annually by human activities (Tubiello et al., 2007).

Increasing demand for land is increasing the pressure on biodiversity and natural resources. Agricultural expansion in the tropics is mainly based on deforestation, since 80% of new tropical croplands are replacing forests (Foley et al., 2007), which affects biodiversity and key ecosystem services. Species-rich regions of the world are under pressure from agriculture conversion, putting animal and plant species at risk, including hundreds of medicinal plants that are the basis for global health care (TEEB, 2010). An estimated one third of the world's cropland is losing topsoil faster than new soil is forming. For example, analysts suggest that Africa has been losing 1% of soil organic matter every year since the 1960s (Clay, 2011), resulting in inefficient use of inputs such as fertilizer and water and reduced grain harvests. These processes result in the spread of hunger and an increased dependence on food imports. Unprecedented water shortages are also increasingly apparent in many parts of the world. Over-extraction of ground water is shrinking the irrigated area, depleting aquifers and degrading soils through salinization.

Future population and economic growth will require a doubling of current food production, including an increase from 2 billion to above 4 billion tonnes of grains annually. Associated with continued dietary shifts from crop-based to livestock-based diets, such trends would require a large agricultural land expansion especially in sub-Saharan Africa and Latin America (Schmidhuber & Tubiello, 2007, AgriMonde, 2010).

THE 21ST CENTURY GLOBAL FOOD CRISIS

Until recently, it was expected that despite climate change and an increasing world population, there would be several decades of food surplus – and low prices – ahead (Easterling et al., 2007). Contrary to this expectation, the volatility of world food prices has increased and two out of the last three years have been characterized by large spikes in international grain prices, with some grains more than doubling in cost. This unforeseen onset of a global food crisis underlines our poor understanding of the complex interactions between food systems, world markets and climate.

UNLESS MAJOR ADAPTATIONS ARE MADE, HIGH SEASONALLY AVERAGED TEMPERATURES WILL CHALLENGE FOOD PRODUCTION IN THE FUTURE.

In 2005-2006, a production shortfall, amplified by export restrictions in major countries such as India and Argentina, led to a dramatic increase in food prices from 2007 to 2008, adding 40 million people to the chronically hungry category (FAO, 2008). Food riots broke out in 48 countries sometimes leading to high levels of political instability (Brinkman and Hendrix, 2010). Subsequently, the summer of 2010 was exceptionally warm in Eastern Europe and large parts of Russia, leading to the loss of 1 million hectares from wildfires, along with crop failures of around 25%, for a total damage of about USD 15 billion. This heat wave prompted Russia to suspend exports, pushing grain prices towards an all time high in February 2011 and contributing to riots and changes of political regimes in the Middle East. In a tight world market, such events are increasing food prices and pushing millions more people into poverty and hunger, thereby contributing to political instability and civil unrest.

Yet, the Earth still provides enough to satisfy every man's need. While a billion people go hungry, half a billion are obese. The global calories deficit reaches only 3% of global supply by agriculture and could be compensated by reducing food over-consumption. Moreover, 40% of the total grain harvest is used to feed livestock, and an additional 6.5% for biofuels. Furthermore, 40% of the totality of global food products is either lost after harvest or wasted (FAO, 2011).

CLIMATE CHANGE EXPOSURE OF AGRICULTURE

Agriculture is intimately linked to climate conditions and therefore highly exposed to climate change. Short-term natural extremes such as storms and floods, inter-annual to decadal climate variability (such as the El Niño) have significant effects on crop and pasture production (Tubiello, 2005).

The climate system is already moving beyond the patterns of natural variability. The extreme drought and heatwave that hit Europe in the summer of 2003 caused a crop yield decline of between 20% to 30% and a green fodder deficit of up to 60% (in France). In Switzerland fodder had to be imported from as far away as Ukraine. In the next 40 years, the risk of summers as warm as 2003 may increase by two orders of magnitude in Europe, and may approach the norm by 2080 under high emission scenarios. Global climate change can therefore be expected to threaten food supply through changing patterns of rainfall and increasing the incidence of extreme weather, leading to greater variability of production, increased price volatility and changes in trade flows (Lobell et al., 2008).

In arid parts of Africa, nomadic herders rely almost entirely on the meat and milk of their livestock to survive. They follow the rains and move their goats, sheep, cattle and camels across vast areas. Between 1980 and 1999, severe droughts have caused mortality rates in national herds of between 20% and 60% in several arid sub-Saharan countries (IPCC, 2007). Again, in 2009-2011, the worst drought in 60 years has triggered a looming humanitarian and food crisis, which would affect more than 10 million people across the region.

Early warning signs suggest that climate change is having an impact on the yields of some major crops. Rice and wheat are the most important sources of calories for humans. Their production increased by around 30% per decade in the 1960s and 1970s. From 1990 to 2009, this rise in production decreased to around 10% per decade for wheat and rice. Of the world's three most important grains, only maize has maintained a rate of increase in the most recent decade that is close to that of the 1970s and 1980s (Long et al., 2010). For two decades, a decline in the growth trend of cereal yields has been observed in many European countries. In France, from 1990 on, heat stress during grain filling and drought during stem elongation have counteracted yield increases achieved through genetic progress (Brisson et al., 2010). This is consistent with statistical modelling showing that global wheat and maize production have declined since 1970 by 5.5% and 3.8%, respectively, compared to a counterfactual without climate trends (Lobell et al., 2011).

It is highly likely (more than a 90% chance) that by the end of the 21st century, growing season temperatures in most of the tropics and subtropics will exceed even the most extreme seasonal temperatures recorded from 1900 to 2006 (Battisti and Naylor, 2009). Unless major adaptations are made, high seasonally averaged temperatures will challenge food production in the future. A further drying of large parts of the subtropics is likely by the end of this century (IPCC, 2007a). Amplification of the hydrological cycle as a consequence of global warming is forecast to lead to more extreme intra-annual precipitation regimes characterized by larger rainfall events and longer intervals between events (IPCC, 2007a). More extreme rainfall regimes are expected to increase the duration and severity of soil water stress as intervals between rainfall events increase (Knapp et al., 2008). In the long run, heat and drought can trigger desertification, as a decline in vegetation cover increases soil erosion and run-off, further reducing soil water storage.

Over the last 30 years, hundreds of carefully managed experiments where soil nutrients were often non-limiting have confirmed that plant biomass and yield tend to increase significantly as CO_2 concentrations increase above current levels. The CO_2 levels that will be reached by 2050-2070 have been shown to increase yields in the range of 10% to 20% for crops of temperate origin (C_3 crops, like wheat and rice) and 0% to 10% for crop species of tropical origin (C_4 crops like maize) (Tubiello et al., 2007). It is important to note though that the effects of elevated CO_2, as measured in experimental settings and subsequently implemented in models, may overestimate actual field and farm-level responses because of interactions with many limiting factors such as high temperatures, low nutrient concentrations, droughts, pests, weeds and air pollutants (Soussana et al., 2010). We still do not know how much of the CO_2 fertilizing effect will remain under these complex conditions, which adds to the many uncertainties for the future supply of food. Moreover, ozone concentrations in future decades, with or without CO_2, with or without climate change, will negatively impact plant production, possibly increasing exposure to pest damage (Tubiello et al., 2007).

AT THE REGIONAL SCALE, CLIMATE CHANGE IMPACTS ARE LIKELY TO VARY WIDELY AND TO LEAD TO CHANGES IN THE DISTRIBUTION OF CROPPING AREAS.

Climate change influences the occurrence, prevalence and severity of plant diseases (Kersebaum et al., 2008). Under future climate conditions, there is a risk that the European corn borer will establish permanent populations in Central Europe therefore extending its climate niche to cover almost the entire area suitable for agriculture by 2040-2075.

Sixty per cent of human epidemics are caused by animal pathogens found in the wild or in domestic flocks. 2010 was the year of the eradication of rinderpest, an infectious disease that has decimated cattle for millennia. Nonetheless, in recent years, several vector-borne, parasitic or zoonotic diseases have (re-)emerged and spread, with major health, ecological, socio-economical and political consequences. Climate change is already affecting the spread of animal diseases such as the bluetongue virus, a sheep disease that is moving north into the temperate zones of Europe (Artz et al. 2010).

Warming also has direct impacts on livestock. The onset of thermal stress in livestock often results in voluntary declines in physical activity with associated

FIGURE 1 WHEAT AND CLIMATE CHANGE

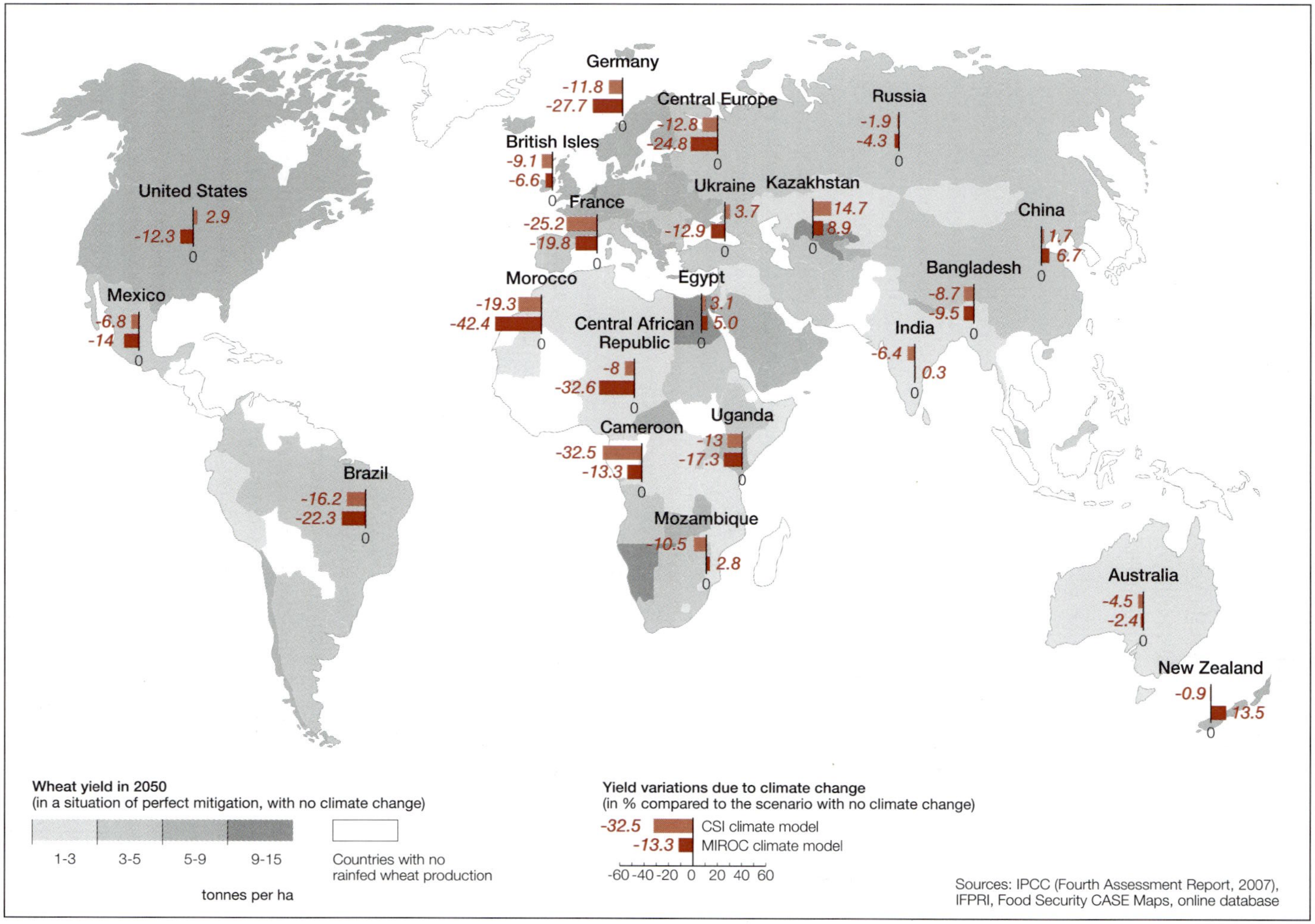

Sources: IPCC (Fourth Assessment Report, 2007), IFPRI, Food Security CASE Maps, online database

Compared to a baseline scenario of an increase in wheat yields due to technological advances, the simulation of the effects of climate change (for the IPCC A1B scenario, where greenhouse gas emissions increase until 2050 and remain significant until 2100), shows convergent results for significant reductions in expected wheat yields in many parts of the world (Europe, North and Sub-Saharan Africa, and Latin America), and increases in a very few situations, such as Central Asia. The magnitude of these variations is still uncertain, as shown by the differences between the two climate models used.

declines in eating and grazing activity. High temperatures put a ceiling on dairy milk yield from feed intake. This can induce an energy deficit and decrease cow fertility, fitness and longevity (King et al., 2005). Increases in air temperature and/or humidity have been shown to affect conception rates of domestic animals that are not adapted to these conditions. This is particularly the case for cattle, in which the primary breeding season occurs in the spring and summer months.

PROJECTIONS OF REGIONAL CLIMATE CHANGE IMPACTS

At the regional scale, climate change impacts are likely to vary widely and to lead to changes in the distribution of cropping areas as shown by the example of Europe. In northern Europe, increases in yield and expansion of climatically suitable areas are expected to dominate, whereas disadvantages from increases in water shortage and extreme weather events (heat, drought, storms) will dominate in southern Europe (Bindi and Olesen, 2011). In the northernmost agricultural areas of Europe and Russia, climate change is projected to result in milder winters, which may enable the cultivation of winter crops to a greater extent. However, fluctuating conditions that currently hamper wheat overwintering, may be exacerbated in the future by increased climatic variability and extreme weather events (Peltonen-Sainio et al., 2010). This could delay the adoption of winter-sown crops (cereals and rapeseed) for several decades.

IN ORDER TO STABILIZE OUTPUTS AND INCOME, PRODUCTION SYSTEMS MUST BECOME MORE RESILIENT, I.E. MORE CAPABLE OF PERFORMING WELL IN THE FACE OF DISRUPTIVE EVENTS.

In addition, climate change is projected to have detrimental impacts on winegrowing in southern Europe, mainly due to increased dryness and cumulative thermal effects during the growing season. Conversely, in cooler areas of western and central Europe, projected future changes will benefit not only wine quality, but might also delineate new potential areas for viticulture, despite some likely threats associated with diseases (Malheiro et al., 2010).

CLIMATE CHANGE ADAPTATION

Agricultural systems need to change in the face of climate risks. In order to stabilize outputs and income, production systems must become more resilient, i.e. more capable of performing well in the face of disruptive events. Farmers already adapt to climate warming through changes in agricultural practices (e.g. sowing and harvest timing, animal stocking density, fertilization...) and changes in plant seed and animal breed mixes. This autonomous adaptation is likely to have substantial benefits under moderate climate change for some agricultural systems. Enhancing the capacity to manage climate risk is also a core adaptation strategy. There are many region- or situation-specific climate risk management options (e.g., crop and livestock diversification) that may also have adaptation value (Howden et al., 2007).

Smallholder, crop, livestock and pastoral subsistence systems, especially those located in marginal environments, are often characterized by livelihood strategies that have evolved (*i*) to reduce overall vulnerability to climate shocks (adaptive strategies), and (*ii*) to manage their impacts *ex-post* (coping strategies) (Morton, 2007). Risk sharing strategies within families and rural communities, as well as insurance mechanisms need to be strengthened to reduce vulnerability.

Under more severe climate changes, there are limits to the effectiveness of autonomous adaptation and planned adaptation is needed (Howden et al., 2007). Planned adaptation in agriculture will require a large coordinated international research effort to develop seeds and breeds adapted to the unchartered climatic conditions of the end of this century and to design resilient and eco-efficient crop and livestock systems, while ensuring the dynamic conservation of soil, water and genetic resources.

Agro-ecological engineering will have to be developed, in particular through the increased use of genetic and species diversity at field and landscape scales, along with eco-technologies to adapt water management by improved water harvesting and increased water use efficiency, to monitor and reduce GHGs and to increase and verify soil carbon stocks. Investments in crop and animal disease protection and in invasive species monitoring will be required to preserve plant, animal and human health. Remote sensing and information technologies, such as seasonal weather forecasting, the geo-monitoring of crops and precision agriculture have the potential to reduce inputs while increasing productivity and resilience. However, the results of adaptation will be a function of both the likely technical effectiveness of adaptations and their adoption rate. Both factors are restricted in some tropical regions by poverty and hunger, lack of financial resources, environmental degradation and conflicts.

CLIMATE CHANGE MITIGATION

GHG emissions due to human activities have grown since pre-industrial times, with an increase of 70% between 1970 and 2004. Without mitigation, the concentration of carbon dioxide in the Earth's atmosphere may reach approximately 1,000 parts per million (ppm) by the end of the present century and remain above this level for thousands of years. In this business as usual worst case scenario, by the end of this century, global average temperatures would rise by more than 5°C, with regional rises of more than 10°C, and will continue to rise for centuries, reaching levels well above tipping points for most ecosystems and agricultural production systems (Schneider, 2009). Sudden and severe declines in crop yields would lead to large numbers of refugees, threaten food security and ignite conflicts in a number of world regions.

Humans have a great influence on the global carbon cycle and utilise about 40% of the Earth's net primary production (Rojstaczer et al., 2001). Globally, about 2,150 gigatonnes of carbon are stored in plants and soil. A fraction of this amount could be released in the next century through climate change and land use change, which would in turn further accelerate climate change. Preserving soil carbon stocks through careful soil management will therefore be required to avoid positive feedbacks on global warming. In addition to reducing its own emissions, carbon sequestration in agricultural soils can play an important role in offsetting emissions from other sectors.

In 2004 agriculture directly contributed about 14% of global anthropogenic GHG emissions in CO_2 equivalents. Land use, land use change and forestry account for a further 17% (IPCC, 2007). In order to limit future global warming to a 2°C temperature increase rail guard, anthropogenic GHG emissions will have to decrease globally by at least 50% by 2050 from 1990 levels. Agriculture is not currently subject to emissions caps, although several countries are already implementing mitigation action plans.

Quantifying GHG emissions from agricultural activities is complex. First, the atomistic nature of production (many individual farmers) in a wide range of geographic and climatic conditions means that emissions are not only highly variable but also difficult and costly to measure precisely. Second, there continues to be a great deal of scientific uncertainty as GHG emissions from agriculture are subject to a complex interplay of many factors such as climate, soil type and production practices.

AS THE POOREST AREAS SLOWLY BECOME MORE INTEGRATED WITH MARKETS, THEY ARE LIKELY TO IMPROVE OVERALL INCOMES AND PRODUCTIVITY, AT THE COST OF BECOMING MORE VULNERABLE TO PRICE SHOCKS.

Many agricultural practices can potentially mitigate GHG emissions, the most prominent of which are improved cropland and grazing land management and restoration of degraded lands and cultivated organic soils (Smith et al., 2008). Reducing excessive nitrogen (N) fertilization and the substitution of mineral N fertilizers by biological N fixation, as well as improved nutrition of domestic ruminants to reduce methane from enteric fermentation and improved manure management can play a significant role. Lower, but still significant mitigation potential is provided by water and rice management and agroforestry (Smith et al., 2008).

The global technical mitigation potential from agriculture (excluding fossil fuel offsets from biomass) by 2030, considering all gases, is estimated to be approximately 5500-6000 Mt CO_2-eq per year, with half of the potential that could be reached in theory for carbon prices up to 50 US$ per tonne of CO_2-eq (IPCC, 2007). In addition, GHG emissions can be reduced by the substitution of fossil fuels for energy production by agricultural feedstocks (e.g. biogas from crop residues and dung) and by dedicated energy crops such as highly productive perennial grasses and short rotation coppices. The economic mitigation potential of biomass energy from agriculture is estimated to be of the same order of magnitude as that of direct GHG mitigation in agriculture (Smith et al., 2008).

However, accounting for the indirect land use changes arising from agricultural production is another important challenge. In particular, the links between production of biofuels from food crops (in many cases subsidized), consequent land use changes, and the rise in food prices demonstrate the importance of foreseeing the range of consequences. Biofuel expansion on arable lands leads to land use changes including deforestation and hence to indirect emissions of CO_2. Some agricultural GHG mitigation options are cost competitive with a number of non-agricultural options in achieving long-term climate objectives. Such options should not reduce agricultural productivity, but rather improve the eco-efficiency of agricultural systems by reducing GHG emissions per unit of crop and animal products.

FOOD SECURITY UNDER CLIMATE CHANGE

The 1996 World Food Summit held under the auspices of the Food and Agriculture Organization (FAO) defined food security as a "situation that exists when all people, at all times, have physical, social, and economic access to sufficient, safe, and nutritious food that meets their dietary needs and food preferences for an active and healthy life" (FAO, 2002). This definition comprises four key dimensions: availability, stability, access, and utilization of food.

The first dimension relates to the overall ability of the agricultural system to meet food demand. The second dimension, stability, relates to the risks for individuals of temporarily or permanently losing their access to the resources needed to consume adequate food. An important cause of unstable access is climate variability. The third dimension, access, covers entitlements by individuals to adequate resources, which encompasses purchasing power, land tenure rights and traditional rights of rural people to a share of common lands.

Millions of people are undernourished and hungry because of a lack of access to sufficient food. They are predominantly located in rural areas in the tropics and their vulnerability is increased by socioeconomic, demographic, and policy trends limiting their capacity to adapt to climate change (Morton, 2007). Among the rural food insecure, most consume more calories than they produce in their fields and are therefore also hurt in the short term by price increases. As the poorest areas slowly become more integrated with markets, they are likely to improve overall incomes and productivity, at the cost of becoming more vulnerable to price shocks.

The final dimension, utilization, encompasses all food safety and quality aspects of nutrition, including the sanitary conditions across the entire food chain. Even net producers of food are often consuming insufficient calories, choosing instead to spend some of their income on sugar, meat, and other more expensive foods (Naylor and Falcon, 2010). Furthermore, food quality may be affected by climate change, cereal and forage crops for example, show lower protein and micronutrients concentrations under elevated CO_2 conditions (Easterling et al., 2007).

By using storylines and climate and crop models, IFPRI (2010) studied a range of plausible futures that suggested that the first half of the 21st century is likely to see increases in real agricultural prices. The rise in food demand driven by population growth and increasing affluence would exceed that of food supply given the negative impacts of climate change. With climate change, price increases would range from 30% for rice in the optimistic scenario (with high income growth and low population growth) to 100% for maize in the baseline scenario (with moderate income and population growth). In the unlikely event that perfect GHG mitigation (a continuation of today's climate in the future) is achieved, the rise in cereal prices would be halved.

Further impacts on food security were also shown in this IFPRI study. Climate change decreases future human well-being, especially among the world's poorest people, increasing the number of malnourished children relative to a world with perfect mitigation. With high per capita income growth and perfect climate mitigation, calorie availability in low-income countries could reach almost 85% of that in developed countries by 2050. In contrast, under the pessimistic scenario, calorie availability – and human well-being more generally – declines in all regions unless climate change is fully mitigated.

TOWARDS CLIMATE SMART FOOD SYSTEMS

Climate smart agriculture has been defined as agriculture that sustainably increases productivity and resilience (adaptation), reduces GHGs (mitigation), and enhances food security and development (FAO, 2010). A sustainable intensification of agriculture, that would allow the closure of yield gaps and an increase in biological efficiencies, especially in developing countries, can enhance food security and contribute to mitigating climate change by stopping deforestation and the expansion of agriculture into sensitive ecosystems. More productive and resilient systems may also lead to beneficial side effects in terms of carbon sequestration and reduction of GHG emissions per unit product. These multiple win-win options require changes in the management of biodiversity and natural resources (e.g. soil conservation and restoration, water harvesting and saving, increased use of biological nitrogen fixation and of integrated systems such as agroforestry and crop-livestock, etc...). However, win-win options are currently limited by gaps in our understanding, as well as by a number of economical and institutional barriers.

Changes in dietary preferences and in bioenergy policy may also help to obtain climate smart food systems. For instance, shifting grain-fed beef consumption to poultry, pork or pasture-fed beef and not using food crops as biofuel feedstocks would significantly enhance food availability and reduce the environmental impacts of agriculture (Foley et al., 2011). Moreover, improving food storage and transport in developing countries to reduce post-harvest losses and, in industrialized countries, the avoidance of waste at the consumer or retail level would together help to reduce the excessively high (ca. 40%) level of food losses and wastage. Climate smart food systems will therefore require a combination of changes at all steps of the planetary food web ranging from a better management of natural resources by resilient farming systems to improved education of consumers, through changes in storage, transport, processing and retailing infrastructures.

Reinvesting in agriculture and food systems will require substantial funding as well as changes in trade and regulatory arrangements to limit price volatility, increase market transparency and encourage international cooperation in agricultural development and rural environment protection.

CONCLUSION

To a large extent, humanity now holds the fate of the global biosphere in its hands, and faces crucial choices concerning its future. Continued development will depend upon our capacity to act as guardians of the biosphere by halting GHG emissions and managing ecosystem services and biodiversity to enable world food systems to operate within a safe space. This will also require a sustained economic growth and welfare redistribution that would lead to a rapid increase of per capita income, to a stabilization of the world population and to an improved global food security. Passing this crucial test of being able to populate the Earth and govern it sustainably, will require the satisfaction of every man's need, but not every man's greed.

Therefore, policies should aim at (IFPRI, 2010): (*i*) increasing the incomes of poor people and their access to food, knowledge and natural resources to achieve sustainable food security and resilience to climate change; (*ii*) investing in climate smart agriculture, consumer education and waste reduction to enhance sustainable food security; (*iii*) revisiting international trade arrangements to compensate for different climate change effects in different locations and (*iv*) cutting GHG emissions and planning adaptation to minimize the harmful effects of climate change on food security. ■

REFERENCES

ARZT, J. et al., 2010, Agricultural Diseases on the Move Early in the Third Millennium. Veterinary Pathology, 47(1), pp.15-27.

BATTISTI D.S., NAYLOR R.L., 2009, Historical warnings of future food insecurity with unprecedented seasonal heat. Science 323, 240–244.

BINDI, M. & OLESEN, J.E., 2011, The responses of agriculture in Europe to climate change. Regional Environmental Change, 11, p.S151-S158.

BRISSON, N. et al., 2010, Why are wheat yields stagnating in Europe? A comprehensive data analysis for France. Field Crops Research, 119(1), pp.201-212.

CLAY J., 2011, Freeze the footprint of food. Nature, 475, 287-289.

EASTERLING W., AGGARWAL P., BATIMA P., BRANDER K., ERDA L., HOWDEN M., KIRILENKO A., MORTON J., SOUSSANA J.-F., SCHMIDHUBER J., TUBIELLO F., 2007, in Climate Change 2007: Impacts, Adaptation and Vulnerability, eds Parry ML, Canziani OF, Palutikof JP, van der Linden PJ, Hanson CE (Cambridge Univ Press, Cambridge, UK), pp 273–313.

FAO, 2002, The State of Food Insecurity in the World 2001, Food and Agriculture Organization, Rome.

FAO, 2008, The State of Food Insecurity in the World 2008 (SOFI): High Food Prices and Food Security – Threats and Opportunities, Rome.

FAO, 2010, "Climate-Smart" Agriculture Policies, Practices and Financing for Food Security, Adaptation and Mitigation.

FOLEY J.A., RAMANKUTTY N., BRAUMAN K.A., CASSIDY E.S., GERBER J.S., JOHNSTON M., MUELLER N.D., O'CONNELL C., RAY D.K., WEST P.C., BALZER C., BENNETT E.M., CARPENTER S.R., HILL J., MONFREDA C., POLASKY S., ROCKSTRÖM J., SHEEHAN J., SIEBERT S., TILMAN D. and ZAKS D.P.M. (2011). Solutions for a cultivated planet. Nature, 337-342.

FOLEY J.A. et al., 2007, Amazonia revealed: forest degradation and loss of ecosystem goods and services in the Amazon Basin. Front. Ecol. Environ. 5, 25–32.

HOWDEN, S.M., SOUSSANA J.-F., TUBIELLO F.N., CHHETRI N., DUNLOP, M. and MEINKE H., 2007, Adapting agriculture to climate change. Proceedings of the National Academy of Sciences of the United States of America, 104(50), 19691-6.

INTERGOVERNMENTAL PANEL ON CLIMATE CHANGE (IPCC), 2007, Climate change: impacts, adaptation and vulnerability. Contribution of Working Group II to the Fourth Assessment Report of the Intergovernmental Panel on Climate Change. Cambridge: Cambridge University Press.

INTERNATIONAL FOOD POLICY RESEARCH INSTITUTE, 2010, Food security and climate change. Challenges to 2050 and beyond. IFPRI Issue Brief 66. December 2010.

KERSEBAUM K C et al., 2008, Simulated effect of climate change on wheat production and nitrogen management at different sites in Germany. Journal of Agrometeorology, 10, pp.266-273.

KING J.M., PARSONS D.J., TURNPENNY J.R., NYANGAGA J., BAKARI P., WATHES C.M., 2005, Anim Sci 82:705–716.

LOBELL David B., BURKE Marshall B., TEBALDI Claudia et al., 2008, Prioritizing climate change adaptation needs for food security in 2030. Science 319, 607-610.

LOBELL, D.B., SCHLENKER, W. and COSTA-ROBERTS, J., 2011, Climate Trends and Global Crop Production Since 1980. Science 333, 616-620.

LONG S. P.; ORT DONALD R. 2010 More than taking the heat: crops and global change. Current Opinion Plant Biol. 13, 241-248.

MALHEIRO A.C. et al., 2010, Climate change scenarios applied to viticultural zoning in Europe. Climate Research, 43(3), pp.163-177.

MORTON et al., 2007, The impact of climate change on smallholder and subsistence agriculture. Proc. Natl Acad. Sci. USA, 104, 19680-19685.

NAYLOR and FALCON, 2010, Food Security in an Era of Economic Volatility. Population and Development Review 36, 4, pp 693–723.

ROCKSTRÖM, J. et al., 2009, A safe operating space for humanity. Nature 461, 472–475.

ROJSTACZER, S. et al., 2001, Human Appropriation of Photosynthesis Products. Science 294, 2549.

SCHMIDHUBER J., TUBIELLO F.N., 2007, Proc. Natl Acad Sci USA 104:19703–19708.

SCHNEIDER S., 2009, The worst-case scenario. Nature 458, 7242, 1104-1105.

SMITH P., MARTINO D., CAI Z. et al., 2008, Greenhouse gas mitigation in agriculture. Phil. Transactions Royal Soc. B-Biological Sciences 363, 789-813.

SOUSSANA J.-F., GRAUX A.I. and TUBIELLO F.N., 2010, Improving the use of modeling for projections of climate change impacts on crops and pastures. Journal of Experimental Botany 61, 2217-2228.

TUBIELLO F.N., SOUSSANA J.-F., HOWDEN S.M., 2007, Proc Natl Acad Sci USA 104:19686–19690.

AGRICULTURE AND CLIMATE CHANGE NEGOTIATIONS

Hayden Montgomery,
First secretary and special adviser at the New Zealand Embassy, Paris, France

Is agriculture discussed within the UNFCCC?

For the last two years the role of agriculture has gained increasing recognition in the UNFCCC, and has been promoted by technical papers, workshops, and other activities by the parties and NGOs. The first formal discussions on agriculture commenced within discussions on mitigation which meant that adaptation was outside the scope of the talks. This created many tensions, particularly for developing countries whose main priority for agriculture in climate change negotiations is adaptation.

Despite this, as this was the only place where agriculture was able to be formally discussed, the parties made the most of it and, for three months at the end of 2009, through negotiations we moved from a narrow view on mitigation to a much broader perspective including adaptation. This led to a draft decision that was to be adopted in the Copenhagen meeting. This did not happen and we've spent the last two years trying to achieve this.

During this period the text on agriculture has changed considerably. While the core of the decision remains the same, an increasing number of issues are being added, including some that had previously been negotiated out of the text. Two years after Copenhagen and a formal work programme on agriculture in the climate change Convention has still not emerged, and the task is becoming more difficult because some parties are calling for specific issues to be addressed.

This creates two possible ways forward: to agree on a simple decision text now, with the possibility to expand its scope during the work programme; or to strive for a complicated text aiming to resolve all issues from the outset, but on which agreement is difficult.

How is agriculture defined in the negotiations?

As yet, the negotiations have not focused on a particular agricultural subsector. Indeed, the definition of agriculture within current negotiations is ambiguous. The Kyoto Protocol defines agriculture as including the following non-CO_2 greenhouse gases (GHGs): nitrous oxide from soil and manure, methane from enteric fermentation, rice production, and manure, and methane and nitrous oxide from biomass burning. What is not included in agriculture is soil carbon which, under the Kyoto Protocol, is classified in the land use, land use change and forestry sector. There are also restrictions preventing parties from accounting for soil carbon to generate credits under the Kyoto Protocol's Clean Development Mechanism. In this current negotiation on agriculture under the Convention we have not yet defined agriculture, but it is likely that most parties will consider it to include all non-CO_2 gases from agricultural production and the soil carbon from these lands.

What are the terms of the debate on agriculture, particularly livestock, within the climate negotiations: is there discussion of a direct reduction of production for some commodities? Of more efficient or more intensive production?

Reduction of agricultural production is not an option, food security is the overriding priority for all and climate negotiations must be seen in this context. The ultimate objective of the Convention mentions the need to ensure that food production is not threatened.

Efficiency and productivity are very important. Indeed, mitigation in agriculture lies mostly in making food production more efficient, thus increasing farm productivity and farmer resilience.

On intensification, I think that the old model of "more input, more output", is no longer accepted. Whether you call it sustainable intensification, ecological intensification, agroecological management or sustainable agriculture, we are moving toward a paradigm of resource-use efficiency that places the focus on a better use of available inputs and an increased use of renewable inputs. It is to move from an extractive production system to one that is sustainable in the long term. It is about being as efficient as possible within the constraints of the farm's environmental characteristics.

Is sustainability limited to carbon efficiency?

Unsurprisingly, there has been a big focus on carbon efficiency because these negotiations are about climate change. For some this raises a concern that sustainability is being reduced to carbon efficiency. Also, because climate change and GHGs specifically, more than other aspects of sustainability, have a financial cost associated with them, countries that are exposed to this cost try to find ways to limit it. Whereas at present other sustainability issues have no such financial cost, despite the fact that real costs exist in terms of long-term viability (e.g. soil erosion, water pollution). But the debate is starting to change and long-term effects are increasingly considered.

I think we are seeing an improvement in the understanding of what is at stake. For example, early in the debate the FAO's 2006 *Livestock's Long Shadow* report generated a lot of attention and some negative reactions from the livestock sector. But despite its limitations, the report raised many important issues related to livestock production not only with regard to climate but also land use, water, etc.

What system of farming is better for climate change adaptation and attenuation?

Formal policy discussions, which are sometimes slow to embrace ideas from the scientific community, are now appreciating that it is not just a question of intensive versus extensive, small holders versus large holders, conventional versus organic, and so on; it is about everything, and everything in between. This is reflected in the negotiations – no single system can be considered as the gold standard, because it depends on the characteristics of the location of that system, the climate, soil, livestock breeds, etc. In some cases more intensive systems will have merits, in other cases extensive systems will. While the negotiations have not seen a debate on extensive versus intensive systems, there has been interest in smallholder farming and the need to consider its vulnerabilities.

We should look for performance improvements in all systems, from rangeland grazing to feedlots. But not only in terms of GHGs because sometimes there is only limited room for improvement here, while there may be considerable potential for improvement in water management, animal welfare, energy use and production, etc. For example, there are many reasons beyond climate change for maintaining grazing systems, such as their importance to the rural economy (particularly in developing countries where agriculture still plays a large role), and for cultural and social benefits.

How do agricultural practices fit between mitigation and adaptation to climate change?

The priorities for farmers are to feed their communities, to make a profit, to be able to grow crops again next year, etc. If a certain practice makes a farmer more efficient, or able to continue to produce crops in a changing climate, whilst also increasing soil carbon or yield etc., then it is difficult to say whether this is mitigation or adaptation (it is really both). Many agricultural practices are difficult to separate into these two, in a way, arbitrary classifications. Agriculture is tied to the land and climate; it is a biological system that we utilize, not an industrial process, and we need to keep this in mind.

Unfortunately, due to the way the negotiations work there is likely to be more finance for what is crudely defined as mitigation. Adaptation seems to be very much integrated into development assistance, whereas mitigation gets more attention from new funds, markets, etc. the challenge is how we can use the momentum around mitigation to deliver adaptation benefits for agriculture.

What opportunities do farmers have, particularly in developing countries, to access carbon credits?

Personally, I think that the pursuit of carbon credits

should not be the first resort for vulnerable farmers. The verification costs for a farmer on a relatively low income would probably outweigh the benefits an individual could receive. Focusing efforts on good agricultural development would probably achieve more for both the climate and the farmer in the long term. Good policies - e.g. land tenure; better soil management; improving animal health and fertility; improving feed digestibility and quality; and optimal fertiliser use - would be good for farmers and climate change, and would help mainstream climate change into agriculture.

If the discussion has not brought carbon credits, what benefits has it yielded?

It is well known that investment in research is too low, even though the returns for society are great. While agricultural research funding has fallen over the last 10 years, the situation is at last starting to change. Although there is frustration that there has not yet been a decision on agriculture in the UNFCCC, one of the benefits of having the on-going discussion in the negotiations is that it has created three years of focused attention on agriculture, which has helped catalyse discussions and other institutions such as development banks, the agricultural rural development community and the research community. A lot of initiatives are now being launched, such as the Global Research Alliance on Agricultural GHGs and the G20 wheat initiative. People are starting to invest in agriculture again and are doing it slightly differently than in the past, with a more global approach, focusing on public good and pre-competitive cooperation between the public and private sectors.

Regarding links between trade and climate change, do you think that free trade would reduce carbon emissions?

Anything that improves the efficiency of food production and its delivery to the consumer, and all the steps in between has to be good. However, removing barriers and inefficiency can be politically and socially difficult.

But is free trade good for mitigation? That's a complicated issue. It is much clearer for adaptation – facilitating the movement of products from a place where they can be grown to a place they cannot – is good for the resilience of the global food system. With regard to climate mitigation and trade: it depends. If you import food from somewhere that produces it less efficiently than your own domestic producers, then free trade would not be good for mitigation. But if the imported food is more efficient, then it would be.

There is no guarantee, however, that the removal of trade barriers will achieve this, or that it would promote more climate-efficient imports. In the short term it will certainly promote more economically-efficient imports, and perhaps more climate-efficient imports would follow. Opening market access to countries currently producing food less efficiently but have the potential for greater efficiency if agricultural investment is increased, would probably encourage investments and thus greater efficiency. So, in the long term it would mean a more climate- and economically-efficient global food system. However, it is difficult to say how rapidly the mitigation benefits could be obtained through the removal of trade barriers.

In the intersection between climate and trade policy, the other factor that should be considered is the nature of the climate policies implemented. If food is produced in a region that has a climate policy that places a cost on emissions, or where those emissions are accounted for by the country of production, e.g. under the Kyoto Protocol, and you move production from that country to a region which does not have a climate policy, then this constitutes "leakage" and would be negative from a climate mitigation perspective.

One last word?

The relatively recent attention on agriculture and climate change should not detract from more fundamental agriculture issues such as sound land tenure policies, subsidies, sufficient investment in agricultural research, etc., that all have profound impacts on food production. We often focus too much on the latest technologies, ignoring existing knowledge about practices that, if implemented, could have a huge impact on the resource efficiency of agricultural systems. I don't consider the problem to be a lack of knowledge, more that we face an implementation gap. Not that we don't need new technologies, but we shouldn't try to run before we can walk. If we can't adopt existing practices to improve the resilience and profitability of farming, leading to more efficient production from a mitigation perspective, how can we hope to adopt new GHG mitigation measures? ■

International agricultural research is at the heart of the debate on the future of agriculture and food systems. Its decision-making bodies are now facing two options: to adapt the agricultural model of the green revolution and the accompanying research principles or to opt for the agroecology model, which would then require an agricultural and scientific revolution.

AGRICULTURAL RESEARCH: A GLOBAL INNOVATION SYSTEM AT THE CORE OF STRATEGIC CHOICES

Benoit Labbouz,
AgroParisTech / Centre International de Recherche sur l'Environnement et le Développement, Paris, France

Sébastien Treyer,
Institute for Sustainable Development and International Relations, Paris, France

Even before the political mobilization embodied in the World Bank's 2007 World Development Report on agriculture and the international initiatives that followed the surge in agricultural prices in 2008 (G8, G20, etc.), the scientific community had already grasped, by the early 2000s, the issue of the future of global agriculture and food. Agriculture foresight exercises such as IAASTD[1] (led by Bob Watson, former chairman of the IPCC) and books such as Michel Griffon's (Feed the Planet, 2006) have raised a typically Malthusian question: in a context of scarcity of natural resources, biodiversity degradation and climate change, by 2050, will the food supply be sufficient to feed the growing world population?

While subject to criticism, the Malthusian perspective has merit in its ability, in situations of scarcity and global change, to open the debate on the various projections that can be made regarding innovation and the future pace of progress in science and techniques. Will these advances enable global agricultural production to increase more rapidly than the rates of resource depletion and growth in demand? Depending on whether one answers yes or no this question, it becomes essential to either liberate the full potential for innovation to increase agricultural productivity, or to aim for a fundamental redesign of the functioning of food chains to act on both the demand and supply (Freibauer et al., 2011).

In both cases, the ability to innovate is central. Technological trajectories in agriculture are closely linked, and will remain so, to the economic and social transformations

1. International Assessment of Agricultural Science and Technology for Development

of production systems and territories. Even if rapid innovation is necessary, it is essential to discuss the technological choices and available options in a transparent way, and their impacts on our societies. While the challenge is global, agricultural situations in different regions of the world are very diverse: innovation systems must therefore reflect this diversity. For this to occur, the research systems in different regions of the world must have sufficient capacity. For some countries, mainly in the South, research and innovation systems are as yet underdeveloped and are largely dependent on other agricultural research structures at the international level: cooperation in public research in the North, private research, coordination of an international research system, etc. This article provides a brief history of the role of this international research system in implementing a type of agricultural modernization, known as the "green revolution", in many developing countries since the 1960s. Decisions regarding the research priorities that govern the innovation capacity of this system are then examined, both for the subject areas for research (what are the technological options?), and for the organization of research practices (how can researchers be encouraged to work alongside those in the field?). The debate over the priorities of the international system for agricultural research for development is one of the strategic environments where decisions on global socio-technical issues are made, which will shape the ability of humanity to feed itself tomorrow.

AGRICULTURAL INNOVATION: HISTORICAL AND EMERGING ACTORS

The international agricultural research institution that crystallizes the debate today is the CGIAR (Consultative Group on International Agricultural Research), which was established in 1971 under the auspices of the World Bank. At the end of World War II, the Ford and Rockefeller foundations were major funders of four agricultural research centres (CIMMYT in Mexico, IRRI in the Philippines, IITA in Nigeria and CIAT in Colombia), which allowed these countries to increase their agricultural production through the breeding of major crops (wheat, maize and rice).

BOX 1 THE GREEN REVOLUTION

The green revolution was based on massive external inputs (chemical fertilizers, pesticides, herbicides, improved seeds, irrigation and, often, mechanization). This modernization was based on a coherent package of technologies and focused solely on maximizing yields, and in doing so allowed the simplification of traditional agricultural systems, which through their great diversity – due to the wide variety of crop and livestock species and their combinations – had provided a way to limit many climate and biological risks. The green revolution enabled a 2.5 multiplication in global agricultural production between 1960 and 2005, while the population "only" doubled during the same period. Its techniques were predominantly accessible only to farmers with sufficient capital and the transfer of these technologies to farmers required accompanying agricultural public policies, subsidies for the purchase of inputs, the construction of infrastructure, changes in land tenure, etc. Although widely criticized for its social and environmental impacts, the green revolution remains a reference model for institutions such as AGRA[1], for the modernisation of agriculture in Africa.

1. Alliance for a Green Revolution in Africa

FIGURE 1 THE 15 SPECIALIZED CENTRES FOR WORLD AGRICULTURE

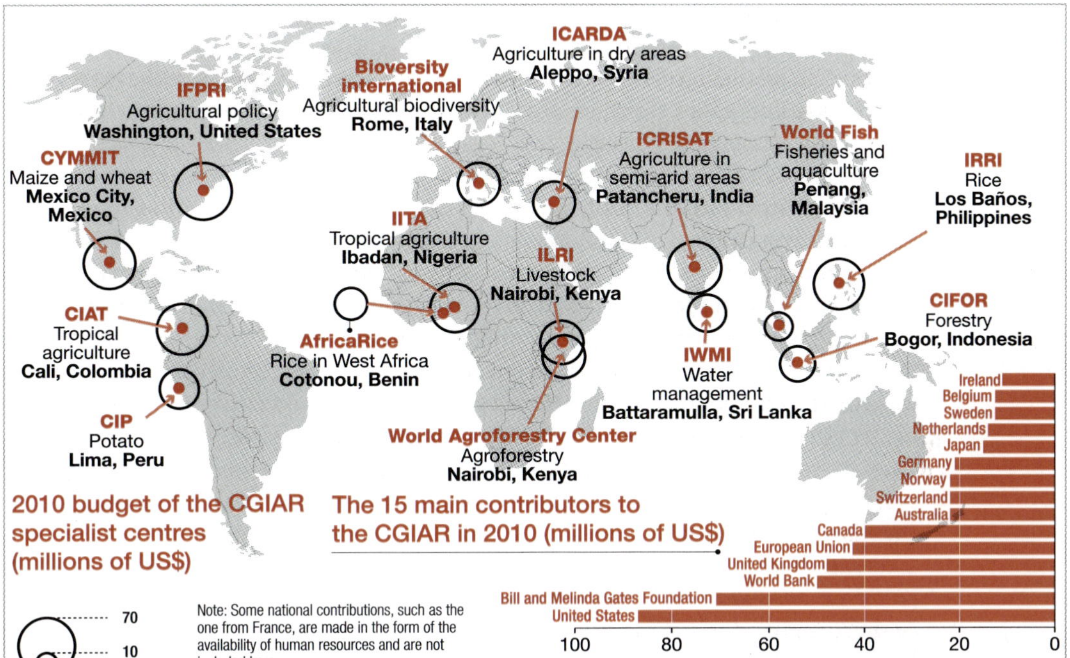

Note: Some national contributions, such as the one from France, are made in the form of the availability of human resources and are not included here.

Source: CGIAR, Financial Report 2010.

These research centres made major contributions to the modernization movement through the dissemination of scientific innovation, that was later to be coined the "green revolution" in Asia and Latin America (see Box 1), and which focused on increasing crop yields to meet the rapid growth in demand. Implicitly, these centres also provided a way to appease the political protests demanding land redistribution. The four centres differed from former colonial centres by their emphasis on genetics and hybridization of high-yielding varieties.

Faced with a need to broaden the funding base of these centres and the on-going risk of hunger, FAO, the World Bank and UNDP[2] set up the CGIAR, a structure for the coordination of research at the four agricultural research centres, to oversee the improvement in the functioning of an international agricultural research system that was gradually becoming established and to mobilize international donors.

Between 1980 and 1990, the number of CGIAR centres expanded from 4 to 18, and there are 15 today (see figure 1). These centres, which are located throughout the world, specialize in specific types of agricultural production: some on particular crops, others on production techniques or systems (irrigation, livestock farming, agroforestry...) or ecosystem types and climatic zones (arid areas, tropical agriculture...).

International cooperation in the agricultural sector has been built around specialized research instruments. All represent crucial issues for global food security. However, despite the support of public and private funds, they do not have equal resources to respond to these issues.

2. United Nations Program for Development

The development of CGIAR centres in countries that lacked efficient national agricultural research centres remains controversial. Although the CGIAR centres have enabled the development of research programmes, thus overcoming the absence of national research systems, they did not however encourage the emergence of such institutions in the least developed countries, rather seeming to substitute them in the long term.

In 2010, CGIAR had 65 members (international organizations, member states and private foundations), which financed its 15 research centres with a total global annual budget of $696 million and rising. For the last 20 years, the CGIAR has been required to provide a better platform to allow the voices of stakeholders to be heard, particularly representatives of farmers from the South who criticize the green revolution in the excessive importance it attaches to technological developments, despite the increasing recognition for the negative social and environmental consequences arising from this strategy. In 2007 this process culminated in a phase of reform for the organization, focusing on its funding and fields of research, which also aims to adapt research to new challenges such as climate change, to further expand the research themes, to ensure the CGIAR operates more transparently and to improve the efficiency of its use of donor funds.

As a response to the CGIAR, the GFAR[3] was launched under the leadership of IFAD[4] in 1996 with a mission to help the national agricultural research systems in the South and to enable producer associations to be heard in the community of international agricultural research. As an international platform for the discussion of strategic options for agricultural research, involving all stakeholders concerned with the future of agriculture, GFAR represents six regional forums. It is financed by the FAO, IFAD and a group of donors that, in 2009, provided the GFAR with a budget of $2.8 million. Its ability to influence the direction of CGIAR research is growing.

In this overview of public agricultural research systems for development, we must not forget the role of research institutes for the development of developed countries (including the former USSR), whose networks of researchers and strategic decisions have contributed to the gradual changes of the CGIAR.

Furthermore, Brazil, India and China are now rapidly entering onto the stage of this international agricultural research system: the extensive development of their national agricultural research systems is attributed to intellectual property rights (patents, the rights of breeders...) and the driving role of private actors and NGOs to guide investment (James et al., 2008). These three countries accounted for 42% of public agricultural research spending in developing countries in 2000 (compared to 25% in 1981).

Private foundations are increasingly becoming engaged in public international agricultural research programmes. For example, private foundations accounted for an 11% share of the budget of the CGIAR in 2010. In particular, in 2010 the Bill and

3. Global Forum on Agricultural Research

4. International Fund for Agricultural Development

Melinda Gates Foundation (in addition to its financial support for AGRA and GAFSP[5], the latter being created in late 2009 in response to the soaring agricultural prices of spring 2008) became an official funding member of the CGIAR, committing to becoming one of its major donors by 2013. In addition, private research by agricultural supply companies is also considerable: in 2005, private agricultural research accounted for nearly 40% of the total expenditure. However, these investments are not homogeneous across the globe, with public research remaining the main form of research in the majority of developing countries (see Figure 2).

The CGIAR reform process constitutes one of the major arenas for all stakeholders to discuss the strategic priorities for agricultural research on a global scale and, in particular, to consider a new model to potentially succeed the green revolution. Today, this debate predominantly forms around two controversial candidate models: a new green revolution or agroecology.

FIGURE 2 PUBLIC AND PRIVATE EXPENDITURE IN AGRICULTURAL RESEARCH IN 2000

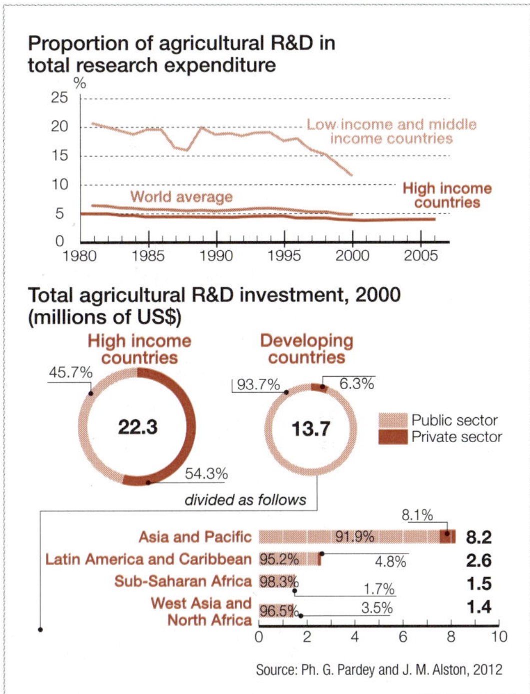

Despite the high stakes, funding for research in agriculture and agricultural science has for decades only represented 7% of government research budgets. While Southern countries have made a greater effort (20% of their research budgets), most funding is provided by Northern countries, which dedicate only 5% of their research budgets to the subject. In recent years, many commitments have been made to reverse this trend, which are awaiting realization.

A "NEW GREEN REVOLUTION" OR AGROECOLOGY: TWO CONTROVERSIAL MODELS

The debate on the direction of CGIAR research is organized around two main reference models, each derived from a particular type of response to the criticism of the green revolution. The models differ in the crop management and agricultural practices they envisage, in the responses they provide to the limited availability of resources, and in the limitations or challenges they raise.

The "new green revolution" theoretically integrates the social criticism that was aimed at the first green revolutions, which bypassed the poorest farmers (Dorward et al., 2004). While the original green revolution was capital intensive (access to inputs, mechanization...), the new green revolution intends to be knowledge intensive. It is mainly based on advances in genetics to achieve similar goals: to improve yields and to improve plant resistance to disease, drought and pests. In addition, it requires the establishment of strong public intervention mechanisms that would focus particularly on the diffusion of technological advances to the poorest farmers.

5. Global Agriculture and Food Security Program

This model promotes major investment in the development of diverse research methods for varietal improvement. By focusing on laboratory-produced knowledge, this "new green revolution" model, similarly to the previous agricultural modernization, positions farmers downstream of research results, agricultural development services and input suppliers, according to a strictly top-down rationale. It seems to ignore the works of the sociology of scientific knowledge, which show the importance of forming associations with the beneficiaries of research, well in advance of the innovation process, for successful implementation (Latour, 1989).

Moreover, while this new modernization project targets small family farms, the productivity gains sought would entail the substitution of labour with capital, and would thus greatly reduce the agricultural workforce, even though this workforce is still important (involving 1 billion people in Asia and 200 million in sub-Saharan Africa), without even accounting for the large population growth that will continue in these rural areas. Unless one anticipates an extraordinary economic boom in another sector of the economy, this path of modernization in Africa may lead to the phenomenon of mass unemployment. The model of the new green revolution should therefore be questioned regarding its ability to develop practices that are sufficiently productive and labour-intensive, and that draw in particular from the modernization trajectories in Asia that had a lesser reliance on mechanization compared to those of Latin America, for example.

WHILE THE ORIGINAL GREEN REVOLUTION WAS CAPITAL INTENSIVE (ACCESS TO INPUTS, MECHANIZATION...), THE NEW GREEN REVOLUTION INTENDS TO BE KNOWLEDGE INTENSIVE.

Finally, though the technologies of the new green revolution are based on the objective to maximize the agricultural productivity from all inputs (land, water and fertilizer) and thus to reduce the environmental impact, they remain largely dependent on fossil fuels and are vulnerable to other types of resource depletion (phosphorus, for example) or socio-political crises and the ecological risks that could be linked the use of biotechnology.

In contrast to this adjustment plan, which represents the continuity of the green revolution paradigm, agroecology offers a completely different perspective for agricultural innovation. A commission chaired by Gordon Conway for the CGIAR in 1996 launched the concept of a "doubly green revolution", but it was actually the IAASTD (McIntyre et al., 2009) that provided the best expertise base to form the foundations of the proposal of an agroecological model as an alternative to the green revolution and its new incarnation. It is based on a fundamental critique of the modernization project that was at the heart of previous technological changes. While the green revolution sought to modernize traditional agriculture by simplifying traditional systems and opening them to the use of external, often synthetic, inputs, agroecology instead puts the ecological functions at the centre of agricultural systems, which are regarded as agro-ecosystems, where the optimization of natural cycles (energy, carbon, nitrogen, phosphorus, water...) would enable the design of systems that are more productive and also more autonomous, that are better able to cope with future environmental issues, benefiting from the complexity and knowledge of traditional systems, and producing not only biomass but also a set of goods and services (protection of water resources, biodiversity, for example). By combining the scientific expertise of ecology and agronomy, agroecology constitutes a separate innovation path, exemplified today by various techniques such as no-till farming (zero tillage), integrated biological control for crop protection, crop rotation, and forming associations between crop species (cereals/legumes, trees/annual crops...) or between cropping and livestock systems. In this innovation model, farmers themselves become producers of knowledge, since innovative agricultural systems must be designed appropriately to suit environmental conditions (as well as socioeconomic ones) that are specific to the local context.

It is likely – with the exception of techniques, such as conservation agriculture, that are fully compatible with large mechanized farms – that innovative agroecologically-designed systems will be more labour intensive, thus providing employment opportunities for the large agricultural labour force (as indicated by the agricultural chapter of the UNEP Green Economy report, 2011): it remains to be seen whether these jobs can be sufficiently remunerated. Indeed, the future success of the agroecology model will be partly based on its ability to design technical devices that enable yields to be maintained at sufficiently high levels (for Northern and emerging countries) and that also provide significant yield increases (for countries where yields are still low, such as in sub-Saharan Africa).

While the green revolution aimed at overcoming environmental conditions through inputs, allowing laboratory performance to be achieved in the field, the agroecology model is much more related to local soil, climatic and ecological conditions. This model, compared to the green revolution, thus leads to a diversification of products, and probably a greater variability of these products. This will be a marketing challenge given the stability requirements of processors, retailers and quality standards, which focus on a small number of primary products.

These two models that polarize the current debate should not obscure the existence of a wide range of proposals that can be attached to one model, to the other, or even to both. The IAASTD presents a set of models (organic, ecological, resilient farming...), that illustrate some trajectories of innovation that do not rely solely on biotechnology and breeding, but on innovation across the entire agricultural ecosystem. Among them, organic farming constitutes a real path of innovation based, like agroecology, on the improvement of the management of natural agroecological cycles and not only on the refusal to use synthetic inputs. This refusal leads to a controversy regarding the ability of organic farming to produce enough food to ensure global food security, which advocates are trying to reframe by highlighting the advantages in terms of local access to food. The agroforestry models presented by Dennis Garrity from the World Agroforestry Centre (Garrity et al., 2011) or the "Evergreen" model developed by MS Swaminathan (Swaminathan, 2004), father of the Indian green revolution, are also based on new ways to connect agronomy and ecology, to link conventional scientific techniques and alternative knowledge. In the field, these alternative models can hybridize to various extents with the new green revolution. But in terms of innovation systems, they lead to a profound questioning of the current mode of research.

PROFOUND CHANGES IN ORGANIZATIONS AND PRACTICES OF AGRICULTURAL RESEARCH AND DEVELOPMENT

In its extreme, the pattern of laboratory research is to give new properties to living organisms (animal or plant) in vitro, to test them in vivo and then to disseminate them to farmers through agricultural advisory and development services, along with inputs enabling real conditions to be brought closer to laboratory ones. In this scheme, farmers are passive recipients of knowledge developed elsewhere (Dockès et al., 2011).

Participatory research practices completely disrupt this pattern, incorporating not only the needs of farmers but farmers themselves at all stages of the research process, using their own empirical knowledge, both traditional and current, and involving:

- analysis of what needs to be improved (yields, the system as a whole, incomes, wellbeing...),
- combined mobilization of scientific knowledge (biology, ecology, chemistry, but also economics, sociology...), of intermediate concepts and knowledge produced in other situations of innovation, and of local knowledge to identify opportunities for improvement and possible solutions,
- combined experimentation within farms, and in interaction with laboratory work,
- validation and dissemination of innovations through farmer network organizations and other local social networks, in partnership with agricultural development services.

In this scheme, agricultural development services have a very large role to play since they must help farmers in expressing their needs, participate in experimentation and must also consider their interface role as a two-way interactive process between researchers and research users, who themselves become innovators. The knowledge of farmers is expressed, acknowledged and shared as much as that of the researchers, thus renewing the role of training systems themselves (Freibauer et al., 2011). This pattern of participatory research is indispensable for agroecological innovation trajectories aiming at the optimal management of agroecosystems in their specific conditions.

The ability to transfer innovations to a large number of different agro-ecosystems is at the heart of the green revolution model, which aims at adapting the ecosystem to improved varieties by benefiting from the contributions of external inputs (James et al., 2008). For agroecology, the diffusion capacity is not based on the transfer of research products themselves, but on the dissemination of knowledge within the research and development network. While economies of scale were a major driver for the operation and dissemination of green revolution innovation, the organization of participatory innovation is necessarily based on other conceptions of the knowledge economy, largely focused on the public nature of the knowledge produced.

In both processes, public and private research are interdependent, but take different forms and methods of organization (partnership between farmers and their organizations with public research, partnerships between private industrial research and public services of research and development). Private research has a place when innovation can be appropriated (intellectual property rights, including on mechanical, chemical, or biotechnological technology). Innovations at the scale of practices, farming systems, agricultural landscapes and their organization, that agroecology seeks to put back at the centre, have the nature of public good and thus reduces the incentives for private research (Pardey et al., 2010; Vanloqueren et al., 2009).

The controversy between the different models of agricultural techniques might seem overly polarized when perhaps we need to allow these different models to coexist. However, it is rooted in profound changes to practices and the organization of research systems, innovation, development and training, all of which require consideration on the ways to control the transition from one model to operate with another.

TRANSITIONS THROUGHOUT THE FOOD SYSTEM AND IN INNOVATION SYSTEMS: THE NECESSARY STRATEGIC CHOICES

The current controversy is not just about emerging farming practices, that could be technically combined in an optimal way with practices from the new green revolution. These models are in fact distinct innovation trajectories and the transition from one to another involves inextricable and profound changes in upstream (agricultural advisory services, farm supply) and downstream (collection, processing, marketing) sectors, and also changes to the organization and functioning of innovation systems, and public policy support for these changes. The strategic choices discussed for the CGIAR reform process will therefore have a wider impact than purely on the 15 research centres involved.

To drive through a transition, the approach of leaving all options open it is not sufficient: such a non-strategic choice ignores the balance of power, the rigidity of structures and institutional resistance. A common representation of transitions in innovation systems is that a dominant regime (a configuration of stakeholders, institutions, with privileged access to resources and funding...) is being challenged by emerging models (Geels et al., 2007). During transition, dominant and emerging models co-evolve through a process of confrontations, until one of these new models becomes dominant again.

The modernization project of the green revolution has proved dominant in agricultural practices, in the research for new technologies, in public policies and private organizations that have been implemented to ensure the effective performance of the project, the structuration of the agri-food sector (elongation of processing lines, concentration of processors...), consumption habits, etc.

For many years this dominance has been challenged, but without it being possible to say that the transition has begun. In the worldwide debate on agricultural research, the agroecological innovation model has become a serious candidate for succession in recent years.

What policies to support innovation are required to drive such transitions? Apart from the need to ensure that niches exist for the emergence of new models, isn't it also important to make choices in support of a particular model to allow a sufficiently rapid transition given the pressing issues of resource scarcity and global change? Or to offer an undifferentiated specific public support to all emerging models, but according to which criteria? Do we have the luxury of time to allow spontaneous change in research and innovation systems to run their course? With no simple answers to these questions, political choices must be made between the contrasting visions of the relationship between the current environmental and social emergencies and the different models of innovation. They make divergent predictions on the speed of transitions in the systems of innovation, and on the proximity in time of environmental, social and economic crises.

Specifically, controlling these transitions entails the enablement of organizational changes in the innovation system, and changes in the distribution of financial, human and symbolic resources. The Global Conference on Agricultural Research for Development (GCARD) in 2010 was a step in the right direction for this "quiet revolution". The first conference of its type, which is to be held every two years, the 2010 GCARD seemed to focus on the capacity of national research systems and farmers in Southern countries (who are the beneficiaries of these research efforts, gathered within the GFAR) to become involved in the upstream determination of priorities of the CGIAR. The outcome of the transition process remains very uncertain however. In Beijing in October 2011, the CGIAR structures held a "Science Forum" to reaffirm their position at the head of, or alongside, GFAR, each claiming the capability to produce the strategic prospective reflections necessary to achieve the required shifts. Moreover, while the GCARD seems to have gathered all public and private initiatives for national and international agricultural research into GFAR and CGIAR, the future influence of foundations such as the Gates Foundation, which currently mainly funds the CGIAR pragmatically aiming above all for "impact", remains unclear. The same applies to national research systems in emerging countries, whose international ambitions are still not well-defined, and also to the expected emergence of national or regional research systems in the least developed countries, which are especially needed for the agro-ecological trajectory of innovation.

THE "NEW GREEN REVOLUTION" THEORETICALLY INTEGRATES THE SOCIAL CRITICISM THAT WAS AIMED AT THE FIRST GREEN REVOLUTIONS, WHICH BYPASSED THE POOREST FARMERS.

This system of the global governance of research is still under construction, in parallel with the reform of the FAO's Committee on World Food Security and its high level panel of experts, or the FAO's Treaty on plant genetic resources for food and agriculture and the Cartagena or Nagoya Protocols of the Convention on biological diversity. However, the same actors circulate between these arenas, disseminating the controversies on the new modalities of the relationship between science and society, be it through the questioning of technology choices, the role of experts or the participatory functioning of innovation systems.

Far from being restricted to the corridors of power where research policies are made, the transformation of this international governance regime for agricultural research for development therefore currently constitutes one of the conflict arenas in which the strategic options for innovation are discussed, which in turn will outline the future trajectories of the world's agricultural and food systems. ■

REFERENCES

Dorward, A. R., Kydd J. G., Morrison J. A. and Urey I., 2004, "A Policy Agenda for Pro-Poor Agricultural Growth." *World Development* 32(1): 73-89.

Freibauer, A., Mathijs, E., Brunori, G., Damianova, Z., Faroult, E., Girona i Gomis, J., O´Brien, L., Treyer, S., 2011, "Sustainable food consumption and production in a resource-constrained world", 3rd Foresight Expert Group Report, European Commission – Standing Committee on Agricultural Research (SCAR)

Dockès, A.-C., Tisenkops, T., Bock, B., 2011, "Agricultural knowledge and innovation systems - Reflection paper", Collaborative Working Group on Agricultural Knowledge and Innovation Systems, Standing Committee on Agricultural Research, European Commission.

Garrity D., Bayala J., Kalinganire A., Tchoundjeu Z., Sinclair F., 2011, Conservation agriculture with trees in the West African Sahel – a review, ICRAF Occasional Paper n°4, 2011, 72 p.

Geels, F., Schot, J., 2007, Typology of Sociotechnical Transition Pathways, *Research Policy*, n° 36, pp. 399-417

James J., Pardey P., Alston J., 2008, Agricultural R&D Policy: A Tragedy of the International Commons. University of Minnesota, Department of Applied Economics Paper Series. InSTePP Paper:08-01

Latour, B., 1989, "La science en action", La Découverte, Paris.

McIntyre, B., Herren, H., Wakhungu, J., Watson, B., 2009, "International assessment of agricultural knowledge, science and technology for development (IAASTD): synthesis report", Island Press, Washington.

Pardey, P., Pingali, P., 2010, "Reassessing International Agricultural Research for Food and Agriculture", Report prepared for the Global Conference on Agricultural Research for Development (GCARD), Montpellier, 28-31 March 2010.

Swaminathan M.S., 2004, Ever-Green Revolution and Sustainable Food Security, National Agricultural Biotechnology Council report 16, 2004, pp. 63 – 76.

UNEP, 2011, Towards a Green Economy: Pathways to Sustainable Development and Poverty Eradication, United Nations Environment Programme, 624p.

Vanloqueren G., Baret P.V., 2009, How agricultural research systems shape a technological regime that develops genetic engineering but locks out agroecological innovations, Research Policy n°38, pp. 971 – 983.

EVALUATING THE GREEN REVOLUTION

Seema Purushothaman,
Ashoka Trust for Research in Ecology and the Environment (ATREE), Bangalore, India

The Indian green revolution provides a reference in debates on new green revolutions elsewhere in the world, particularly Sub-Saharan Africa, serving as an example from which valuable lessons can be drawn. What is the state of the debate in India regarding the evaluation of the green revolution? Do people talk about adapting or changing models?

Among the large body of literature evaluating India's green revolution, there is a consensus on three positive aspects. First, it is generally agreed that the green revolution played an important role in the avoidance of famines in the 1970s and 1980s, even though it did not completely alleviate hunger. Second, the green revolution has improved infrastructure (irrigation, marketing, transport) in many rural areas (with a similar caveat that improvements were not uniform). Third, it led to a greater investment in agricultural education and research in most regions.

However, there are also many criticisms. The first revolves around the disparities (*i.e.* rural/urban, agricultural/non-agricultural) that were exacerbated by the uneven distribution of the technology package in India. The second criticism concerns the focus on a limited number of crops, thus creating an imbalance in the supply of food grains that subsequently led to an imbalance in rural nutrition: there was a dramatic decrease of calories, protein and vitamin-rich foods. A third problem is the focus on a small number of technologies, namely chemical fertilizers, pesticides and irrigation. This narrow focus on both crops and technologies may not be the exact focus of the green revolution package as envisaged by M.S. Swaminathan (who states that early green revolution literature actually refers to the precautionary principle), but could be an unintended outcome of the dissemination of the technology change. Even though the green revolution was technology-driven, its impact and focus were affected by many other factors such as infrastructure, the availability of inputs and other policies.

The green revolution's requirement for technological advancement meant that incentives had to be created, i.e. that the necessary technologies were subsidized, giving rise to a fourth problem: input and output subsidies became integral to Indian agriculture. Subsidies are not only a budgetary burden but they also benefit farming industries and big farmers disproportionately and have negative impacts on the sustainability of farming systems.

Is the evaluation of the green revolution consensual or controversial?

During the 1970s and 1980s, much of the scientific literature highlighted the achievements of the green revolution in terms of the adoption of new varieties, a production increase and a decrease in rural poverty. In the 1990s, the amount of literature on the negative impacts started to increase, opening a debate between four stakeholder groups: (i) farmers, (ii) NGOs, (iii) researchers and (iv) policy makers.

Regarding the first stakeholder group, sixty per cent of India's agricultural population is dominated by small farmers with less than two hectares, for whom farming is important as both consumers and producers. However, India's most influential farmer organisations are focused on specific crops (sugar-cane, onion, etc.) and as small farmers tend to cultivate more than one crop, their voices are not generally heard in such organisations, losing out to powerful lobbies of bigger farmers. Thus, small farmers are insufficiently represented in national agricultural debates. Furthermore, small farmers in many of India's agroclimate regions still use traditional technologies. Although not strictly classifiable as organic or agro-ecological, they do not adopt high input intensive practices as an entire package, but on a need-based approach.

The second stakeholder group is NGOs, all of which are critical of current agricultural models and policies, mainly on the grounds that they are cost and credit intensive. On this matter, both social and environmental NGOs frequently overlap. Voluntary agencies play an important role in agricultural discussions in the media, but less so in academia or in policy corridors where they remain unheard.

A third group is researchers. They can be subdivided into governmental researchers (that work mainly in agricultural universities), corporate researchers, and independent researchers such as members of the Ashoka Trust for Research in Ecology and the Environment or the M.S. Swaminathan Foundation. Increasingly, the work of government researchers has moved closer to that of corporate researchers who generally publish in the same disciplinary journals. Independent research groups though still a minority, generally try to incorporate intersectoral and interdisciplinary perspectives.

The fourth group of stakeholders is policy-makers. While state policies (or provinces) are moving away from green revolution techniques, federal level policies remain unchanged and maintain subsidies for fertilisers and irrigation as it was in the 1960s. Constitutionally, agriculture is a state-run affair. States can formulate agricultural policy and allocate budgets, but this conflicts with centrally implemented policies of fertilizer subsidisation etc. Hence, in states such as Karnataka, Andhra Pradesh or Kerala, which invest in sustainable agriculture, there is both coexistence and competition between policies in two directions. It is not possible for a state to exclude itself from fertiliser policies, because all fertiliser outlets within a state sell fertilisers at a subsidised price. Thus the only solution is for states to subsidise organic fertilisers and sustainable inputs, which is increasingly happening.

These four stakeholder groups have different approaches and advocate different development paths for Indian agriculture, but the contours are moving. For example, the national commission on farmers has called for a new green revolution, qualifying their aim by stating that we must first learn from the mistakes of the previous one. This represents a real concept change.

How does the discussion between mainstream and ecological agriculture play out in India? How does it relate to employment issues?

In India, the mainstream approach to farming is much more dominant as it receives federal and commercial support, while the agroecological school mainly involves the voluntary sector and independent researchers.

In India's rural areas, off-farm employment possibilities are limited. Agriculture is often the only employment source, thus favouring labour intensive agriculture in the context of a low opportunity cost. Conversely, productivity for small farmers is considered in terms of productivity per labour unit, and not per hectare. This does not mean that small-scale farmers are averse to technology, but that they need appropriate technologies that take their circumstances into account.

Moreover, India cannot afford to move millions of smallholders and their families away from farming. Indian cities can no longer cope with an influx of unskilled workers, given that urban development is taking place at a gradual pace. Unskilled workers will not find stable employment in the service or manufacturing sectors.

To move forward, what should be done in terms of regulation and legislation?

As an academic I cannot make recommendations, but only suggestions that are based on extrapolations from my research. It is possible for India to meet the challenge of food and agricultural sustainability, but to do so requires policies that target the millions of smallholder farmers that want to engage as both consumers and food producers. They do not go against the grain of development. They want their children to be educated and, if they so choose, to move away from the sector. But such ambitions should not be motivated by distress, but happen in due course in line with the acquisition of new non-agricultural skills.

Policies should not ignore the traditional skills and cultural peculiarities of these smallholder farmers. Furthermore, small-scale farming has the lowest ecological footprint; therefore an alternative development path for this sector could be an opportunity for India to reduce its carbon emissions.

So, how can policies be better designed to account for smallholders? There are at least three areas that must be addressed. The first two concern the marketing of inputs and outputs. Small farmers are at a constant disadvantage when they depend completely on markets since they are subjected to uncertainties, especially for outputs. It is also a question of getting good prices. In earlier times, Indian farmers had to gamble only with the monsoon, now they must also gamble in world markets. Furthermore, because small farmers lack storage and marketing infrastructure, they cannot get good prices on export markets and cannot therefore depend on them. This is an issue that needs to be addressed by policies, especially since the Indian domestic market is large. If the government could mitigate the risk linked to the connection to local markets, it would be an important contribution to the sustainability of the small farm sector. The second point to tackle is to reduce the dependence on market dependent inputs, in order to mitigate cost escalation risks; and the third point is the need for climatic risk management, by providing these farmers with insurance against climate variations.

It is only if these three areas can be successfully tackled that a fourth factor can be improved, namely the capacity to mobilise credits and loans. The extension of loan facilities is necessary to make small farming a sustainable enterprise, but this must be accompanied with a good output market and sufficient climate insurance. If not, the extension of loan facilities to farmers without upper limits would be a recipe for disaster, as shown by the large numbers of farmers that commit suicide in India. Although it must be said that farmer suicides are not always linked to farming activities and loans.

The much needed policy package should therefore be more than just a list of measures, but must be designed strategically. Furthermore, the issue of how policies are made is also critical. The agroecological school in agricultural management adopts a very decentralised vision in both policy implementation and formulation, and even the research agenda. In many cases centrally decided priorities do not fit with local realities. For example, the green revolution package extended irrigation to masses of farmers without assessing the costs and benefits to the specific ecosystem, sometimes leading to over-irrigation among other problems. This was due to a top down imposition of technologies: where the extension process does not take specific local aspects into account, which may be crucial for the sustainability of a technology, or at least for it to have a neutral impact. Hence, in summary, policies should focus on smallholders, prioritise the mitigation of the three risk types and operate in a decentralised manner. ■

The market is not only a place for setting the price, nor is it merely a location for commodity exchange. It is a power field where farmers, as relatively passive agents of trade, encounter the "production" of price. This article seeks to analyze the interaction between fields and markets from the vantage point of cotton farmers.

AN ETHNOGRAPHY OF COTTON PRODUCTION AND EXCHANGE IN A TURKISH VILLAGE

Koray Caliskan,
Bogazici University, Istanbul, Turkey

This ethnography describes the universe of cotton production and exchange in Pamukköy, the largest cotton-producing village in the Söke Plain of western Turkey. It documents how farmers mobilize resources, interact with agricultural workers, obtain credit and sell their product. In contrast to the cotton field, where farmers perceive themselves as active and formative agents in the rural political economy, cotton growers view the market as dominated by traders and various mercantile tools that weaken their agency.

STUDYING AGRICULTURAL MARKETS: A HISTORICAL CONTEXT

Late 19th and early 20th century perspectives on agriculture, especially those inspired by Marx, regarded farmers as a transitory class doomed to disappear as a result of rapid industrialization and modernization. In the post-World War II period, this perception of agriculture expanded to encompass the economic motivations of the peasantry, modes of production, relations between farmers and the state and other classes, and their role in rural transformation and development. A vast literature gradually accumulated, which included: *articulation theories*, looking at the ways in which urban actors (such as the state and capitalists) extract surpluses from farmers; *persistence theses*, explaining the survival strategies of farmers against worldwide market expansion or capitalism; and *differentiation theses* that focus on the pace of proletariatization and internal variation among rural producers. In these works, capitalist relations and markets are, in general, implicitly considered to be outside factors that impact on the lives of farmers, and it is the task of the social scientist to study the process and consequence of that impact.

Polanyi's *Great Transformation* reinterpreted the meaning of markets, emphasizing the role of the state and politics in the making of markets in the modern period, and the formative role of social relations in pre-modern societies (Polanyi, 1944). The challenge now is to carry out research that empirically analyzes the processes through which prices are made in concrete market settings.

There is no commodity other than cotton that gives us a better vantage point to study the interaction between the field and the market. Cotton is located at the intersection of industrial, financial and agricultural relations of exchange/production that connect more than a billion people to each other through agriculture, trade and textile manufacture. Every year, more than 50 million farmers from 81 countries produce around 90 million bales of cotton. Compared to other cash crops that farmers rely on – excluding crops for the farmers' direct consumption, such as wheat – cotton covers the largest production area in the world, followed by sugar cane, sunflowers, coffee and tobacco. In terms of trade volume, no other agricultural commodity comes close to the circulation of cotton. Every year, more than one third of cotton produced all over the world crosses national boundaries and is consumed in a country other than where it was originally produced. This is the largest share of any agricultural market in the world. Historically, too, cotton trade has put its mark on world trade, with a total value of twice as much as gold and silver combined in the late nineteenth century.

COTTON PRODUCTION IN PAMUKKÖY

This essay investigates cotton growing and exchange in and around the Turkish village of Pamukköy. It examines the ways in which farmers encounter the cotton market as they sell their produce to private merchants or TARIS (Union of Agricultural Co-Operatives for the Sale of Figs, Raisins, Cotton, Olives and Olive Oil), the only cooperative that deals with cotton trade in western Turkey. Farmers do not see the cooperative as a merchant. Because they have an organic relationship with the cooperative, its cotton purchase is seen as endogenous to their production process. In contrast, cotton traders and their offices are considered exogenous to farmers' fields. As one farmer put it, "things happen there," referring to the trading tools and manners that traders deploy to gain powerful bargaining positions. The article analyses this power relationship and argues that prices should not be seen as a mere exchange tool set by the supply and demand of cotton. Serving as a social and financial interface between the field and the market, the price is also a tool utilised by merchants, which – together with other specific ways of trading – limits the farmers' agency.

As farmers produce and exchange cotton, they are affected by several markets, but the two in which they participate directly are the labour and cotton exchange markets. As demonstrated below, there is no single price for cotton, as it is reflected in the index of world cotton prices. Rather, different price forms are experienced, produced and deployed by farmers, merchants, the state and the world market's big players. Hence, prices are made in fields of power, and the market is one such power field, where farmers encounter price production as relatively powerless agents of trade.

TABLE 1 LAND OWNERSHIP IN PAMUKKÖY

hectares	Number of household	Percentage of households (%)	Cumulative percentage of household (%)
5 et plus	5	2	2
4-5	2	0.80	2.80
3-4	4	1.60	4.40
2-3	20	8	12.40
1-2	44	17.60	30
0.5-1	49	19.60	49.60
0-0.5	54	21.60	71.20
None	72	28.80	100

Slightly more than 10% of Turkey's total production comes from the Söke Plain, Turkey's second-largest cotton-growing plain. Located in the province of Aydın, the plain is in the administrative boundaries of the town of Söke, an agro-industrial city 155 km west of Izmir. Cotton is the single most important cash crop of the plain. Looking down from the hills of the Samsun Mountains in August, the plain looks like a green sea of cotton, encircled by Lake Bafa, the Beflparmak Mountains and the Aegean Sea. Depending on the source quoted, between 4,010 and 6,152 farmers grow cotton in the plain and live in villages and towns encircling this green sea which provides cash revenue not only to cotton growers, but also to thousands of migrant agricultural workers who arrive in the plain every year to work in the cotton fields.

According to the village headman, until 20 years ago the population of Pamukköy was around 1,700. The neoliberal reforms of the early 1980s increased migration to cities, thus decreasing Pamukköy's population to 734 by 2001. The 250 households of Pamukköy own 249.5 hectares of land, an average of about one hectare per household. 71% of households own land in the village (see Table 1). Half of the remaining households rent land for cotton production, usually no more than half a hectare. The rest do not grow cotton, but have household members who work in cotton-related jobs; they work in others' fields and drive their tractors. Those who do not reside in the village own fourteen percent of the village land. Farmers who reside in Pamukköy use almost all of this land in exchange for either rent or cotton. Land use patterns follow land distribution patterns in the village.

THE HARVEST

The sale of cotton is the only major cash revenue for Pamukköy farmers, the harvest of which occurs between late September and mid-November. Harvesting is the most labour-intensive part of the cotton-growing process and there are three distinct groups of farmers that deploy three distinct ways of concluding the growing seasons:

- those who have a large enough labour supply in their households and relatively less land to work on, by and large draw on non-monetary forms of labour exchange during the harvest. It is safe to assume that a family can rely on household labour if it has less than 0.2 hectare of land per working household member. Even then, these families usually borrow labour. However, a land/household member ratio above this level requires some sort of monetary or non-monetary labour exchange in the village.
- those families who own more than 0.2 hectares per working household member, but do not have enough money to hire labourers, tend to employ labour-pooling where a few families come together to work on each other's land. Farmers use

complex methods of labour control in such arrangements, known locally as an *Ödek*. Families of different sizes come together to pool their resources but because all have different capacities to contribute, the surplus or shortage is covered by cash. These complex cooperatives are common in many developing countries where the majority of farmers rely on unpaid domestic labour for survival. Their flexibility is striking in the sense that these informal cooperatives can form a workshop in a couple of hours and dissolve it even quicker.

- those that depend solely on hired labour. Compared to the first and second groups, these farmers use less complex means of labour control in the field. Workers are paid according to the weight of cotton they pick. The poverty of workers and their lack of union power facilitate labour control for those who have enough means to hire them. If a worker loses her job, she cannot make it to the next harvest without borrowing money.

The speed of picking cotton depends on dexterity, determination and wealth. Those who work on their own fields can work at a gentler pace. Those who work for friends and relatives can choose not to work in difficult muddy conditions. The neediest and most impoverished workers, who are mostly Kurdish seasonal migrants, are obliged to work in any conditions.

A family of four working members can hand-pick their cotton in two months or less for a field not exceeding 2 hectares. With four seasonal workers, the same amount can be hand-picked in 12 to 15 working days. On average, workers pick 85 kg of seed cotton per day.

Most Pamukköy cotton farmers barter labour and settle the difference with cash or cotton in various labour-pooling mechanisms. Those with larger fields hire workers to carry out production. Whether they use money or not, all these labour exchange processes include market performances composed of various resources and tactics. These performances are shaped through power contestations and social positions. The amount of land one has, the degree of dispossession of a worker, the farmers' and workers' relations with labour gang leaders, the farmers' indebtedness and the power asymmetries between farmers and workers at the moment of exchange, all play a dynamic part in shaping the labour market. As the size of land increases, farmers tend to rely more on monetary exchange. As the level of dispossession increases, farmers find themselves in a weaker position vis-à-vis those who hire them as workers. Yet, even the most powerful farmers in these village-level exchanges experience disempowerment and poverty once they enter the market for cotton exchange. To analyze this market and the ways in which fields and cotton markets connect in Pamukköy, we need to examine the dynamics of commodity exchange on the ground.

THE MARKET: EXCHANGING COTTON IN PAMUKKÖY

Neo-classical economists and sociologists describe markets as price-making contexts and explain prices as being made in markets. Aiming to break out of this circular reasoning, other researchers have sought to show that prices and markets are embedded and constructed in the very social and cultural contexts in which they operate. Recently, however, new research in economic and social anthropology has begun to focus on markets as socio-technical universes from the vantage point of price-making. This literature shows that prices are produced in the specific geography of the markets where they emerge within relations of power. Despite a globally accepted price form (such as US dollars), each specific price is made in a multiplicity of power relations that go beyond a mere coming together of supply and demand (Callon, 1998; Muniesa, 2000). Each price is a tool produced and deployed in the specific context of an exchange relationship. The making of prices in the cotton markets of the Söke Plain is in line with the findings of the literature on the anthropology of markets. From the harvest to the final exchange of cotton, categorically different price forms emerge.

FROM THE HARVEST TO THE EMERGENCE OF PRICES

The end of the harvest marks the last stage of the cotton's life cycle in villages like Pamukköy. Farmers sell their entire crop before the last day of October. There are two main buyers of cotton in the plain: the agricultural sales cooperative TARIS, and private merchants who also own ginning factories. Since cotton has to be ginned before it can be sold in the institutionalized market settings – such as the cotton-trading pit of the Izmir Mercantile Exchange (IME) – a vast majority of farmers (except for a few large landowners) are excluded from lint markets. For the farmers of Pamukköy, the exchange of their cotton takes place outside the exchange building. It is possible to keep their seed cotton and sell it to a trader later, but this possibility is realized only in a few marginal instances.

There are many reasons for the immediate selling of the crop. The primary reason is that growing cotton requires farmers to borrow heavily, thus creating an urgent need for cash after the harvest. Farmers who own more than two hectares hire labour and pay the daily wages in cash, usually by borrowing. Farmers have different means of finding money, for example Ibrahim, a farmer owning 1.8 hectares, borrowed money from a merchant who had a ginning factory in Söke. Although the terms were similar, he chose not to borrow from the state-owned Agricultural Bank (*Ziraat Bankasi*) for several reasons. First, the bank required the farmer to have land under his name as collateral. Ibrahim's land was not registered under his name. He owned the land, yet the land was formally registered as village property. Secondly, banks use formal written means of communication that can be seen by anyone who frequents the *kahve* (coffee house). To be known as someone with inadequate financial resources is shameful for many farmers and implies a decline in status.

Informal bankers generally apply high interest rates, at least five percent higher than bank rates. They manage to buy their clients' cotton for less money, thus contributing to the temporary depression of prices during and immediately after the harvest. Moreover, when it comes to buying the cotton from the indebted farmer, it is not uncommon for them to overweigh the produce and downgrade the ginning outturn of the cotton. Farmers cannot resist selling their cotton to merchants, for they always sign a contract with these cotton-money merchants to pay their debt back on 31st October, immediately following the harvest. Without financial backing, it is not possible for growers to keep their crop until the time comes to sell it under relatively better terms. The abundance of cotton and the urgent need for cash depress the prices between late September and early November. It would not make any difference, if they chose to sell to TARIS, since the cooperative pays the price of the day when cotton farmers bring their crop.

TRADERS' PRICE

An analysis of monthly cotton prices reveals the large price differential between prices in October compared to other months. Between September 1993 and August 2004, the average of the Cotlook A Index, a price index commonly referred to as the world spot price of cotton, was 66.78 cents/lb. The October average was 63.78 cents/lb. This is a deviation of exactly 300 points, a difference vast enough for the New York Board of Trade (NYBOT) to formally enforce a halt in trading, because in NYBOT's globally recognized cotton futures market a 300-cent deviation in a single trading session is an institutionally recognized sign of market crisis.

PRICES ARE MADE IN FIELDS OF POWER, AND THE MARKET IS ONE SUCH POWER FIELD, WHERE FARMERS ENCOUNTER PRICE PRODUCTION AS RELATIVELY POWERLESS AGENTS OF TRADE.

The Cotlook A Index only represents price deviation to a limited extent and possibly under-represents the variance as it incorporates traders' price perceptions. The Izmir Mercantile Exchange reveals a more alarming difference. The average October price between 1993 and 2004, including the price of 2001, is 65.23 cents/lb, while the average for all other months is 73.19 cents/lb, making a vast difference of 796 points. This is approximately eight cents for each pound of cotton, easily enough to be considered as a market crisis if occurring in the world of merchants.

Farmers are excluded from lint markets because they do not have warehouses to store their cotton, the financial means to insure it, the means to follow prices on a daily basis or the political power to participate in price-making. These prices belong to the world of traders, merchants, brokers and a few large landowners. For example, cotton prices in the IME, Turkey's cotton market centre, are produced in three main forms in four different market places (see box 1). Each of these price forms is produced in specific market geographies, almost all of which are designed to keep the farmers outside the exchange (Çaliskan, forthcoming). *The market price* of the IME is a price indicator used to set individual prices; *the pit trading price* is a rehearsal device that merchants produce to probe the market to see whether it is bullish or bearish; and *postpit prices* are the actual prices of cotton exchanged in the market, but because they are not posted, no one has certain knowledge of them. In short, the market as it takes shape in Izmir has multiple prices that work as "prosthetic" devices used by traders to pursue their mercantile interests. Prosthetic prices are developed to shape actual transaction prices, but cannot be used to exchange actual cotton; rather, they are used to *set* actual prices for cotton exchange.

Apart from the production and deployment of various price forms, traders also affect the making of prices by negotiating each year's supply and demand estimates. These estimates are more effective in creating prices than their real levels, for the actual levels are known only after all the cotton is bought and sold on the market in late October.

Three main groups with three diverse motivations – merchants, the TARIS cooperative, and government representatives – attend the meeting of the Permanent Working Group on Cotton where supply and demand are estimated each year. Merchants dominate the meeting as they host the caucus in the trader-run exchange and also because they attend en masse, creating the absolute majority in the meeting room. They have an interest in depressed prices because lower prices make it possible for them to trade larger volumes. As a result, they work to prevent the meeting from underestimating the production level, since a forecast of decreasing supply pushes up prices. The cooperative representatives have a contrasting motivation as they benefit more from increasing prices as this translates into higher profit margins for the cooperative. Thus, they aim to prevent overestimation of the production level, because the expectation of a higher supply depresses prices. Government representatives, however, have a different focus. Their main purpose is to record the cotton supply levels as accurately as possible for the Minister of Agriculture.

The author's fieldwork suggests that merchants have the largest say in these estimates. Every year, traders manage to influence the meeting to produce a higher estimate than the actual levels and, thus, successfully depress prices before the harvest. These price forms disappear as the traders approach the countryside as they mean little to the vast majority of farmers, for a number of reasons:

First, the terms "merchant" or "trader" are used only in the singular form and in a pejorative sense. The widespread belief is that they protect the interests of yarn producers and themselves. Therefore, *any* price in any *form* associated with merchant houses is regarded with suspicion, as just another tool to obtain the farmer's produce at the lowest possible price. Second, whatever the price, a great majority of cotton farmers are either locked into debt relations with merchants, or pledged into selling their cotton to TARIS, as this is the only way they can get the credit needed to continue production. TARIS extends indirect credits – such as fertilizer, seed or pesticide – during the planting season and deducts their cost when the farmers sell their produce back to the cooperative.

If the merchants depress prices, they can buy more cotton and sell for less. Not only does the mark-up make the difference, but when the cotton price is depressed, more and more yarn factories switch from cheap, oil-based polyester inputs to cotton; they buy more cotton and thus contribute more to the merchants' income. Finally, as the price decreases in the countryside, merchants need less capital to purchase the commodity, so their carry-over increases in real terms. Paradoxically, the reason for the increase in merchant income can lead to the eventual loss of all profit. The more they depress prices, the more difficult it becomes for small farmers to grow cotton - not only because of the decreasing contribution of cotton to their income, but also because the farmers are selling their land to pay off their debt, thus losing the chance

BOX 1 MAPPING THE COTTON PRICES: COTTON TRADING IN IZMIR

The IME cotton price has three forms: *Rehearsal Price, Transaction Price* and *Market Price*. These forms are produced in four temporally and spatially specific places of the market: *the Pit, during Post-Pit Trading, the Closing Price Committee Meeting and the Permanent Working Group on Cotton Meetings.* Prices are best viewed as prosthetic devices to further various trading objectives. They are made, produced and challenged by multiple actors in a market process.

Making such processes transparent can enable those affected by prices to participate more actively in price formation. Market intervention and maintenance are constant. Therefore, *laissez-faireism* acquires a new meaning since allowing some but not all to be involved in price formation, directly affects the livelihoods of those who make and consume the commodity, the price of which is produced amid power relationships.

Pit trading, the entry point for studying cotton trading in Izmir, is framed in temporal and geographical limits, bringing traders together for ten minutes between 12:20 and 12:30. It is here that "the theatre of the market" takes place. Market players in this theatre produce *rehearsal prices*, which are made to exchange only a very limited quantity of the commodity, serving as a rehearsal for the actual volume of trade that has yet to be reached during post-pit trading. The pit price is a market device deployed and produced by merchants to strengthen their positions when making actual transaction prices. The rehearsal price is also a device because it is simultaneously an investment for post-pit trading and also an actual transaction price since cotton is exchanged by its acceptance.

Once this first market boundary has been crossed, literally by leaving the pit at 12:30 and entering the second location - the post-pit, traders begin to use rehearsal prices to negotiate actual transaction prices. However, both the traders and their prices are scattered in this second market location. Hundreds of registered and unregistered prices are made in this dynamic situation. What is particularly interesting is that this is not the place where *the market price* emerges.

Following post-pit trading, the market price is formed in the Closing Price Committee meeting. The committee members remove the highest and the lowest market prices from the list they receive from IME administration, and take the weighted arithmetic average of individual registered transaction prices. Hence, the market price is not set by the coming together of demand and supply, as neo-classical price theory suggests, but is produced by the application of a mathematical formula to a political process of deliberation and bargaining, that is affected on a daily basis by the market power and positions of traders.

Demand and supply also play a role, but again in a surprising setting. Before traders perceive their effects, supply and demand impacts are negotiated by specially-formed committees of Ministry of Agriculture bureaucrats, traders, economists, agricultural engineers, a few large landowners and officials from TARIS, the cotton sales cooperative that represents farmers that produce more than a quarter of the cotton in the country. As the market process progresses, actual supply and demand figures cannot be known for sure, only becoming clear once all the cotton has been sold. Prices in their multiple forms are made *before* these levels are known. The perceptions of "what the market will be" are discussed and negotiated, and figures such as supply volume or production costs are produced, thus further informing the ways in which prices are made.

Three main groups attend meetings; with three diverse sets of motives that prefigure their approach to cotton supply estimates. Traders, merchants and brokers comprise the first group, all of whom have an interest in depressed prices because lower prices enable larger volumes to be traded. As a result they strive to prevent underestimation of the production level, because a lower supply forecast pushes prices up. Large landowners and cooperative representatives have the opposite motivation. Obviously, they benefit from increasing prices and thus have an incentive to prevent overestimation of the production level, because the higher the supply prediction, the lower the price. Ministerial representatives, however, do not share the same concerns for price levels as the previous two groups. Their main purpose is to obtain the highest possible accuracy in the estimate of cotton supply levels, since this figure informs the allocation of agricultural subsidies in the annual budget. An underestimate can lead to budget deficit, while an overestimate can make the government look unsuccessful.

to grow cotton altogether. If this process continues for two more decades, the land will be consolidated in fewer hands and prosthetic prices may then become more relevant in the countryside, not because the farmers will have understood how to use derivative markets, but because there will be no small farmers left.

THE FARMERS' PRICE

The cotton price for a farmer is never the formal posted price. For them, formal prices – whether posted by merchant houses or TARIS – are only indicative of what they can get for their cotton. "The price is what I carry in my pocket after I sell my crop," said Numan, a cotton grower who had to sell his cotton to a ginning factory owner from whom he'd borrowed cash. Numan received his money in less than an hour on the condition that he "gave" his cotton to this private trader, settling the difference after the debt had been deducted.

The price that farmers receive is always lower than the price they are offered. To obtain the cotton, traders begin with illegally lending money, yet creating a legal context for it. When Numan made his deal with the ginning factory owner, the trader told him that he had some bank credit that he did not need at the time, and "since he liked Numan a lot," he offered Numan that money to use. The interest rate he applied was even one point lower than the private bank rate. He sold his money, but pretended that he served as an intermediary between the bank and the farmer. "Usually, they never give money directly to us. We go and get it from the bank. It looks as if the bank gives us the money," explains Numan.

"I know that it doesn't come from the bank. He [the money lender] sends a note to his friends in the bank, authorizing me to withdraw money from his account, and it is his money that I get. Other villagers don't know. They think it is the bank that gives them the money and that the *fabrikacı* is the intermediary. When there is a bank, there is paperwork, the state is involved, courts, etc. They send letters and stuff. Everybody knows that you borrow money in the village, so other villagers don't want to work with you.

WITHOUT FINANCIAL BACKING, IT IS NOT POSSIBLE FOR GROWERS TO KEEP THEIR CROP UNTIL THE TIME COMES TO SELL IT UNDER RELATIVELY BETTER TERMS.

"The first thing he [the *fabrikacı*] asks is not the amount of money we need, but the field size we have. Because he will not get the cash back, he will ask for my cotton instead.

"You cannot have a farmer sell his crop in advance without having TARIS's price. The *fabrikacı* also offers the TARIS price. But the amount he pays is always lower. For example, when you bring your cotton, he never likes it. The *randıman* [ginning outturn] is everything in cotton. If the cotton has more seeds, it has less lint. So the money I get goes down if the *randıman* decreases. The *fabrikacı* always mess with these roller-gins. They fix their scissors so that they pick a little bit less lint. So they pay less. They also play with their scales. Farmers' cotton always weighs less. They make money in this way too."

A former ginning factory owner confirmed Numan's description of how the cotton is sold in the market. He commented:

"This business of debt is making the rich richer and the poor poorer. But what can farmers really do? The Agricultural Bank does not give credit anymore. TARIS's bank is taken away. Farmers are left alone. They need money, and usually very rapidly. Who has money? The ginners. They don't like the poorest or the richest farmers. The poor have nothing to lose; the rich have power to resist. They want the middle farmers who produce enough, yet don't have money to finance themselves. These farmers also need them, because they want to borrow fast, and they want to borrow it secretly. Ginners keep it secret, and sometimes the farmers lose their land secretly if they cannot pay the debt."

Not all farmers are indebted, yet they still work with private traders. Scales and sampling roller-gins are usually more balanced when it comes to working with farmers who are not indebted. A discussion on the formation of farmers' prices must include mention of the "performances" put on by traders, which weigh heavily against the farmers. A farmer with better financial resources wanted to sell his cotton to a private trader. He talked to a few ginners before the harvest about his intention to sell. "Initially all of them thought that I wanted to borrow money. They were nice to me, wanting to help like a father. Attitudes changed however when he brought them samples of his cotton: "They looked at it, didn't like it, yet still wanted to buy it. They are real actors. Like those on TV."

When a rich trader and a poor farmer come together within an exchange context, the latter finds himself in a field of power that works against him. Traders use an idiom that farmers are unaccustomed to; they are experienced in buying and selling and thus outperform the farmers; and they have a financial cushion to rely on if they do not win the bargaining. This is hardly the case for farmers, who sell small amounts compared to the quantities the traders buy everyday.

Knowing that trading performances are "only acting" does not help the Pamukköy farmers. Farmer motivations in the marketplace are dominated by their urgency to sell their produce to pay their debts. They cannot get their cotton ginned and wait for the best opportunity to sell their commodity. They cannot produce and deploy various prices to empower their trading position, for they do not have one. They cannot travel with their cotton between various potential buyers, nor can they shop around to find the best merchant. The market as seen by Pamukköy farmers is not a neutral place where supply and demand set the price. It is a place where the farmers are relatively powerless agents of the exchange relationship that turns their cotton into money. This conclusion suggests a rethinking of the relationship between the field and the market.

CONCLUSION

For farmers, contact with the market involves bargaining and agreeing a price in the trader's office, usually located in a ginning factory owned by the merchant himself. Three factors prefigure this bargaining process.

First, the previous relationship between the trader and the farmer has a strong effect on negotiations. If the farmer owes money to the trader, he has to pay it back, usually before 31st October. Traders informally cooperate not to buy from a farmer who is indebted to another trader. Thus, farmers can only sell to the trader from whom they borrow. According to the author's calculations, in 2001 the Pamukköy farmers who borrowed from private traders sold their cotton for eleven percent less than the farmers who borrowed from family members or the cooperative.

The second factor relates to the specific processes that take place in the IME. All active traders in the Söke Plain have representatives in the Exchange. Well before the cotton hits the market, traders begin to prepare the ground. Various prosthetic prices are probed and produced during pit and post-pit trading. Traders also work hard to affect the market process. For instance, they lobby agricultural engineers, statisticians, researchers, government officials and cooperative representatives to influence their perception of the coming harvest. The objective being to produce an estimate of a high yield and low production costs. In caucuses dominated by traders, such as the Permanent Working Group on Cotton, the very supply of cotton is determined even before farmers harvest their crop. These estimates are always lower than the actual levels.

FARMERS ARE EXCLUDED FROM LINT MARKETS BECAUSE THEY DO NOT HAVE WAREHOUSES TO STORE THEIR COTTON, THE FINANCIAL MEANS TO INSURE IT, THE MEANS TO FOLLOW PRICES ON A DAILY BASIS OR THE POLITICAL POWER TO PARTICIPATE IN PRICE-MAKING.

The third relates to the trading performances and politics that merchants forge on the ground. Traders attempt to undermine a farmer's produce to strengthen their trading position. Furthermore, merchants cooperate to ensure that they fix their sellers even before the harvest. Most cotton buyers in the plain lend money to farmers, then buy the farmers' cotton and settle the difference.

The market price offered to farmers, who are relatively passive agents of trade, is produced prior to the moment of exchange. Merchants devote much time to building networks to affect prices, while farmers spend all their available time growing cotton, hiring or pooling labour, hoeing fields or picking cotton. Farmers do not have time to simultaneously maintain market platforms and grow cotton.

The price that serves as the interface between the field and the market can be regarded as a summary of all the power contestations that take place before and during the moment of exchange. Studies of rural relations of exchange and production should take into account the specific nature of price-making on the ground and consider the asymmetries in the way prices are produced. Treating prices as things that are set in the market is itself a political investment in shaping markets. Prices are not set through actual demand and supply; they are produced as prostheses to be used by a select group of market participants only. Academic research on agricultural production and exchange must take these factors into account, or risk being a formative part of the politico-economic universe it studies.

The Pamukköy farmers see the markets and fields as places where various trading performances take place. The farmers' capacity to improvise and exploit these performances increases in direct proportion to the amount of land they own. Yet, for most of these cotton farmers, such relative autonomy vanishes when they enter the cotton market.

In conclusion, economic activities in the countryside involve fields of power where webs of asymmetrical relations are forged and maintained by a multiplicity of actors. These actors operate in a cascading relationship of domination, resistance and negotiation, structured in an essentially political topography of encounter. Such a political context can only be understood if one pays ethnographic attention to the specific predicaments of different actors involved in commodity production and exchange. ■

REFERENCES

Çaliskan Koray, forthcoming, "Price as a Market Device: Cotton Trading in Izmir Mercantile Exchange", in *Market Devices*, ed. M Callon, Millo Y., and F. Muniesa, London: Blackwell

Callon Michel, 1998, ed., *The Laws of the Markets*, London: Blackwell

Muniesa, 2000, "Performing Prices: The Case of Price Discovery Automation in the Financial Markets", in *Oekonomie and Gesellschaft*, ed. H. Kalthoff, R. Rottenburg, and H.-J. Wagener, Marburg: Metropolis

Polanyi Karl, 1944, *The Great Transformation*, New York: Farrar & Rinehart

RICE PRICE FORMATION IN THE SHORT AND LONG RUN: THE ROLE OF MARKET STRUCTURE IN VOLATILITY

C. Peter Timmer,
Center for Global Development, Washington, United States of America

Nearly half the world's population consumes rice as a staple food, 90% of which is produced in Asia. Of the 440 million metric tons (mmt) of milled rice equivalent produced globally, most is consumed within a short distance of where it was produced. Only 7-8% crosses an international border. Nevertheless, the world rice market is important – providing essential supplies and sending out global price signals. It is therefore a major concern when such market signals are highly volatile. This paper looks at the unique causes of rice price volatility, addressing questions on price formation trends and the influence of speculation on the rice price spikes.

CAUSES OF RICE PRICE VOLATILITY

The variation coefficient of world rice prices has often been double that of wheat or maize. Understanding this volatility is difficult due to the nature of the market, where countries stabilize internal rice prices by the disposal of surpluses onto the market or by importing more to meet deficits. Thus, the political decisions of key Asian countries directly influence worldwide supply and demand (Timmer and Falcon, 1975).

Rice price volatility is also driven by rice production structure and by marketing and consumption trends in Asia. Millions of farmers, traders, processors and retailers, along with billions of consumers, all handle a commodity that can be stored for over a year in a consumable form[1]. The price expectations of these market participants are critical to their decisions on how much to grow, sell, store and consume.

ANALYZING FOOD PRICE FORMATION

Understanding causation in food price formation implies an empirically refutable analytical model based on supply and demand mechanisms, with equilibrium prices derived from basic competitive forces[2]. Gradual changes in supply and demand can cause gradual increases in prices but the food price explosion of 2007 and 2008 clearly requires additional explanation beyond a simple supply and demand model.

The "supply of storage" model provides the basis for interpreting short-run price behaviour for storable commodities (Houthakker, 1987). It stresses the interrelated behaviour of speculators and hedgers as they judge inventory levels in relation to use. To make this model operational in the short run, it is used to gain

1.The oxidation of traces of bran on milled rice can significantly reduce the shelf life of rice stored in the tropics.

2. Such statistical analyses were carried out and the results can be viewed here: http://ideas.repec.org/p/cgd/wpaper/172.html.

FIGURE 1 RICE, A MARKET THAT IS NOT PREDOMINANTLY GLOBAL

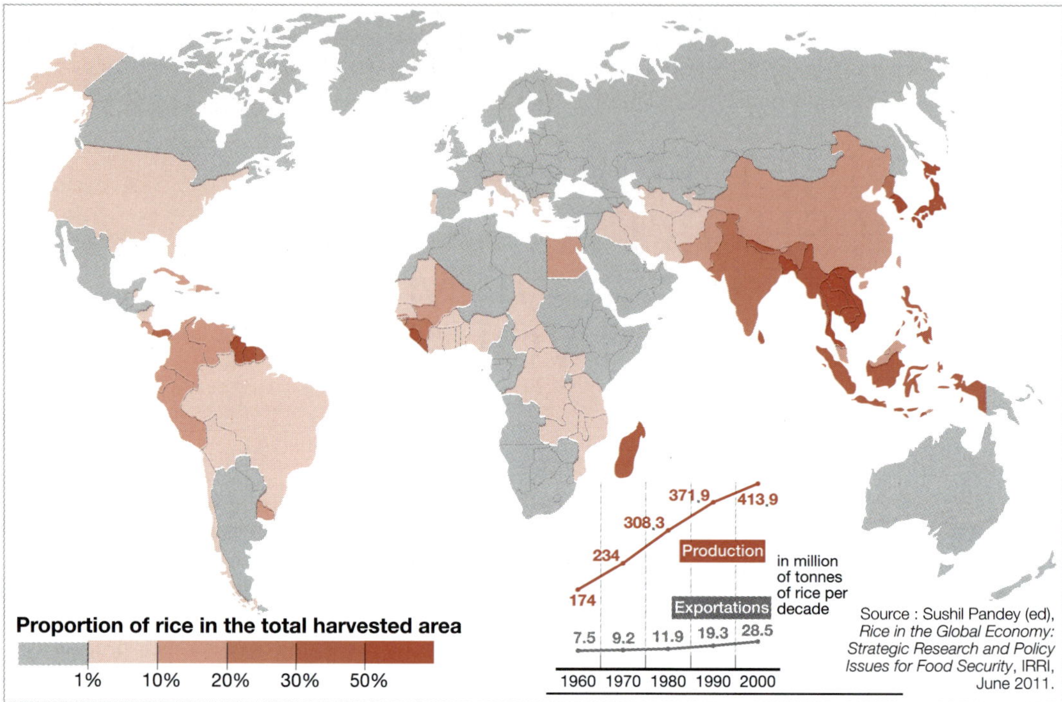

Unlike other major cereals, such as wheat and maize, rice production is mainly consumed in the countries that produce it. Trade thus represents only 7% of world production.

insight on the formation of price expectations. These expectations seem to be driven by a combination of commodity price behaviours and the specifics of individual commodities. Broad commodity price trends are captured by various indices, such as the IMF commodity price index. Thus, traders operating in any one commodity market will track the broader price movements for all commodities (Sanders and Irwin, 2008). These broad price movements seem to be driven by basic macroeconomic forces such as economic growth rates, the value of international currencies and relative inflation rates (Timmer, 2008a).

Traders also follow their specific commodity closely. Here inventories determine price formation, once a crop's harvest/supply situation has been established. Typically, commodities with reasonably reliable inventory data tend to have short-run prices driven by unexpected supply behaviour, whereas commodities with poor inventory data, especially where significant inventories can be in the hands of millions of small agents (farmers, traders, consumers, etc.), tend to have extremes in price behaviour generated by rapidly changing price expectations, and consequent hoarding or dis-hoarding of the commodity. The short-run price dynamics for rice thus look significantly different to wheat or maize.

THE ROLE OF MARKET STRUCTURE IN RICE PRICES

The actual production/consumption balance for rice has been relatively favourable since 2005. Short-run substitutions in both production and consumption between rice and other food commodities are limited, and until late 2007 it seemed that the rice market might avoid the price spikes seen in wheat, maize and vegetable oil. The limited futures market for rice also made financial speculation less attractive.

However, with only 7 to 8% of global production traded internationally, the world rice market is very thin, leaving it vulnerable to large price moves. The market is also relatively concentrated, with Thailand, Vietnam, India, the US and Pakistan routinely providing nearly 80% of available supplies.

Does this difference in market structure also account for the difference in long-run price trends between rice and the other two staple food grains? Only to a limited extent. The political economy of high variance in world rice prices leads countries to retreat into autarky, dumping their own instability into a smaller world rice market. One consequence of this self-sufficiency drive among rice importers is larger overall production than expected, which contributes to a long-run decline in world prices. But other factors (faster technological change, slower population growth in rice-consuming countries, changing consumer tastes…) could account for less tension between supply and demand for rice, than for wheat and maize.

THE 2007 – 2008 COMMODITY PRICES SPIKE

The scope for extreme volatility is clear in light of the political and market structure dimensions for rice. Understanding how these factors contribute to the formation of price expectations on the part of market participants is essential to grasp the proximate causes of unstable rice prices. These expectations can drive "destabilizing speculative behaviour" among millions or billions of market participants, such that price formation seems to have a large, destabilizing, speculative component.

In 2007, concerns grew that world food supplies were limited and prices for wheat, maize and vegetable oils were rising. Several Asian countries increased domestic rice stocks to protect against future shortages[3]. The fear of shortage spread and a cumulative price spiral began.

In India, the 2007 wheat harvest was damaged by drought and disease, so the food authority announced that it would retain more domestically-produced rice rather than import more wheat. An outright ban on non-Basmati exports followed in April 2008 (India is usually the world's second largest rice exporter).

Other rice exporting countries followed suit as rice prices spiked. The commerce minister of Thailand's new government openly discussed export restrictions from Thailand – the world's largest rice exporter (9.5 mmt in 2007). By March 2008, rice prices in Thailand jumped $75 per mt, continuing to skyrocket until April, reaching $1100 per mt.

Poor consumers in many rice importing countries, including Vietnam and the Philippines, were seriously affected by this price volatility; while some countries, including India, China and Indonesia, were able to protect domestic populations from the rice shock through export restrictions. Leaders in these countries argued that domestic stability was more important than international stability[4], reflecting Asia's longstanding view that rice price stability is essential for political stability and economic growth (Timmer and Dawe, 2007).

This is a vicious circle: world market instability causes countries to withdraw from importing and exporting rice, thus thinning the market, which increases instability and causes ad hoc policy reactions. Breaking this cycle requires long term binding agreements between rice importers and exporters to rebuild market confidence. National or international pressure for such agreements is, however, virtually non-existent[5].

Hoarding behaviour, a "rational" response to panicky governments, was a critical component of the 2007 rice price rises. Financial speculation, however, seems to have played only a small role (partly because futures markets for rice are limited). Instead, decisions by millions of households, farmers, traders and some governments sparked a sudden surge in rice demand and changed a gradual price increase into an explosion.

3.See Slayton (2009) for a detailed analysis and chronology

4. The overwhelming victory by the Congress Party in India in May 2009 gives credence to these political arguments.

5. One way forward might be for an ASEAN-led initiative to create a free trade zone for rice in the region. If such a zone included south China, Bangladesh and India, most of the world's rice producers and consumers would be exposed to more stable prices, and the entire world rice market would also be more stable. Perhaps Indonesia, with its newfound self-sufficiency in rice, could host discussions on such an initiative.

CONCLUSION

Market structure plays an important role in short-run rice price formation, stemming from the highly unusual organization of the world's rice economy, with many small producers, traders, retailers and consumers handling a product that is storable at each stage.

The supply of storage model shows that this highly decentralized storage capacity is subject to changes in price expectations throughout the supply chain. As data on the rice stocks held by market participants is non-existent, their impact on rice price formation is unpredictable.

Most analysts think that long-run linkages exist among the staple grains, based on commodity substitutions in production and consumption (Timmer, Falcon and Pearson, 1983). Commodity markets for rice, wheat and maize are similar in that they are influenced by basic supply and demand factors, although the long-run fall in rice prices has been faster than for wheat and maize. Wheat and maize are more closely linked to each other than to rice markets, while any price impact caused by bio-fuels would have been mediated through the maize market and its financial linkages to other commodity markets.

Financial speculators were almost certainly responsible for some of the increase in wheat and maize prices. Conversely, speculation did not cause the rice price rise, due to the different industrial organization of domestic and international rice markets, along with government intervention.

The unwinding of speculative positions leads to rapid price falls, until perceived surplus stocks are consumed. With the exception of India, this process of short-run de-stocking seemed to end in late 2008, well before world prices began to rise again. The world financial crisis will probably prevent more rapid rises in rice prices, but the supply/demand balance for the next few years suggests export prices will be closer to $500 per ton rather than $300 per ton in the pre-crisis period.

These conclusions are reached mostly by eliminating the other explanations and by logical reasoning. There is no specific quantitative test of the hypothesis on offer, and indeed, one may not be possible. ■

REFERENCES

HOUTHAKKER, Hendrik, 1987, "Futures Trading", in Eatwell, John, Murray Milgate, and Peter Newman, eds., *The New Palgrave: A Dictionary of Economics, Volume 2.* MacMillan Press Ltd., London, UK, pp. 447-449.

SANDERS, Dwight H., and SCOTT H. Irwin, 2008, "Futures Imperfect", *New York Times*, July 20, p.12.

SLAYTON, Tom, 2009, "Rice Crisis Forensics: How Asian Governments Carelessly Set the World Rice Market on Fire", CGD Working Paper 163. Washington, DC: Center for Global Development.

TIMMER, C. Peter and FALCON Walter P., 1975, "The Political Economy of Rice Production and Trade in Asia" in Lloyd Reynolds, ed., *Agriculture in Development Theory* (New Haven: Yale University Press), pp. 373-408.

TIMMER, C. Peter, FALCON Walter P. and PEARSON Scott R., 1983, *Food Policy Analysis*. Johns Hopkins University Press for the World Bank, Baltimore, MD.

TIMMER, C. Peter and DAWE David, 2007, "Managing Food Price Instability in Asia: A Macro Food Security Perspective", *Asian Economic Journal*. Vol. 21, No. 1, pp. 1-18.

TIMMER, C. Peter, 2008, "Causes of High Food Prices", Chapter 2.3 in *Asian Development Outlook Update,* September, Asian Development Bank, Manila.

CHAPTER 11

The economic efficiency of the agro-industrial model is often at the expense of the natural environment and generates social injustices between countries and among chain actors. Sustainability will only be ensured by its coexistence with an alternative model based on proximity and networks of small businesses that value the natural and cultural heritage of specific territories.

THE AGRI-FOOD INDUSTRY AT THE HEART OF THE GLOBAL FOOD SYSTEM

Jean-Louis Rastoin,
Montpellier Supagro, Montpellier, France

On a historical scale, the processing of agricultural raw materials and storable perishable commodities used for food preparation, that we define today as the "food industry", is an ancient activity. Its origins probably date back to Neolithic agriculture and settlement, which means that processes such as the manufacture of flour and meal from cereals, cheese making, fermentation to make beverages and the drying and smoking of meat and fish, have been practiced for about 11,000 years.

For a long time this activity remained integrated with agricultural production and was carried out on an artisanal or family scale. The agri-food industry, in the contemporary sense of the term, did not arise until relatively recently during the Industrial Revolution of the nineteenth century. As occurred in other sectors (such as textiles and metallurgy), the industrialization of food processing originated through technical innovations, such as: the beet sugar extraction process (Chaptal and Delessert, 1811), the heat sterilization method for canning (Appert, 1802) and chocolate making (Meunier, 1824). New ways of organizing production accompanied the profusion of innovations at this time, with a movement away from artisanal production towards industrial factories. Large agribusinesses (Nestlé in Switzerland, Unilever in the Netherlands and the United Kingdom, Liebig in Germany, etc.) started to appear at the end of the nineteenth and early twentieth centuries. However, the market structure did not change significantly until the Second World War (Rastoin, 2000).

Today, the agri-food industry is situated at the centre of a very important economic complex that we call the "food system", the purpose of which is to feed populations,

FIGURE 1 HIGHLY CONTRASTING FOOD INDUSTRIES

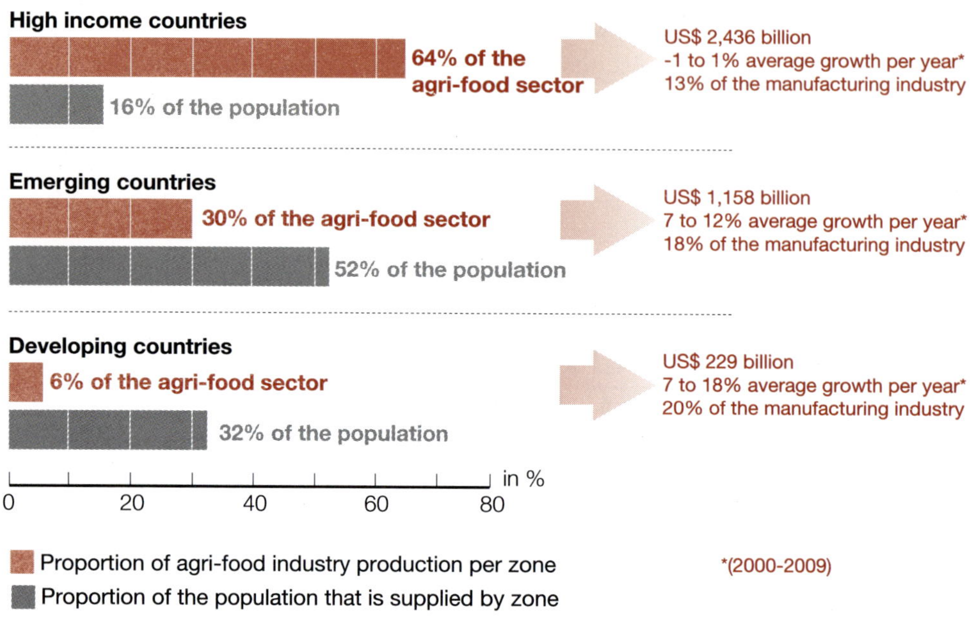

Sources: author's estimations, based on CIAA, *Data & Trends of the European Food and Drink Industry*, 2010 and Unido Databases, 2011.

The importance of the production of the agri-food industry does not correlate to the size of the supplied population, but rather to the purchasing power and investment capacity of a country. It remains primarily concentrated in rich countries.

mostly through a system of market relationships (Rastoin and Ghersi, 2010). The position occupied by the agri-food industry within the food system varies by country. However, there is convergence towards a global agro-industrial service sector, a model in which the agri-food industry plays a pivotal role, that is described in the first part of this chapter. Following these foundations, the second part of the article discusses the trends of globalization and financialization inherent to this model, before the conclusion outlines three long-term scenarios for the agri-food industry.

HETEROGENEITY AND CONVERGENCE OF THE AGRI-FOOD INDUSTRY

In a majority of countries, the agri-food industry is primarily the largest subsection of the manufacturing industries, representing 10% to 30% of this sector. With an output of nearly $4000 billion and more than 25 million employees in 2009, the importance of the industry relates to its technical and economic proximity to agriculture (utilization of basic raw materials) and by its role to satisfy an inexhaustible demand – the sustenance of humankind.

The agri-food industry is highly heterogeneous in terms of its structure and geographic space. Indeed, it is composed of a large number of sectors (for example, the processing of cereals, oilseeds and animal products). Each of these sectors has

FIGURE 2 WORLD LEADERS IN THE AGRI-FOOD INDUSTRY, 2009

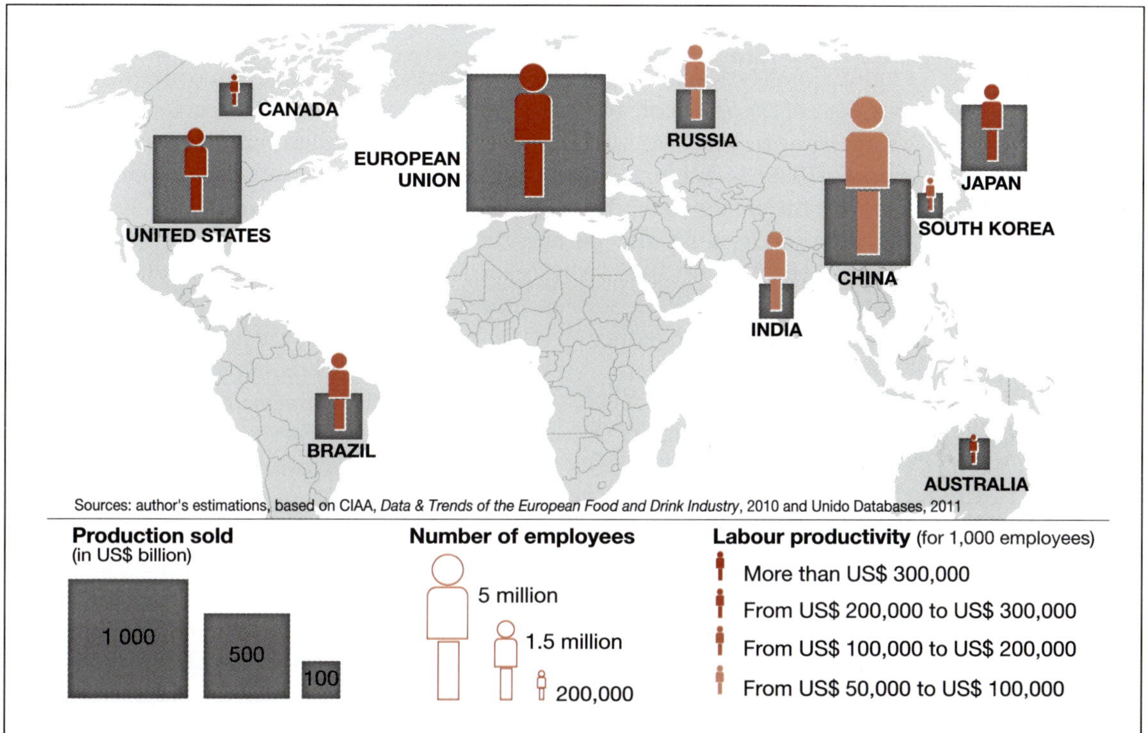

Sources: author's estimations, based on CIAA, *Data & Trends of the European Food and Drink Industry*, 2010 and Unido Databases, 2011

distinct technological, economic and managerial characteristics. In addition, agri-food sector profiles differ according to the agro-climatic constraints, markets and economic levels that vary from one country to another. In 2009, high-income economies (HIE), which contain about 16% of the world's population, accounted for 64% of the industry's global production in terms of economic value (72% of global GDP), while developing countries account for 32% of the global population and only 6% of production.

Growth in the agri-food industry has "plateaued" in rich countries, is steady in emerging countries and rapid in developing countries, which confirms the basic and demographic nature of food markets: in the process of economic growth, the agri-food industry serves as a kick-start sector which is able to mobilize small amounts of capital and large workforces to accompany, through the relative decline in food production costs, the rise of other consumer goods. However, most governments neglect to stimulate the sector. In HIEs, this observation remains valid, because if current needs become saturated and demand therefore stagnates (i.e. a "mature" market develops), the emergence of new demands (environment, health...) potentially has the knock-on effect of generating economic growth in the agri-food industry.

The heterogeneity of the agri-food industry remains very high, but there are signs of its erosion, as is also occurring in most other capital-intensive activities that are

Alongside the historical players in the agri-food industry - from Europe, the United States and Japan - are emerging countries, which are currently developing large production capacities. The performance of these new competitors is particularly impressive because they have not yet achieved the level of productivity of Northern countries.

under pressure from three powerful and interrelated forces: concentration, globalization, and financialization.

A RAPID CONCENTRATION DICTATED BY ECONOMIES OF SCALE Ten countries concentrate 85% of the agri-food industry's global production and 70% of its workforce. It is interesting to note that, in the early 2000s the absolute supremacy of major rich countries (USA, Germany, France and Japan) was superseded by a shared leadership with the emerging world: China, Russia, Brazil and India.

Concentration must also be appreciated at the company level. According to calculations (Rastoin, 2008) with regard to the most advanced sectors in terms of technology and marketing, such as ultra-fresh dairy products, soft drinks and snack foods, an oligopolistic reduction has taken place, leaving a handful of companies in control of nearly three-quarters of the market. The average profile of the top ten agri-food companies – which in 2007 was a revenue of nearly $47 billion, 119,000 employees and a profit of about $3.7 billion – demonstrates the size and therefore the considerable economic power of these companies. In comparison, Danone, the industry's highest-ranking French company, is at number 12 in the rankings with a revenue of $17 billion (five times less than Nestlé, the biggest company) and profits of $1.9 billion (six times less than Nestlé). It is worth noting here that US dominance of the industry remains a reality: nine of the top 15 companies are American. Some 8000 foreign branches of the world's top 100 agri-food companies are widely present in emerging and developing countries, which increases the concentration of agri-food capital in HIEs.

THE MOST LIKELY EVOLUTION OF THE FOOD SYSTEM IS THE COEXISTENCE BETWEEN TWO MODELS: THE SERVICE SECTOR-BASED AGRO-INDUSTRIAL MODEL AND AN ALTERNATIVE MODEL BASED ON PROXIMITY.

At the "fringe" of this oligopoly are over 600,000 businesses worldwide, the majority of which are small and very small businesses that form a dense "mesh" in rural and suburban areas, with "territorialized" products which often trade on the strength of their organoleptic qualities and local cultures (Noronha Vaz et al., 2008). In the EU in 2009, very small businesses (0 to 9 employees) and small and average businesses (10 to 249 employees) accounted for 86% of the 360,000 businesses in the agri-food sector, 63% of employment and 48% of revenue (CIAA, 2011).

HIGH ECONOMIC AND TECHNOLOGICAL PERFORMANCE Concentration is a result of the drive for competitiveness. In industrialized food systems, large-scale food retailers dominate the market of the agri-food sector, which is itself highly concentrated (in Western Europe, households make more than 80% of their food purchases in such locations). Competition between large-scale retailers occurs mainly through consumer pricing and thus through the net prices paid to agri-food suppliers. The agri-food industry is therefore firmly focused on productivity gains, continually pushing to reduce fixed unit costs through an increase in production capacity (economies of scale, substitution of labour by mechanization and automation).

Technological advances have enabled both the significant reduction of losses along with the improvement of quality control regarding health and safety issues. This *Saint-Simonian* achievement has two drawbacks: the significant loss of employment (caused by a three quarter reduction in the number of industrial sites)[1] and negative effects on the environment and food security. The process of concentration is observed throughout agriculture and the agri-food industry once changes start to occur in the marketing structure with the arrival of large distribution chain stores (Reardon and Minten, 2011). We estimate that in 2010, 55% of the world's population shopped in supermarkets, becoming part of an agro-industrial food system.

There is a second explanation behind concentration in the food industry, namely, the need for food safety (non-toxic food). The BSE (mad cow) crisis of the mid-1990s and those that followed (salmonellosis, etc.) prompted large-scale food retailers to establish procedures for monitoring quality and traceability that are very restrictive in terms of the agri-food industry. Today the food system operates under the influence of standards (IFS, BRC, GlobalGap, etc.), which have been partly developed by these retailers. It is obligatory for agri-food companies seeking supermarket endorsement to adhere to these standards, which involves a considerable expense.

Finally, such concentration ensures the financing of intangible investments (innovation and communication), which enables companies to gain market share due to the leverage effect derived from size. These investments are considerable. Thus, in 2008 R&D represented around 1.5% of the revenue of major agri-food companies in the OECD, while advertising accounted for between 5% and 15% of the final price depending on the product. Again, the effects of scale come into play and work in favour of large companies.

The historical trajectory of the market economy is such that today in HIEs, the agri-food industry spends more on services (mainly logistics and marketing) than on agricultural commodities. In European countries, there has been a transition in the average final price structure of food products from "40/A-40/I-20/S[2]" to "20/A-30/I-50/S" in just over half a century: the agri-food sector has thus become a tertiary industry (Nefussi, 2004), which is certainly a strange paradox! In reality, the cost structure reveals firstly the development of consumer behaviour, and secondly a "sharing" of the economic value of goods that reflects the market power of the different stakeholders involved in the chain.

THE INESCAPABLE LAW OF GLOBALIZATION AND FINANCIALIZATION

Despite its many specificities related to the exploitation of living materials, the agri-food industry shows a tendency to become a mainstream and purely consumer

1. The stability of the workforce in the agri-food industry is due to small businesses (0-19 employees) which, through job creation, offset the significant job losses caused by large companies of more than 500 employees.

2. The share of agricultural commodities (A), industrial processing (I) and services (S) in a food product with a price of 100.

FIGURE 3 GLOBAL AGRI-FOOD TRADE

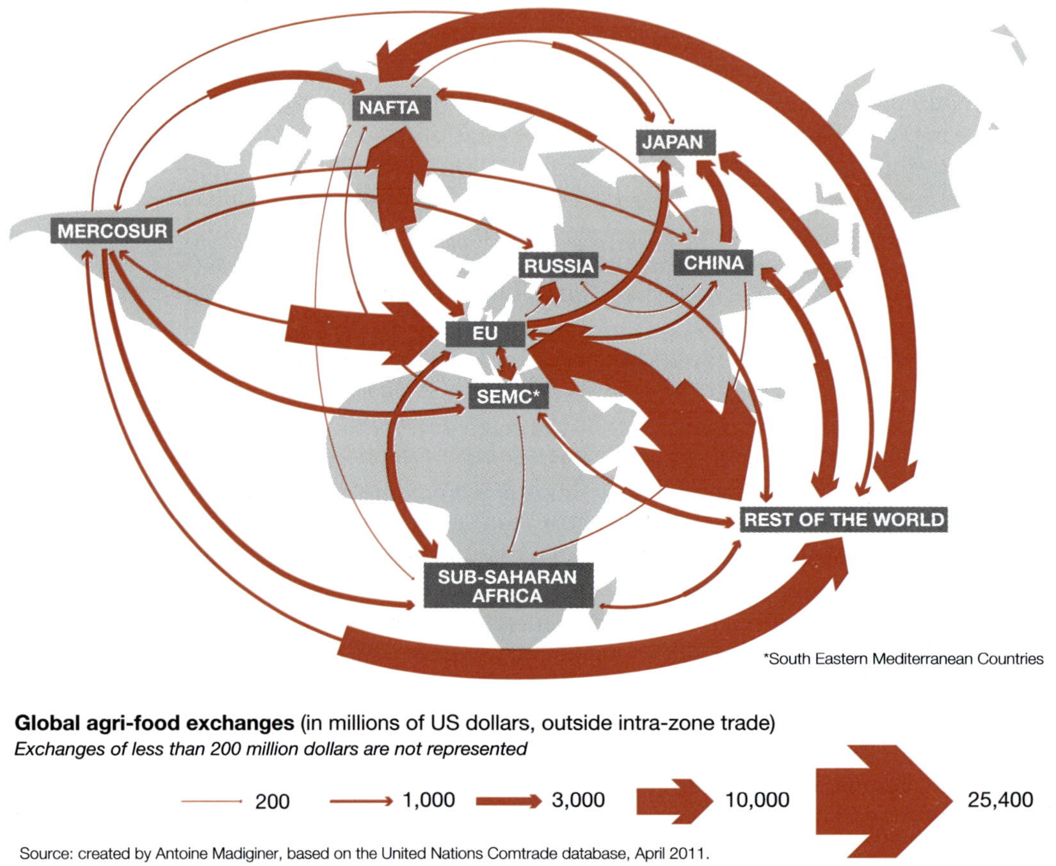

Source: created by Antoine Madiginer, based on the United Nations Comtrade database, April 2011.

While agri-food exports are growing rapidly in emerging countries, including Brazil, global inter-zone agri-food trade is still dominated by stakeholders from the North: EU and NAFTA.

goods industry, both globalized and driven by financial markets, without managing to provide food security.

RAPID GLOBALIZATION OF AGRI-FOOD MARKETS Globalization is characterized by an intensification of all types of international trade, a new geopolitics for such exchanges, and by in depth modification at the managerial level, providing the impulse for new modes of governance. The agri-food industry has adjusted to this movement with much greater ease than agriculture because of its capacity to transform perishable products (agricultural commodities) into storable and transportable goods.

The most visible manifestation of globalization is that of the sustained, although irregular, growth of international trade. Global exports of agricultural and food products have thus increased by a factor of 4.4 over the last 30 years to around $1000 billion in 2009. However, agricultural commodities (basic raw materials) have experienced much slower growth in international trade than the processed products of the agri-food industry (the exports of which in 2009 represented 40% of the total trade of agricultural and food products). The EU is the world's leading agri-food power, accounting for 54% of exports and 49% of imports of processed products globally. Three-quarters of this latter figure consist of intra-community trade flows, but even disregarding this amount, European supremacy remains clear with 31% of exports and 22% of imports globally, which is well ahead of Mercosur (20% and 2% respectively) and NAFTA (10% and 17%). However, the rapid growth of the total exports of raw and processed food products from Brazil, "the world's farm", and from other emerging countries, inexorably remodels the global leadership landscape.

This dynamic, however, is conditioned by the completion of the Doha Round, the aim of which is to dismantle tariff protection and technical barriers to trade for agricultural and agri-food products. In a context of stalled negotiations, regional and bilateral agreements are multiplying, often exacerbating international distortions. The geopolitical issue that then arises is the construction of a Euro-Mediterranean and African political strategy in order to avoid a hegemonic US/China duopoly. The current problems of economic management in the Eurozone make such an objective seem doubtful in the medium term.

Globalization also has an impact on the organization of productive activities. A new phenomenon has developed in recent years, where large companies distribute their functional activities - i.e. research, human resource management, information technology and finance departments - throughout the globe according to cost/benefit criteria. This geo-strategy is motivated by two factors: the search for growing markets and to obtain advantages in terms of production costs, which accounts for the two recent investment waves: central and eastern Europe and Asia.

The most controversial issue here is that of the delocalization of industrial units on the basis of competitive advantages. However, this trend does not currently benefit developing countries, which only account for 0.06% (agriculture) and 2.8%

BOX 1 FOOD MILES, LCA AND SUSTAINABLE DEVELOPMENT

A British study (Smith et al., 2005) calculated the negative externalities of food miles (£9 billion in 2002, i.e. 13% of the added value of the food system), with a high proportion of the total cost attributable to congestion (57%) and a significant impact in terms of public health (accidents, pollution and noise: 30%).

According to an American study, 84% of greenhouse gas (GHG) emissions in 1997 were due to agricultural and agri-food production and only 6% were due to transport within the food system (Weber et al., 2008). Therefore, buying locally would enable maximum savings of 4% to 5% in terms of GHG emissions. A change in eating habits, such as a one-day per week substitution of red meat and dairy products with an alternative source of animal or plant protein, would have a similar impact.

Finally, environmental and social life cycle analyses (LCA) appear much more relevant than food miles for assessing the impacts of different food production and marketing channels.

(agri-food) of direct investment abroad. On the contrary, these countries continue to pay the price of this strategy which entails the location of activities on sites that offer advantages in terms of fixed per unit costs and, above all, organization and technology, which are located in rich or emerging countries. From these sites products are exported worldwide, as shown for example by the flooding of African markets by American and Brazilian mass-produced frozen chicken, destroying local industries and in the long term eradicating regional culinary heritage.

Globalization also impacts upon health risks (increased vulnerability to pandemics due to animal production concentration), environmental damage (pollution and GHG emissions), professional organizations and NGOs. Multi-national protest movements are becoming increasingly frequent and large scale, due to the technologies of the digital society. In the agri-food sector, crises provide the trigger for international reactions which are coordinated with varying degrees of effectiveness, but all of which are becoming increasingly visible, as observed during the 2008 price spike.

GOVERNANCE: THE IMPACT OF THE FINANCIALIZATION OF MAJOR AGRI-FOOD BUSINESSES

The agri-food industry has not been spared from the financialization of the global economy over the last thirty years: most companies have resorted to the stock market to finance themselves. Consequently, investment funds, which are based on short-termism and the demand for high returns, make up a part of their capital[3]. These two indicators justify often brutal decisions to buy or sell shares, generating chronic instability which is incompatible with agri-food activities. Indeed, this activity is subject to the vagaries of biological and climatic hazards and thus to a much longer temporal horizon than the electronics or clothing industries. In addition the intervention of hedge funds on international exchanges dealing with commodities, such as Chicago, New York or London, amplifies the effects of volatility (thus in 2007 and 2008, the prices of cereals and oilseeds were first tripled and then halved). These upheavals pose major problems for farmers as well as for the agri-food industry, especially small businesses.

WHAT ABOUT CONSUMERS? Agriculture and the agri-food industry have a vital role to play in consumer food safety. Indeed, together they constitute the "productive heart" of this food safety, with a desirable contribution to the objective defined by the FAO in 1996, which is to ensure food for all that is of good quality, in sufficient quantities and in accordance with local cultures. The global challenge of feeding a growing population has been met or even surpassed in the last fifty years, in quantitative terms, illustrated by the fact that the current global per capita food availability exceeds the standard requirements set by nutritionists (2800 kcal/day/person in 2007, i.e. 600kcal/day/person above the requirement). The growth in food availability provided by agriculture and the agri-food industry has been higher than

3. For example, Danone's "stable core" of shareholders is estimated to account for 5% of the capital.

demographic growth. In addition, at the global scale, microbial toxicity risks have been greatly reduced by progress in science, technology and organization.

However, these observations must be qualified: significant disparities exist between countries and between regions within countries, some of which are marked by major food deficits, by deficiencies and/or nutritional excesses. In addition to the one billion people that the FAO estimates as suffering from calorific deficit, deficiencies of micronutrients (especially iron) and vitamins affect two billion people worldwide, causing often irreversible health problems particularly among children (Delpeuch et al., 2005). In 2008, at the other end of the scale, 1.5 billion people were overweight[4] and obesity[5] affected more than 500 million people aged over 20 (WHO, 2011). Furthermore, scientists suspect that severe pathological effects may result from the accumulation of heavy metals or residues from agrochemicals or synthetic food ingredients (Schlosser, 2005).

The conclusions that can be drawn from the above review cast the global agri-food industry in a mixed light, which must of course be adjusted according to different countries and sectors. The activity, which is oriented towards the very basic food market, continues to occupy a prominent place in modern economies. This position is supported by economic and social ripple effects, both upstream (agricultural supply and agriculture itself) and downstream (marketing), as well as on the periphery (industries and services such as equipment and logistics): it is estimated that a single job in the agri-food industry "generates" an average of four jobs in the surrounding sectors. The model of mass production based on economies of scale, to which most countries converge, allows a significant reduction in real food prices and leads to relative food and health security. However, it is accompanied by a relocation of industrial sites, which induces a redistribution of agricultural production areas and aggravates international and regional asymmetries. There is a further, controversial, consequence: the convergence of dietary

4. As defined by a Body Mass Index (BMI) that is equal to or above 25.

5. BMI = / > 30

BOX 2 THE FOOD ISSUE IN DEVELOPING COUNTRIES

In developing countries, and particularly LDCs, the situation is radically different. Indeed, over time, in most such countries a dual, if not schizophrenic, food system emerges: one sub-sector aims at the middle and upper classes of major urban areas, focusing on export and reproducing the agro-industrial pattern; while another traditional sub-sector exists to cater for the majority of rural areas. Generally, these countries have large agricultural populations, food prices are relatively high (accounting for the bulk of household income), women spend a considerable amount of time involved in food preparation since food products are at a basic development level and international trade integration remains low, except for a few rare commodities. The priority for these countries is obviously to escape from poverty (Sen, 1981) through agricultural modernization and the diversification of activities. It is therefore very important to combine agricultural policies and food policies (Raoult-Wack and Bricas, 2002). These countries must avoid the mistake of repeating a model that is today showing its limitations, and should instead integrate objectives of sustainable development into their policies.

trends caused by standardized global products that are promoted to consumers via expensive and often excessive packaging and marketing. Purchase incentives on food products and a lack of education on the subject lead to the inevitable consequence of considerable wastage in rich countries, where close to one third of food purchased is wasted (Ventour, 2008). Finally, negative externalities from concentration in the agri-food industry generate significant costs in terms of public health and the environment, which are not currently integrated into company responsibilities because - to varying degrees - these costs are met by the community. The agri-food industry now faces a huge challenge: given its position at the interface between insufficient agricultural production (in developing countries) or production that is weakened by specialization and intensification (in HIEs) and consumers that are often poorly nourished, it could make a major contribution to global food security.

TOWARDS A "HYBRID" FOOD TRANSITION?

The combination of pressure from changing variables (population increase, decline in natural resources and biodiversity, climate change, socio-economic crises) together with variation in political and strategic frameworks, justify the development of two prospective scenarios for the global food system leading to 2050. The first is the generalization of the service sector-based agro-industrial model (AIM), in a context of the continuance of "liberal" capitalism. The second scenario envisages the consolidation of an alternative model based on proximity (AMP) and networks of small and very small businesses, assuming the strengthening of policies for a balanced local development.

Within the framework of the AIM, it can be estimated that by 2050 only 500,000 farms of 4000 ha[6] (agribusinesses), and around a hundred giant corporations in the sectors of agricultural supply, the agri-food industry, logistics and distribution, will be responsible for the bulk of food production and marketing. The combination of a market that is shaped by huge advertising budgets and intense collusive lobbying of multinational corporations on cross-cutting issues such as product quality standards, consumer information, taxation, etc., will provide these companies with control over the global governance of the food system.

By contrast, in the AMP the reduced dimensions of companies (micro-businesses and small and medium businesses) reduces their capital requirements, so that they do not have to resort to financial markets and are able to adopt a more transparent legal status than the anonymous companies that form the majority under the AIM, and can also allow greater stakeholder involvement (direct contact with shareholders and employees). In agriculture, the family business predominates (about 50 million). Company size in the AMP leads to customized technologies and a reduction in the size of industrial units, while forms of organization and business management are based on the sharing of resources and expertise across business networks, enabling

6. Theoretical average does not account for the remaining international disparities.

the reduction of costs that would have benefitted from economies of scale under the AIM, allowing the creation of synergies between stakeholders.

Projections for the food system built around these two contrasting scenarios, shows that the AIM is unable to respond adequately to the requirements of sustainable development. Indeed, while this model generally manages to provide food at low prices (economic efficiency), it is often at the expense of the natural environment (negative externalities) and the generation of social inequalities between and within countries and among stakeholders in different sectors - whether they are producers, traders or consumers. The alternative scenario on the other hand, while better satisfying the three requirements of sustainable development (equity, environment and participatory governance), raises problems regarding economic competitiveness and the ability to provide cheap food[7].

Given the inertia of producers and consumers, and the limitations inherent in the democratic system (politicians have very limited horizons when making future plans, a period usually dependent on the date of the next election), the most likely evolution of the food system is a coexistence between the two models presented, with uncertainties remaining regarding the consolidation and growth of the alternative scheme. Indeed, an adjustment of the agro-industrial model to take into account some of the constraints mentioned above is already underway.

It is therefore essential to consider how to organize the transition towards a new model of "sustainable" development for food. Over a period of two generations, this model is likely to take on a hybrid form, combining attitudes and behaviours according to geographical areas, integrating modern configurations (based on globalization) as well as post-modern ideas (based on territorial roots), due to the extreme diversity of the observed situations.

To move towards sustainable food, we cannot rely on self-regulation by the market alone. A proper food policy must be established, the like of which has not been seen in any country in the world to date.

A food policy must be an effective incentive to improve the nutritional diet. It should be based on a change in consumer behaviour through education that must be started at an early age. It requires reflection on the allocation of budgetary resources (revaluation of the food budget) and of time (increasing the domestic time spent on meal preparation). It must also guide agricultural and industrial policy in the direction of improving the nutritional quality of products, and it must also help transform the production-commercialization model through diversification and shorter cycles. Finally, such a policy should include R&D effort into relevant models, especially those aimed at new crop management, food baskets and the sizes of businesses.

Such a food policy requires regional coordination and international consultation. Given the excesses caused by the globalization of agricultural markets and agri-food

7. The Agrimonde quantitative model leads to an important conclusion: in two contrasting scenarios akin to the models discussed here, the resources of agricultural biomass in the world that can potentially be mobilized would enable the food demands of nine billion people to be met by 2050 (Paillard et al., 2010).

markets, a form of "regionalized globalization" may be a viable solution (i.e. to envisage specific deterrents and incentives at borders and within macro-regions such as the Euro-Mediterranean) to achieve the "re-localisation" of food systems. This would reduce the distances between places of production and consumption. Some of the main benefits of the re-establishment of short food chains would be the maintenance (if it is not already too late) of diverse consumption patterns (by moving them towards better nutritional adequacy); the stabilization or creation of activities and therefore jobs in rural areas in most countries of the world; and to restore a meaningful relationship between producers and consumers[8] (Winter, 2003).

The alternative scenario invites us to invent a new agri-food model that values the specific historical heritage of individual societies and territories, while integrating the scientific and technical knowledge of the present century. ■

8. Such as the one envisaged by the Tikei movement in Japan or the association of community-supported agriculture (AMAP) in France

REFERENCES

CIAA, 2010, *Data & Trends of the European Food and Drink Industry*, Brussels

Delpeuch F., Le Bihan G., Maire B., 2005, Les malnutritions dans le monde: de la sous-alimentation à l'obésité, In: Ghersi G., dir., *Nourrir 9 milliards d'hommes*, ADPF, Paris, 32-37.

European Commission, JRC/DG RTD, 2009, *The 2009 EU Industrial R&D Investment Scoreboard*, Brussels: 126 p.

FAO, 2009, *Food insecurity in the world*, Rome.

Nefussi J., 2004, La tertiarisation des filières agroalimentaires, *Economies et Sociétés*, Tome XXXVIII, n°3, série AG, n°26, Paris: 613-629.

Noronha Vaz de T., Nijkamp P., and Rastoin J.L., Editors, 2008, *Traditional Food Production and Sustainable Development: A European Challenge*, Ashgate, London, 285 p.

Paillard Sandrine, Treyer Sébastien and Dorin Bruno (ed.), 2010. *Agrimonde, Scenarios and Challenges for Feeding the World in 2050*. Quæ, Versailles: 296 p.

Raoult-Wack A.L. and Bricas N., 2002, Ethical issues related to food sector evolution in developing countries: about sustainability and equity, *Journal of Agricultural & Environnemental Ethics*, 15, 325-334.

Rastoin J.-L, 2000, Une brève histoire économique de l'industrie alimentaire, *Economie Rurale*, n° 255-256, Paris, 61-85.

Rastoin J.L., 2008, Les multinationales dans le système alimentaire, *Projet*, n° 306, La Plaine St Denis: 61-69.

Rastoin J.L., Ghersi G., 2010, *Le système alimentaire mondial, Concepts et méthodes, analyses et dynamiques*, Editions Quae, Paris: 581 p. Preface by Olivier de Schutter, Special Rapporteur of the United Nations on the Right to Food.

Reardon T., Minten B., 2011, Surprised by Supermarkets: Diffusion of Modern Food Retail in India, *Journal of Agribusiness in Developing and Emerging Economies*, 1(2)

Schlosser E., 2002, *Fast Food Nation, the Dark Side of the all-American Meal*, Perennial/HarpersCollins Publisher: 276 p.

Sen A. K., 1981, *Poverty and Famines*, Clarendon Press, Oxford.

Smith A., Watkiss P., Tweddle G., Mc Kinnon A., Browne M., Hunt A., Trevelen C., Nash C., Cross S., 2005, *The Validity of Food Miles as an Indicator of Sustainable Development*, AEA Technology Environment/DEPRA, Oxon, UK: 117 p.

Ventour L., 2008, *The food we waste, Food waste report v2*, Wastes and Resources Action Programme, WRAP, Banbury: 237 p.

Weber C.L., Matthews H.S., 2008, Food-Miles and the Relative Climate Impacts of Food Choices in the United States, *Environmental Science and Technology*, 42(10), American Chemical Society: 3508-3513

WHO, 2011, *Obesity and overweight*, Fact sheet n° 311, Geneva.

Winter M., (2003), Embeddedness, the New Food Economy and Defensive Localism, *Journal of Rural Studies*, 19, 23-32.

SMALL FARMERS AND MARKETS IN LATIN AMERICA: IN SEARCH OF GREATER SOCIAL INTEGRATION

Octavio Sotomayor and Javier Meneses,
ECLAC, Santiago, Chile.

Similarly to their counterparts in other regions, small farmers in Latin America and the Caribbean have had to face the market in isolation, either through intermediaries or by selling their products directly, often in very poor conditions. However, in recent years, a number of public and private strategies have emerged to address this problem.

COOPERATIVE ORGANIZATION

The traditional approach has been to form associations so that producers are unified on entering the market. Many cooperatives were created as part of the land reform process initiated in the 1960s, but difficult economic and political conditions experienced in the region in the 1980s and 1990s have prevented their consolidation, with a few exceptions such as the dairy cooperative CONAPROLE (Uruguay), the National Federation of Coffee Growers of Colombia or the hundreds of agricultural cooperatives promoted by the Landless Workers' Movement (MST) in Brazil. While this approach remains effective under certain conditions, especially in more isolated rural areas, it has serious limitations: small farmers have little capital, it can be difficult for them to form organizations, and they face considerable risks when competing in open and changing markets.

"PRODUCTIVE CHAINS" ALONG THE LINE

Alongside the traditional approach where producers organize themselves independently (with support from the state) to obtain new technologies and sell their produce, an alternative strategy is to link agro-industries, that are already well-established in the market, with small supplier businesses (Sotomayor et al., 2011), thus generating a sales volume to benefit both low income communities and agro-industries (Figure 1).

Supplier Network Programmes (Programas de Desarrollo de Proveedores - PDP) are an example of the "productive chain". Established in Mexico (in 1997) and Chile (1998) to provide technical assistance to small and medium-sized supplier businesses, these programmes are in operation in all production sectors and are particularly well suited to the reality of a highly fragmented agricultural sector, and to the supply requirements of the agro-industrial sector.

In Chile, PDPs are implemented by the Corporación de Fomento[1] (CORFO), a public agency that provides grants to agro-industries to finance such agreements. In addition to playing an active role in the technology transfer process, agro-industries also provide a market for commodities, which enables

1. "Development corporation"

improvement in the competitiveness of production chains. Stable contractual links between agro-industrial companies and their suppliers create trusting relationships that facilitate the processes of specialization and productive complementarity, which is beneficial to all those in the chain. Between 2002 and 2007, CORFO financed 150 PDPs, involving 3,800 small and medium suppliers. Today, another public agency, the Institute for Agricultural Development (Instituto de Dessarrollo Agropecuario, INDAP) continues this strategy through the implementation of productive alliances, a similar instrument to the PDP, for an additional 4,000 small producers. In Mexico, PDPs incorporated 1,600 supplier micro-businesses between 2003 and 2008. The Mexican experience has been replicated in El Salvador, Honduras and Colombia, under the coordination of ministries, NGOs and chambers of commerce and industry.

A further variation of this approach is the "inclusive business" (negocios inclusivos - NI) model, which is currently being developed by the Ecuadorian Ministry of Agriculture, as part of its strategy for food sovereignty. This approach explicitly seeks to foster win-win relationships between agro-industries ("anchor companies") and small producers, to insert low-income rural families into new productive activities. The goal is to go beyond philanthropy and corporate social responsibility, to create economic activities that can be self-funded and sustainable over time. Further experiments are being developed in Peru, Colombia and Central America (Van Haeringen and De Jongh, 2010).

In PDPs, relationships are defined according to the needs of the anchor companies (the supply of commodities), while the inclusive business concept defines relationships symmetrically with producers, to create a win-win situation. In PDPs, the creation of value is focused on the product, whereas inclusive business seeks economic co-creation, centred on the identification of opportunities arising in the community (Simanis and Hart, 2009) and leading to the realization of joint ventures. At the operational level, PDPs focus on technical assistance since they establish a bilateral individual relationship between the agricultural advisory services and the producers; inclusive business encourages association through the establishment of a relationship between groups of producers and a company. Furthermore, inclusive business extends as far as the signing of contracts for commodity purchases by industries, in addition to credit and technical assistance.

After more than ten years in operation, PDPs have demonstrated a capacity to promote the processes of technical innovation and small farm development. The inclusive businesses awaiting realization have a great potential because they target a more comprehensive type of chain. However, it is conceivable that price variations may generate tensions between anchor businesses and producers, revealing relationships of dependence and domination between stakeholders. Since inclusive business has emerged as one of the axes of corporate social responsibility, the value of this approach lies in the fact that both parties are defined as strategic partners, thus formalizing explicit commitment from anchor businesses towards small producers. Even so, there remains a need for relationships of trust between stakeholders, along with a number of well-designed regulations, either through contracts (annual and multi-annual), transparent systems of quality analysis and mechanisms for dispute resolution.

SHORT CHAINS

The promotion of short chains is another strategy used to improve marketing, primarily used by NGOs, agro-ecology networks, schools and universities, and organisations representing farmers, consumers and indigenous people.

Short chains remain an emerging trend, materializing mainly through the establishment of agricultural fairs and ecological or organic markets, such as in Loja in Ecuador or Jalisco in Mexico. However, the politicization of the food sovereignty issue, resulting from the increases in food prices in 2008 and 2011, has mobilized governments. In this context, three new trends have emerged:

– the promotion of urban and peri-urban agricultural projects: successful examples of such projects can be found in Argentina, Bolivia, Colombia, Cuba, Haiti and Uruguay, among others. In Argentina, for example, the Pro-Huerta Program, which has been funded by the Ministry of Social Development for several years, has supported 624,000 family, school and community vegetable gardens, as well as 125,000 peri-urban farms. In Cuba, urban agriculture has played a key role in ensuring public access to food, with 3,000 ha of vegetable gardens for auto-consumption, as well as more than 1000 ha of plots for intensive vegetable production, either under glass or in low input systems. In 2005, 384,000 jobs were created through such systems (Herrera, 2009);

– the creation of public programmes for food procurement from family farms: the most emblematic example of this type of project is the "Zero Hunger" programme in Brazil, which seeks to establish food procurement from family farms. The programme has enabled the supply of food from family farms to sufficiently match state school demand. Between 2003 and 2008, the federal government spent 1.18 billion dollars in acquiring nearly 2 million tonnes of food, with the participation of 118,900 small farmers in 2008 (Veiga Aranha, 2010). This programme directly depends on the political commitment of the Brazilian government. The second generation of this programme, known as *Brazil sem Miseria* (Brazil without misery), which aims to eradicate extreme poverty within four years, maintains the public programmes of food procurement from family farms, and seeks to involve supermarkets and the private sector in general;

– the development of closed local production models: some of the cooperatives created by the Brazilian MST have formed small centres for ethanol production from sugar cane and sunflower oil, for direct use in their own farm machinery, for travelling short distances, as well as to sell it through short chains to provide an income. This model allows the completion of the commodities cycle (the residue is reused as cattle feed) and provides an energy supply, a level of economic autonomy and reduces costs, as well as providing a small income from the sale of surplus ethanol to fuel local vehicles.

The assertion of the state's power of control is a modality that has been very successful in Brazil, and is beginning to be examined in other countries in the region. It is very important to note that despite these trends and other emerging initiatives, the issue of short chains has not been subject to public policies.

CONCLUSION

The diversity of these trials illustrates the extent of the search for new relationships between small producers and markets, which remains the main challenge for family farming. Consumers themselves are still virtually absent from this process. Although there are a number of experiments underway (local fairs, direct farm sales, internet sales, quality labels, among others), the direct relationship between producers and consumers could in future experience a new development. ■

FIGURE 1 EVOLUTION OF RELATIONS BETWEEN PRODUCERS AND THE MARKET

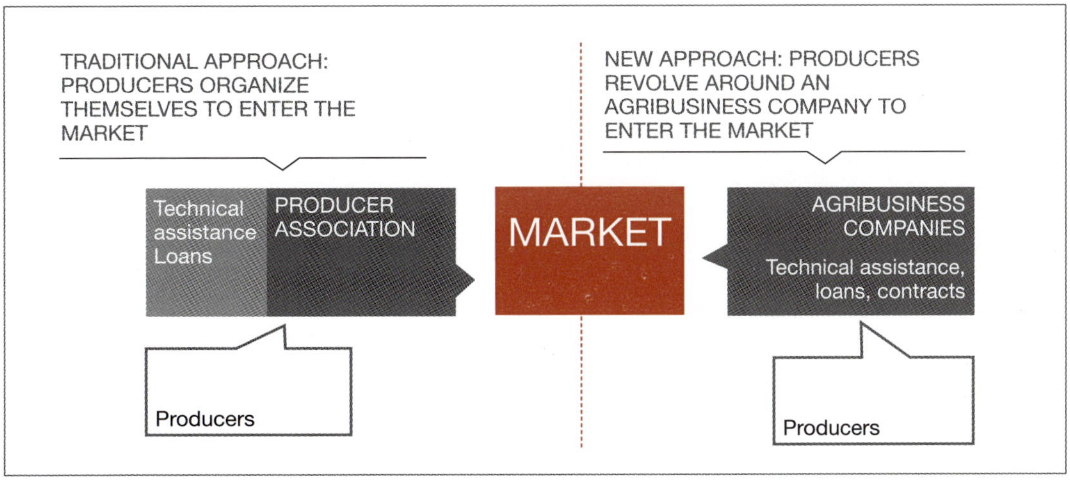

REFERENCES

Herrera, A., 2009, Impacto de la agricultura urbana en Cuba. Revue Novedades en Población Année 5, N°9.

Simanis, E. and Hart S., 2008, Beyond Selling to the Poor: Building Bussiness Intimacy Trough Embedded Innovation, Cornell University, April.

Sotomayor, O., Rodríguez A. and Rodrigues M., 2011. Competitividad, sostenibilidad e inclusión social en la agricultura. Nuevas direcciones para el diseño de políticas en América Latina y el Caribe. CEPAL, Santiago de Chili.

Van Haeringen, R. and de Jongh W. R., 2010, Los negocios inclusivos en el sector agropecuario: práctica y desafíos. Revue Estudios Agrarios, N° 44, Procaduría Agraria, México, pp. 63-74.

Veiga Aranha, A., 2010, Fome Zero. Um projeto transformado em estratégia de governo. En: Da Silva, J. G., M. E. del Grossi y C. Galvão de França (Org.), Fome Zero. A experiência brasileira. MDA, Brasilia, pp. 85-109.

CHAPTER 12

Agricultural price volatility has returned to centre stage since 2008. Since then, economists and politicians have debated its causes, with biofuels and speculators as two prime suspects. But rather than addressing the root of the problem, the political response, from the G20 in particular, seems focused only on treating its effects.

INSTABILITY IN AGRICULTURAL PRICES: A COMFORTABLE TRUTH

Tancrède Voituriez,
Institute for Sustainable Development and International Relations, Paris, France

The commodity price rises that occurred between 2006 and 2008, and subsequently during the winter of 2010 to 2011, have attained a profile in media and political debates that is higher than that reached by these products since the price increases of 1974 and 1979, which were concurrent with the two oil crises. Agricultural price movements on international markets were usually only subject to media attention because of apparent relentless decline (Figure 1). After an optimistic pause in the 1970s, at the turn of the century the world, and especially the developing world, rediscovered the tribulations of excess. Oil itself did not escape the curse of abundance, prompting the still famous "Drowning in Oil" article in *The Economist*, concerning the inexorable decline of oil's market price[1].

In this context, the 2006-2008 and 2010-2011 crises have a special resonance. Both have surprised by their magnitude and abruptness. The IMF's commodity price index was set at 100 in 2005, in July 2006 it was raised to 130, in November 2007 it was more than 157, and by July 2008 it stood at 219; i.e. more than doubling within three years. After a decline during the financial crisis, the index rebounded and in March 2011 it reached 210. For comparison, the average figure for the period from 1992, the year the index was established, until 2004, would be just 60. How is it possible that such a phenomenon was not anticipated by politicians, who were wholly unprepared for its impact on disposable income, purchasing power and

1. See "Drowning in oil" and "The next shock" in *The Economist*, 4 March 1999, in which the British newspaper reported that the price of oil could go under $10.

inflation; and why was the market itself incapable of anticipating the scarcity? Is it because the fine adjustment of supply and demand is such an elusive and mysterious process, that prices are suddenly able to double or triple in value within a few months, even though in many cases markets have been established for more than a century, with prices remaining stable or on average declining for periods of almost 20 years?

In a collective response, the G20 countries have taken onboard these two issues and focused not on identifying the causes of instability, but on the treatment of its effects (Box 1). This approach, to which the realists will give a free rein, is motivated firstly, at least officially, by the need for visible action, and by the observation that the causes are uncertain and controversial, and therefore difficult to target. In short, we are facing new crises and the key is to limit their intensity and effects.

The conclusions of the G20 agriculture ministers were met with a favourable, even rapturous, reception. Only a few dissenters stressed that key issues had not been dealt with and that the delay in addressing the causes would certainly allow a deal to be reached, but that it would be an agreement without any real scope. For the dissenters, it was of particular significance that trade liberalization had been dismissed as potentially culpable, that the role of biofuels had not even been

BOX 1 G20 AGRICULTURE MINISTERS' RESPONSES TO MARKET INSTABILITY

At meetings in Paris on 22 and 23 June 2011, G20 agriculture ministers adopted an action plan on food price volatility and agriculture. The action plan is to be submitted to G20 heads of state at the November 2011 summit in Cannes. The action plan has five major elements:

Increase of food production – There are commitments to implement "a broad scope of actions to boost agricultural growth", and the strengthening of research and innovation with, as a first step, the launch of an "International Research Initiative for Wheat Improvement (IRIWI)" to coordinate research efforts on this crop; and to encourage and increase public and private investment in agriculture.

Information and market transparency - The G20 will launch, within the FAO, the "Agricultural Market Information System (AMIS)" to encourage major players in agri-food markets to share data, to enhance existing information systems, to promote greater shared understanding of food price developments and to promote political dialogue and cooperation. This database will be implemented "as soon as possible" and will be complemented by a satellite observing system that will share satellite images of agricultural production to enable the effects of climate change on world agricultural production to be accounted for.

International policy coordination - The action plan recommends the implementation of a "Rapid Response Forum" placed within the FAO. This mechanism for immediate action, composed of senior G20 officials, will help make decisions in the shortest possible timescales in the event of a drop in supply from a major producer, allowing countries to consult and respond collectively. Any decisions made during a crisis will be taken within the Rapid Response Forum.

Reduce the effects of price volatility for the most vulnerable - Ministers agreed on the objectives, principles, modalities and timing of an agriculture and food security risk management toolbox, and supported the development of a proposal for a targeted emergency humanitarian food reserves system (to complement existing regional and national food reserves).

Financial regulation – G20 Finance Ministers and Central Bank Governors are "strongly encouraged" to make appropriate decisions for "a better regulation and supervision of agricultural financial markets".

mentioned, and that the commitments against financial speculation were vague and inconsistent[2].

This article balances these two views by first examining the potential causes that have been put forward by relevant institutions and stakeholder groups since 2006, and then reviewing the state of scientific knowledge on these issues. We start with trivial questions: 'agricultural prices are unstable, so what?' before examining the potential liability of trade liberalization, biofuels and speculation. In conclusion, we find that there was no reason for the G20 to exclude or defer the inclusion of the latter two factors in their restricted agricultural governance plan.

AGRICULTURAL PRICES ARE UNSTABLE, SO WHAT?

Two terms are used alternately to describe market price movements of agricultural commodities: instability (of markets) and volatility (of prices). The same measurement method is generally used, which is derived from the financial world, to provide information on price dispersion around the average. The greater this dispersion, the more frequent the major variations ("major" relative to the average) will be, and the greater the market instability and price volatility (Box 2).

The fact that a market is unstable with volatile prices poses no economic problems itself. Such phenomena only become problematic when they are a source of uncertainty and risk for economic operators (producers, investors, buyers...) – a risk so significant that these agents modify their behaviour and distance themselves from what is both in their own interest and that of the community. Unstable prices may not necessarily create risks, either because operators do not perceive the risk in question, or because there are means to guard against it – such as through insurance. The state of the world referred to when measuring economic gains and losses related to instability is therefore not a world where prices are stable, but a world where all risks are insurable. Operators are free to insure themselves or not.

2. This viewpoint was espoused by La Confédération paysanne (Peasant Confederation), la Coordination rurale (Rural Coordination) and MODEF.

BOX 2 MEASUREMENT OF PRICE VOLATILITY

The most common measure of long-term volatility or "historical annualized" volatility is obtained by calculating the standard deviation of price variations (usually expressed as a logarithm) over a given period, multiplied by the square root of the frequency of observations:

$$\text{Volatility} = \sigma\sqrt{T} \text{ with } \sigma = \sqrt{\sum_{i=1}^{n} \left(r_i - \mu\right)^2 / \left(n-1\right)}$$

and $r_t = \ln(P_t) - \ln(P_{t-1})$, n is the number of observations, μ is the mean variation of the logarithms of prices (r_t) and T is the frequency of observations (252 for daily data, 12 for monthly data etc).

When price fluctuations represent uninsurable risks, producers produce less (than would otherwise have been produced if the risk was null or insurable), investors invest less and consumers consume less (if not necessarily the product in question, then other products may be eschewed to meet the need for "products with unstable prices"). Ultimately, the consumer is the first loser. Indeed, the risk (uninsured) linked to instability reduces production and investment (and therefore long-term production) compared to a "risk-free" situation. With a constant demand, which is the case in agriculture where demand is assumed to be inelastic, i.e. low price sensitivity, a lower production will result in higher prices. The consumer therefore bears the cost of the risk perceived by the producer by paying more for the product (than would have been paid in risk-free situations or with an insured risk). The market transfers a part of consumer wellbeing to the producer, via prices that on average are higher (compared to a risk-free universe). Instability is therefore a redistribution mechanism. It should be noted that there is another loser: poor producers. Such producers are, empirically, more risk-averse than rich producers. Rich producers see profit opportunity in instability, while poor producers feel the threat of bankruptcy and ruin. Poor producers are therefore more likely to avoid high-risk crops than rich producers, thus guarding against the negative effects of instability – lower prices in this case – but will not be able to take advantage of the benefits – record-breaking revenues when prices soar. Within the production sphere, instability again serves as a redistribution mechanism. Redistribution towards the rich in this case.

AGRO-EXPORTING COUNTRIES CAN GAIN FROM INSTABILITY, WHILE AGRO-IMPORTING COUNTRIES WILL LOSE OUT THROUGH BEARING THE COSTS OF RISK.

The significance of this setup is that the risk (not covered) linked to instability creates two types of problems: it is a source of inefficiency (higher prices are paid on average for products than in a risk-free world) and inequity (it is to the detriment of poor households). But the risk is not universal. Insurance mechanisms are being developed in some regions (in brief, OECD and some emerging countries) and less in others (LDCs). However, poor rural households occur in higher proportions in areas such as LDCs than in others (OECD). What is true for households is also true at the national scale. Agro-exporting countries can gain from instability, while agro-importing countries will lose out through bearing the costs of risk, the importation of inflation and by having to divert resources and savings towards the purchase of food at the expense of other imported products. Thus, it appears that market instability is primarily an economical problem that affects the poor, which may be its only universality.

DOES TRADE LIBERALIZATION MAKE MARKETS MORE UNSTABLE?

Price rises (as distinct from an increase in volatility, which is itself only very relative) between 2006 and 2008, and then in 2011, occurred in a context of globalization and market liberalization, suggesting that the two phenomena could be related[3]. This idea is not absurd, gaining support from theories that were formulated even before the events in question occurred. The problem is that these theories are incompatible.

The first "theory" emphasizes the benefits of agricultural trade liberalization on instability. By considering instability as a result of external shocks on the supply or demand (weather shocks, but also political shocks or "externalities", for example through the introduction of unilateral measures such as subsidies or export tax), the gradual integration of domestic agricultural economies into a large global market attenuates price instability through compensation of weather shocks (a good harvest somewhere in the world compensates a poor harvest somewhere else) and through mitigation of political shocks or "externalities" (a progressive ban on subsidies and taxes for example). It could be said that liberalization takes the edge off the shocks.

This does not mean the resulting prices are stable; in theory, prices are expected to be less unstable, in other words, that their distribution becomes closer to established laws of probability and that they become insurable under the law of large numbers. They are therefore more "stable" in the narrow sense that they are "more insurable". In fact, global prices are generally more "insurable" under classical financial techniques than domestic prices.

A competing theory is that the beneficial effect of shock compensation that globalization brings is, on average, offset by the negative consequences of the dismantling of public policies for the stabilization of domestic prices, which is a consequence of globalization. The disappearance of guaranteed price policies that were developed in OECD countries after the Second World War (or shortly beforehand in the United States) has increased uncertainty and risk for the producer, who is left exposed to market price fluctuations. Without the availability of an insurance mechanism (or without wanting to seek recourse to insurance since it requires payment, while public policies cost nothing) the producer makes erroneous production decisions by producing too much or too little, and so prices are never in equilibrium but adjust constantly according to the whim of the producer. Instability is endogenous. It creates risk, which creates error, which creates instability.

Both theories are valid. Neither can be proven superior by empirical testing.

However, empirical observations do suggest that, in retrospect, long-term international prices of most agricultural commodities have never been as unstable as during the 1970s (Figure 2), which was the golden age of agricultural policies. In contrast, most prices have never been as stable as in the 1990s, which was a major era of domestic and trade liberalization (LDCs undergoing structural adjustment and MacSharry's 1992 CAP reforms). Moreover, the observed relationship between stability and agricultural policies and trade liberalization has remained true for a long time. Periods of increased instability (as opposed to price rises) have since 1700 been associated with periods of restrictions in international trade (Jacks et al. 2010).

3. This argument is particularly salient in the opinions of organizations such as la Confédération paysanne or Via Campesina.

Furthermore, we observe that on average agricultural prices are unstable during increases and stable when heading downward. Depression is long and steady while increases are sudden and condensed. This is a constant of agricultural markets. The reason being that when surplus occurs, farmers do not uproot their vines or cocoa trees, nor do they put their land into fallow (they hope that their neighbours will), thus the downturn spirals. When prices are high, even if hectares are available to convert, it takes time for production to react, which according to a producer's risk aversion may be one or two seasons – a precise period overlooked by the market, hence leading to instability. The recent phenomenon is not extraordinary when considered over the long term; it is in fact quite marginal when compared to events in the 1970s.

Can we not find political grounds for the present instability – regardless of its relative historical importance? It is worth recalling that in 1994 WTO member countries agreed to abandon policies of domestic price stabilization. The objective being to stabilize revenues, rather than domestic markets. A gradual redirection of public intervention was intended, away from the setting of a price floor towards direct revenue aid – handouts. The producer was left alone to manage savings and investment choices, like any entrepreneur (in reality not exactly like any entrepreneur because small businesses do not receive the equivalent of 20,000 euros in subsidies for a full-time equivalent worker, which is the average for OECD farmers). Two consequences were expected from liberalization: that domestic prices would be more unstable (due to the elimination of domestic stabilization policies); and that global prices would be more stable in the sense defined above, i.e. more insurable (due to the increase in the geographic scope of trade which then reduces "exported" instability caused by domestic stabilization policies that lead to surpluses or irregular and unpredictable imports onto global markets). Basically, it was expected that instability would be levelled out and equalized, and that financial instruments (futures contracts and options) could be used to limit it.

WHEN PRICES ARE HIGH, EVEN IF HECTARES ARE AVAILABLE TO CONVERT, IT TAKES TIME FOR PRODUCTION TO REACT, WHICH ACCORDING TO A PRODUCER'S RISK AVERSION MAY BE ONE OR TWO SEASONS.

The WTO agricultural policies "package" therefore combines trade liberalization, subsidies to producers to facilitate the transition, and the use of financial instruments for risk protection. The package is inaccessible to LDCs, which is to be expected since the WTO agricultural agreement was neither written by nor aimed at LDCs.

The management of instability through the natural effect of market expansion and of the dismantling of public price stabilization policies in fact discriminates "against" households and poor countries that lack access to direct aid or financial instruments for risk management. They only have access to external markets, an access that lacks sufficient power to stabilize their revenues, which the other two elements of the "package" provide. As mentioned above, instability poses a fundamental problem of inequality, as does its contemporary management if we consider the "package" of measures that was institutionalized by the WTO agricultural agreement. Even if liberalization does not increase the instability of agricultural markets, it does mean that poor households are more consistently exposed.

ARE BIOFUELS RESPONSIBLE?

The new link between agricultural commodity markets and energy markets is difficult to contest. The rising global need for maize imports was concurrent with that of the increase in US domestic demand for maize for ethanol production, which currently stands at about 29% of maize consumption in this country. As Collins (2008) emphasized, expansion of maize usage for ethanol production has truly accelerated from 2004/05, with the enactment in the US of the Energy Policy Act during a period of a sharp increase in oil prices, which had the explicit objective to increase the share of biofuels in the country's energy "mix". Agricultural markets and energy markets are therefore linked due to the anticipation of high oil barrel prices along with the US need to secure energy supplies.

Similar concerns occur in Europe in the "climate and energy package" adopted by the European Parliament on 17 December 2008, which stipulates that each Member State of the Union must meet the 10% target of biofuels in the final energy consumption in transport by 2020. This objective implies that, by the stated date, 60% of Europe's consumption of vegetable oil will be for biofuel production (European Commission, 2007). Retrospective and prospective data collected by Goldman Sachs (Currie, 2007) show that in 2004 there was a "spike" in the demand for agricultural products for energy, and the legitimate expectation that an increase in biofuel supply would continue.

Considering cereal production as a whole, USDA estimates show that the US maize demand for ethanol production accounts for one third of the growth in global demand for these commodities (compared to one quarter for animal consumption and about 45% for food). For some, these figures put US maize ethanol demand into perspective regarding its culpability for the rise in global prices of wheat and other cereals – and ultimately in the rise of agricultural prices as a whole.

It is true that aggregate data seems to clear biofuels of blame. According to Agrimonde (a prospective project led by CIRAD and INRA) estimations, less than 5% of the total worldwide production of calories of plant origin is for non-food uses, including agrofuels. According to the International Energy Agency, in 2005 1% of cultivated land was used to produce agrofuels, replacing 1% of the global consumption of fossil fuels. However, the announced rise in the use of cereals for agrofuels has increased the attractiveness of the market for financial speculators interested in agricultural commodities so that, according to Bricas and Bru (2008), "it is more the anticipation of an increase in demand, rather than an actual increase, that accounts for the price surge". Furthermore, the use of globally aggregated data consistently underestimates the statistical share of North American maize energy consumption out of all produced and consumed cereals. According to the same authors: "It is not the current volumes of agricultural commodities used for agrofuels

TABLE 1: ESTIMATIONS OF THE SCOPE OF THE EFFECT OF INCREASED US BIOFUEL PRODUCTION ON CEREAL PRICES

Author	Proportion of increase accounted for	Product
Lipsky (2008)	70%	Maize
	40%	Soy
Collins (2008)	60%	Maize
Rosegrant et al. (2008)	47%	Maize
	26%	Wheat
	25%	Rice

that explain the rise in prices, except at a local level for US maize providers, who are the main producers of these fuels." The fact that US maize prices "surge" due to an increase in energetic demand may be sufficient to align maize prices in various places of export or import across the world, with the price of US maize remaining the reference in international trade. If one accepts the idea that prices are established at the marginal value of the quantities traded, the price of maize may be determined by the price of (US) ethanol, even though it only accounts for a small share of the aggregate or average consumption globally.

While rising US maize and cereal prices can be explained by the rapid increase of domestic demand for maize for biofuel production, there are discrepancies between the different analyses on the contamination effect of rising US prices that reaches across the world and spreads to other products. Even though the figures are questionable (especially given the absence of firm counterfactuals), it is important to consider the orders of magnitude of the quantitative simulations conducted to determine the contribution made by the increased supply of US biofuel maize towards generating the global agricultural price surge of 2006-2007 (Table 1).

It is also important to consider that the tension caused by the actual or anticipated scarcity of food maize on the North American market has had direct impacts on soy and wheat prices, by their replacement in crop rotations, which has contributed to the increase and exceptional amplitude of price; even if the exact measurement of this contribution is impossible, as is its relative measurement, in the available global models. Biofuel policies seem to be the last instrument of national sovereignty in agriculture. In the case of "large countries", their effects go beyond national borders and they become objects of multilateral coordination, thus necessitating inclusion on the G20 agenda.

IS SPECULATION IRRESPONSIBLE?

In the final version of its Communication on commodities and raw materials on 2 February 2011, the European Commission acknowledged the existence of a "strong correlation between positions on derivative markets and spot prices". This statement replaces one contained in a previous version that read: "There is no conclusive evidence on the causality between speculation in derivatives markets [on the one hand] and increased volatility and price increases in the underlying physical markets [on the other hand]." To balance Nicolas Sarkozy's demands to regulate speculation with Angela Merkel's reluctance on this issue, the Commission has used the term "strong correlations" as a skilful and prudent compromise, demonstrating the extent to which the role of speculation in commodity price instability is both politically sensitive and scientifically controversial. So, is there any empirical evidence for the liability, or neutrality, of speculation in terms of causing the spike in commodity prices? Here we present the most salient reasoning on this subject, produced by economics experts since the 2006-2008 spike.

Three observations emerge. The first is that economics does not provide a complete and coherent explanatory model of the observed price developments on a given market, whether it is "spot" or "futures". The second observation is that to fill this gap, the empirical evidence to blame or exonerate speculation is sparse and often contradictory. Finally, considering the precautions imposed by the two previous points, the temporary and fragile idea seems to prevail that derivatives markets are at least partly responsible for the increase of instability.

WHAT HAVE WE LEARNED FROM THE 2006-2008 CRISIS? Speculation is only one factor among several that researchers and international institutions have identified as causal agents of the increase in prices and the volatility of commodity markets after 2006. It is also the focus of the media today. The rising profile of the subject is due firstly to the hesitancy of economists themselves who, all things being equal, are unable to definitively measure the contribution of one variable in isolation from others, so that today we are still no closer to establishing the hierarchy of causes that has led to the tripling of prices for some products[4]. Secondly, the general denunciation of "speculation" following the bursting of the US financial bubble, has mean that both subprime speculation and the very ordinary type on wheat or soy futures prices have been tarred with the same brush. Finally, the uproar created by the July 2008 testimony of an investment fund manager hasn't helped the reputation of speculation: in "The Accidental Hunt Brothers – How Institutional Investors Are Driving Up Food and Energy Prices", Michael Masters and his co-author Adam White firmly put the blame on systematic long-only institutional investors (those that always "*go long*"), that speculate on commodity indices such as the Goldman Sachs Commodity Index or the Dow Jones-AIG Commodity Index.

At the heart of the Masters and White argument is the fact that the proportion of positions taken by index speculators on futures markets has experienced an unprecedented rise between 2005 and 2008, during the "spike" in prices. In 1998, physical hedgers held 77% of the positions in commodities futures markets, traditional speculators ("non commercial") accounted for 16% and index speculators 7%. In 2008, the proportions were reversed. The physical hedgers (or "commercial") held 31% of positions, traditional speculators 28% and index speculators 41%. This proportion approaches 70% on certain products, in particular wheat and livestock (Figure 3). There were 2.7 million commodities futures contracts bought by index speculators

4. Credit should be given to Gohin and Dronne (2008) in their attempt to integrate all the potential causes of the price rise into one model: "Les principaux déterminants de l'évolution des prix agricoles internationaux," mimeo INRA. Their model, however, fails to explain, as noted by the authors themselves, nearly half of the price movements observed.

FIGURE 3 THE GROWING IMPORTANCE OF INDEX SPECULATORS

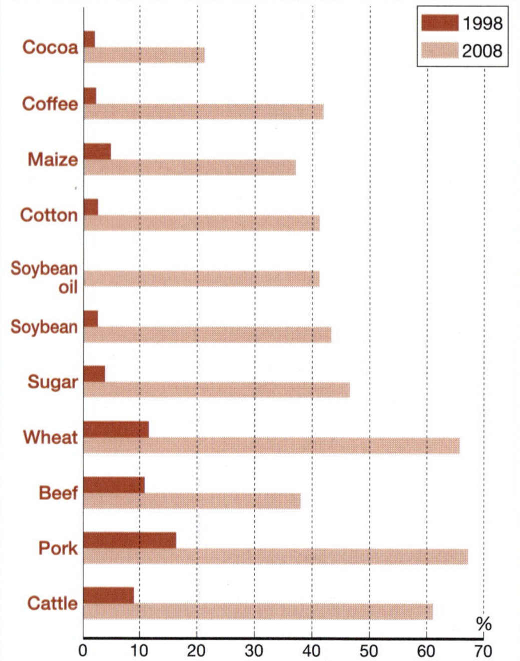

Source: From Masters and White data (2008)

Depending on the product, between one third and two thirds of futures purchases were made by index speculators in 2008, against an average of less than 10% in the ten previous years. A growing role that can only enhance the rise in agricultural prices.

between 2003 and 2008, compared to 1.4 million by traditional speculators and less than 900,000 for hedgers. According to Masters and White, there has clearly been a bubble.

CORRELATION DOESN'T PROVE CAUSATION All is not straightforward though. To paraphrase the cautionary language of the Commission, the findings of Masters and White demonstrate correlation, but not causality. Work of the French Ministry of Agriculture has stressed in particular that the arrival of index speculators precedes the rise in prices by nearly two years, and also that the number of open positions (unsettled purchase contracts) remained unchanged during the increase. Some econometric work on the futures market of the Chicago Board of Trade suggest a causal relationship between index speculation and price inflation in the case of soy – but only for soy, thus concluding overall that there is only limited evidence that index speculation has caused changes in commodity prices between 2006 and 2008.

An objective weakness of these works is to associate "speculation" with a particular market player – the "non commercial" speculator, as opposed to "physical hedgers" that supposedly buy and sell futures without "speculating". Buying or selling futures, whether these operations are accompanied by a transaction in the physical market or not, are inherently "speculative", reflecting a bet or a belief about the future development of prices. Thus, systematically "long-only" investor positions have met systematically "short-only" investor positions, the fact that the latter are "hedgers" or "commercial" does not exclude them from having the intention to speculate. We can just about infer that a systematic purchase position, such as that taken by the index investors between 2005 and 2008, makes the demand for contracts insensitive to price – in economist jargon, demand becomes "rigid". It is known that a rigid demand has a particular ability to make the market overreact, because adjustment is no longer made through quantities (demand for quantities does not vary, whatever the price) but through prices.

A last, more intuitive, element can help comprehend the role of speculation. In the current episode of rising commodity prices, a number of causal (or supposedly causal) factors that were present in 2006-2008 are absent or seriously attenuated, such as: there is presently no particular tension regarding oil, there remains no drastic increase in demand from emerging countries or for biofuels, and there are no particular fears that a speculative bubble in the assets markets will burst. If there is only one causal factor present again this time, it is "speculation", and perhaps this is indeed the culprit. Instead of reaching this conclusion through intuition, it would have been better if we could rely on more consistent evidence. Ahead of the next crisis, it is vital that economic science, which was caught out this time, makes itself useful in a debate that completely evades it at present.

CONCLUSION

Are the causes of instability really so elusive that we can justify only focusing on its effects? There exists an economic challenge to deal with the causes – according to the efficiency argument, any "problem" must be treated at its root. The ability to deal with the effects is also morally justified, because the poorest are most at risk. It therefore seems that the G20 agriculture ministers choose to favour coordination for moral reasons of justice, more than for efficiency. Our literature review shows that there is a realistic preconception behind such a choice. Indeed, no studies have isolated one single factor from amongst several, as responsible for the bursts of instability experienced by the agricultural markets.

However, there is no reason to exclude the political "disturbances" that the support policies for biofuels constitute, especially in North America, or the deregulation policies of the agricultural financial markets since the 1990s, especially when the rationale behind the current governance of agricultural trade (i.e. the WTO rules) is based on the elimination of disturbances in exterior markets, which is created by various public policies. The provisional result of the agricultural G20 is inconsistent in this respect.

Regarding biofuels, as for the agricultural futures markets, the systematic purchase of positions has created the conditions for price instability: anticipation errors were compounded by a demand that has somehow become more rigid. The globalization of trade and the financialisation of markets should have made this demand more elastic to the price than it was previously – the agent benefiting from increased opportunities for arbitrage, buying or selling according to market variation. However, index speculators have systematically been buyers. As were the ethanol producers due to the subsidies allocated to the sector. The rigidity of demand has amplified the few imbalances in production, which in retrospect, were not exceptional and did not justify such price movements. ■

REFERENCES

BRICAS N., BRU E., 2008, *La hausse des prix alimentaires au Sud: causes, conséquences, propositions* (disponible sur: http://www.cirad.fr/fr/actualite/communique.php?id=919).

COURLEUX F., 2008, "De la hausse à la baisse des prix: impacts de la crise économique sur l'agriculture et les industries agro-alimentaires", Working Paper, Ministry of Agriculture and Fisheries, Prospective et évaluation, n°1, November 2008.

COLLINS K., 2008, "The Role of Biofuels and Other Factors in Increasing Farm and Food Prices", supporting material for a review conducted by Kraft Food Global, Inc.

COURLEUX, Frédéric and LECOQ, Pierre-Emmanuel, 2011, "Que peut-on dire de la financiarisation des marches agricole ? Constats et réponses", Sfer Seminar, 14 January 2011.

CURRIE, Jeffrey, 2007, "Food, Feed, and Fuels: An outlook on the Agriculture, livestock and biofuel markets", Goldman Sachs International, March.

EUROPEAN COMMISSION, 2007, *The impact of a minimum 10% obligation for biofuel use in the EU-27 in 2020 on agricultural markets*, AGRI G-2/WM D(2007).

GILBERT, Christopher, 2008a, "Commodity Speculation and Commodity Investment", Discussion Paper 20, Universita Degli Studi Di Trento.

GILBERT, Christopher, 2008b, "How to Understand High Food Prices", Discussion Paper 23, Universita Degli Studi Di Trento.

GOHIN, Alexandre, and DRONNE, Yves, 2008, "Les principaux déterminants de l'évolution des prix agricoles internationaux", mimeo Inra.

IRWIN Scott, and MERRIN Robert, 2008, "The Adequacy of Speculation in Agricultural Futures Markets: Too Much of a Good Thing?", Marketing & Outlook Research Reports, Univ. Illinois.

JACKS, David, O'ROURKE, Kevin and WILLIAMSON, Jeffrey, 2010, "Commodity price volatility and world market integration since 1700", NBER working paper 14748.

TENSIONS AND ASYMMETRY ON AGRICULTURAL MARKETS POTENTIAL REGULATIONS AND FUTURE PROSPECTS

Michel Petit,
Institut Agronomique Méditerranéen, Montpellier, France

Current debate on the subject of international markets for agricultural products is indicative of the major issues affecting global agriculture today. While the media focuses on price volatility, there are more significant issues from a long-term perspective: how can global food production meet the needs of nine billion people by 2050 without destroying the natural resources – soil, water and biodiversity – upon which this production is based? Under what conditions will everyone have access to food? And with what social and political consequences? What will be the sources of supply? What role for the hundreds of millions of small farmers in poor countries?

There is no doubt that the answers to these questions will be greatly influenced by both domestic and international agricultural markets, and much will depend on how these markets are regulated. However, we are witnessing a paradoxical decrease of regulations in this area of international trade because the consensus for progressive trade liberalization, embodied in the GATT and by successive WTO Rounds until the 1994 Marrakech Agreement, is now seriously eroded. Yet, it is on the basis of this consensus that a multilateral regulation of national public policies affecting the agricultural market was established after 1994. Does the failure or paralysis of the Doha Round mark the end of such regulation? Are other regulations feasible?

PARADOXICAL GROWTH IN INTERNATIONAL TRADE AND PARALYSIS AT THE WTO

The multilateral trade negotiations launched in 2001 by the WTO in Doha seem to be ending in failure. During the last few months, some participants have pushed for at least a minimal agreement to be reached by the end of 2011; however, indications are that even this limited objective is unlikely to be achieved. It appears that agricultural negotiations are the central cause for the breakdown in the Doha Round talks. Yet, the main components of an agreement in this sector (which would mark a significant step towards trade liberalization or, more precisely, an increase in restrictions to limit market distortions resulting from unilateral policies) have been on the negotiation table for many years. Such components include the elimination of export subsidies, a substantial reduction of tariffs and other barriers to domestic markets that are the most attractive for global exporters, and significant reductions in domestic subsidies for agriculture in rich countries. Thus, the inability to reach a consensus on this matter is paradoxical: an agreement on agriculture cannot be obtained even though there is general accord on the potential outline of a final agreement.

A second paradox is that trade in agricultural and food products continues to increase despite the paralysis of the WTO negotiations. In such circumstances, it is difficult not to conclude that the WTO negotiations are secondary compared to the main determinants of trade; a notion that is supported by the fact that stakeholders, such as large multinational food companies, have not recently engaged in pressurizing WTO negotiators into reaching an agreement.

EROSION OF THE CONSENSUS FOR TRADE LIBERALIZATION

The breakdown of the WTO negotiations reflects the erosion of the international consensus, of Western origin, on the progressive liberalization of trade. This consensus was responsible for the creation of the GATT at the end of World War II and the WTO in 1994, and also the continuous advancement of liberalization in the successive Rounds of the GATT until 1994. WTO negotiations in the late 1990s and the launch of the Doha Round in 2001, despite the failure of the ministerial meeting in Seattle in 1999, were in the same vein. The consensus resulted from genuine remorse at the end of World War II: those responsible for the economic policies of the inter-war period acknowledged the mistakes of such policies in Western countries, and that insufficient consideration of the effects of unilateral national decisions on other countries had in fact delayed the end of the Great Depression of the 1930s, thus fostering, unintentionally, the Nazi and fascist regimes. For example, the race for protectionism was initiated in 1930 in the United States by the adoption of exorbitant border tariffs through the Smoot-Hawley Act.

There are several common errors in the interpretation of the consensus for trade liberalization. Contrary to the beliefs of many liberals and anti-globalists, the consensus was never about the full establishment of free trade. Today, more than sixty years after the creation of the GATT, trade liberalization is far from complete and such total liberalization is not even a consideration. This is why, for example, significant concessions in terms of access to agricultural markets for the poorest countries are not on the agenda of the Doha Round. Moreover, these negotiations reflect a considerable amount of pragmatism, based on a clear awareness of the political obstacles to liberalization. It is widely acknowledged that trade liberalization creates losers: such as the producers of protected products, who are typically few but well identified and easy to mobilize politically since they have a full awareness of what they stand to lose; and also winners, such as the consumers of a previously protected product, who are often numerous but not easily mobilized. In consequence, negotiations are characterised by long adjustment processes and are of a multi-sectoral and multilateral nature, traits that tend to increase the number of possible coalitions and avoid overly polarized debates.

All these procedural tricks worked well for sixty years and led to a significant liberalization of trade, or rather to practices that limited government leeway, in order to avoid unilateral decisions that would cause the greatest distortions in global markets. However, agriculture has only belatedly become involved in this general trend: it was only through the adoption of a major European agricultural policy reform in 1992 (the “MacSharry reform”) that the Uruguay GATT Round was subsequently able to reach a conclusion with the Marrakech Accords, including the first substantial commitments on agriculture.

Since then, however, the consensus obtained has been much eroded. In agriculture, the global anti-globalization movement, which grew around the Via Campesina organization, has put the fight against free trade at the centre of its demands. Moreover, and probably more significantly, there are few governments at the WTO that seem willing to make substantial concessions to reach a general agreement. The United States, for example, is paralyzed by the conflict between its export sectors (cereals and soybeans for example) that are eager to win new markets, and its producers of protected products (cotton, sugar, milk, etc.) that reject any proposals to withdraw benefits (subsidies or other assistance). Emerging and poor countries are united by a shared belief that the Marrakech Accords are unfair because they allow rich countries to provide massive support to sustain their agriculture, causing large distortions of competition. Ultimately, multilateral liberalization of agricultural trade does not seem to be a priority for any of these countries.

FIGURE 1 THE INTERNATIONAL REGULATION OF TRADE, AN UNFINISHED STORY

1999 Seattle conference. An attempt to launch a new round fails and governments, which are subject to conflicting pressures from civil society, cannot agree

2001 Launch of the ninth round in Doha (Qatar), known as the 'development round', reflecting the increased number of developing countries among the members of the WTO

2003 Ministerial Conference in Cancun (Mexico), a dispute, mainly on agriculture, between developing countries (both emerging and least developed countries) and the rich countries that support their agriculture. Paradoxical result: the United States and the European Union, whose opposition in agriculture had dominated the discussions and negotiations for decades, come together on the defensive

2004 Agreement in Geneva, between delegations of member countries, without the ministers, on how to continue negotiations

2005 Ministerial Conference in Hong Kong, which had been scheduled to conclude the round: no agreement despite the commitment of the European Union to eventually eliminate agricultural export subsidies in the event of a global agreement

2008 Ministerial conference in Geneva: failure. Ostensibly, because of a conflict between India and the United States on a clause allowing developing countries to curb agricultural imports when they increase sharply. More profoundly, the blockage is due to the fact that no governments were prepared to make significant concessions to meet the demands of any other government on agriculture or other areas under negotiation

Since 2008, no substantive progress.

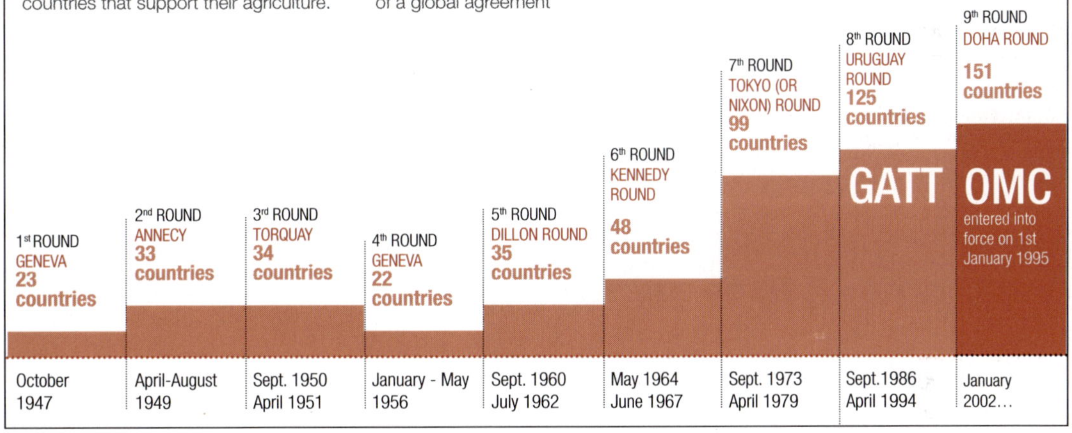

Primarily sectoral, private standards have expanded in terms of scope and the number of stakeholders involved in their creation and implementation. Public actors are now trying to regain control of these standards by becoming strongly involved in the generation of knowledge on these standards.

OUTLOOK FOR THE REGULATION OF WORLD AGRICULTURE IN THE COMING DECADES

Firstly, it must be emphasized that the stalling of the Doha Round does not mean the end of the WTO, as the existing rules and procedures, such as the Marrakech Accords and the decisions of the Dispute Settlement Body (DSB), are still very much required. While the implementation of DSB judgements is far from perfect: a verdict against American cotton during a dispute with Brazil in 2005 has still not been executed; in other instances WTO pressure has had a significant impact, such as forcing the EU to radically reform its sugar policy following its dispute with Brazil. Furthermore, even if the Doha Round fails, further negotiations will probably take place within the WTO to adjust trade rules to reach a possible multilateral agreement on climate change, and also to harmonize the results of multiple bilateral and regional negotiations on preferential agreements, which in recent years have proliferated – the OECD counted 297 in place or under negotiation!

In addition, recent discussions and decisions of

the agricultural G20 can be interpreted as new attempts to forge intergovernmental regulations beyond the WTO negotiations, but with significant areas of overlap such as possible rules on export restrictions. It is too early to assess the impact of a tentative agreement, which would aim among other things at the organization of greater market transparency by sharing information on stocks (a lot will depend on the modality of such an agreement's application). This is in fact an attempt to overcome the paralysis in WTO negotiations and to move beyond its strict framework.

Finally, any discussion on the future of the international regulation of agricultural markets must take into account the recent emergence of mechanisms, involving the participation of a variety of agents from civil society and private and public sectors, which are engaged in common objectives, particularly in terms of environmental protection such as the "Roundtable on Sustainable Palm Oil". Such mechanisms can indeed be considered as a new form of regulation since private companies are bound to codes of conduct that set defined limits to their pursuit of profit for the benefit of the greater good, even if this approach has well known limitations. ■

REFERENCES

G20 France 2011, Ministerial Declaration Action plan on food price volatility and agriculture, Meeting of Agriculture Ministers of the G20, Paris, 22 and 23 June 2011

Petit M., 2005, "Les négociations agricoles à l'OMC: Où en sont-elles? Où vont-elles? *Cahiers Agricultures,* 14,4, 399-403, July-August 2005.

CHAPTER 13

"Organic farming", "fair trade", "responsible" products... private standards such as these have multiplied to encourage farmers to participate in voluntary quality approaches. Combined with retailer standards, such voluntary approaches are increasingly necessary to ensure market access and are used as reference by public authorities.

SUSTAINABLE VOLUNTARY STANDARDS: TOWARDS PRIVATIZED REGULATION IN THE FOOD AND FARM SECTOR?

Eve Fouilleux,
CNRS and Centre International de Recherche Agronomique pour le Développement, UMR Moisa, France

In a global context of widespread liberalization and a growing political importance given to issues of so-called sustainable development, agriculture has not proved the exception to the transversal phenomenon of the privatization of standards. In response to the growing industrialization of agricultural production methods, to the proliferation of large-scale agricultural pollution and to a growing awareness of the working conditions of farmers and agricultural workers from the "South", the theme of sustainable development has become increasingly visible. In parallel to the development of public policy and other international agreements, private standards have increased, which have a declared intention of encouraging farmers to minimize their negative impacts from an environmental and social point of view.

While corresponding to a variety of objectives and/or secondary aims in the field of sustainable development (organic agriculture, fair trade, food safety issues, responsible production, ethics, etc.), these standards refer however to similar institutional and procedural schemes. Based on specifications of environmental and/or social "best practices", such standards are often – but not always[1] – associated with specific labels or statements on products, which enable consumers to identify them. Producers adopt these specifications voluntarily (hence they are known as voluntary standards) and "third party certification" procedures are employed to monitor compliance – where an auditor ensures that farm practices conform to specifications. Producers

1. We distinguish "business to business" standards, which are not visible to end consumers but serve as traceability insurance and a guarantee between the various operators in the food chain, from "business to consumer" standards, which have an associated label aimed at the final consumer.

are awarded with the corresponding label if standards are met, which can be used on their products to promote sales or marketing opportunities.

As already demonstrated (Fouilleux, 2010), various devices are gathered under the term private voluntary standards[2]. Using examples from the agri-food sector, the first part of this paper shows that despite this variety, standards tend towards convergence and transnationalization. The second part reviews the debate that exists between public and private spheres that is generated by the use of private standards as an increasingly powerful form of control.

A SHORT HISTORY OF VOLUNTARY STANDARDS: TRANSNATIONALIZATION OF ORGANIZATIONS AND INSTRUMENTS

ORGANIC AGRICULTURE STANDARDS Among the different types of private standards, some are managed by politically and/or socially-oriented organizations. This is the case for organic agriculture, an area where alternative producer groups identified certain consumer concerns and implemented relevant standards. Such standards were originally private, to be followed by government regulation in France (1981), the European Union (1992), and the rest of the world (more than 60 governments have now established public regulations in organic farming). In parallel, hundreds of private organic standards also exist throughout the world, which vary considerably in terms of the number of farmers and volumes of products involved. While initially organic farming was structured primarily within the European Union, and more generally in industrialized countries, it has gradually extended to include so-called "Southern" or "developing" countries, mainly as product exporters to European and US markets. Organic standards were the first to be organized internationally, which was carried out through the International Federation of Organic Agriculture Movement (IFOAM), founded in 1972 at the initiative of the French Nature and Progress association, the British Soil Association, a Danish group of biodynamic farmers and the Rodale Institute, a leading advocate of organic farming in the United States. IFOAM has played an important role in the dissemination of organic agriculture throughout the world.

As producer associations, consumers, government policies, businesses and other economic agents were mobilized, a real "global organic agriculture sector" was gradually institutionalized (Raynolds, 2004). The first formal international specifications – although initially vague – defining organic production were established in 1980, the "voluntary standard" instrument and its associated "third party certification" dimension undoubtedly had a role in the structuring, institutionalization and consolidation of a global market for certified organic products. Initially, the producer groups that had set up the initial specifications developed the certification; farmers inspected each other on a voluntary basis involving mutual knowledge and trust. Then, a sharp

2. Many public voluntary standards also exist, such as geographical indications in Europe, the Label Rouge in France.

increase in trade flows and an increase in the amount of organic sales by supermarkets (which are now the main retailers of such products, selling far greater quantities than speciality shops (Daviron and Vagneron, 2011)) led to an increase in the number of standards and practices that took on a more market-oriented significance, creating conflicts and political contradictions all the way up to the IFOAM level.

Indeed, the increasing formalization of specifications – linked to the need to provide criteria and indicators to enable certifiers to make objective and straightforward measurements during audits – has gradually diluted the initial objectives of the founders of the organic movement, who emphasized the need for a systemic, holistic and ad hoc approach to agriculture and the interactions between humans, nature and agriculture. While IFOAM still tolerates less formalized specifications and alternative control systems, it now gives particular emphasis to third party certification, which it justifies by the demands of markets and states. In addition to IFOAM standards that producers can adopt directly, IFOAM is now proposing a multilateral system of equivalence (a kind of organic "meta-standard"), which could accompany other organic standards, whether public or private, and allow producers to use the IFOAM logo next to those of other standards.

FAIR TRADE STANDARDS Fair trade provides another example of a voluntary standard initially supported by private citizens engaged in explicitly political projects. This movement was originally structured around militant organizations based mainly in Europe (and the US), who purchased products directly from producer organizations in developing countries, which they then sold to consumers via a network of speciality shops. These shops – part of the *"Magasins du Monde"* organization – sell products under their own name (*Traidcraft, Oxfam, Solidaridad, Equal Exchange, Artisans du Monde*), particularly local crafts (although including some agricultural products, especially coffee). In this model, rather than a shared common standard and a certified label, product equity is guaranteed to the consumer by long-established relationships of trust between producers, buyers and the place of purchase (Daviron and Vagneron, 2011). This model is now embodied in the comprehensive plan by the World Fair Trade Organization (WFTO) that endorses an approach based on trust and peer review, known as an "integrated system", which it considers as a more suitable method than third party certification.

However, the speciality stores model has faced strong competition since the late 1980s from the development of a second generation of fair trade initiatives, based on standardization and third party certification. The Max Haavelar Foundation, established in the Netherlands in 1988 by a Dutch priest and an executive from the Dutch development NGO Solidaridad, is considered the first initiative to standardize fair trade. With labelling and third party certification, the aim was to develop a larger market by selling products in supermarkets, generating increased support to producer incomes in developing countries (Raynolds et al., 2007). The immediate success of Max Haavelar with consumers induced a proliferation of standardization initiatives and the considerable diversification of stakeholders that supported the standards. Given this explosion of initiatives in 1997, the Fair Trade Labelling Organisation (FLO) was created to unite fair trade labelling organizations, and it today gathers 24 national members, including 19 standards organisations or "labelling initiatives" (covering 23 countries) and three networks of certified producers (Africa, 2004; Latin America, 2004; and Asia 2005). In addition to representing and promoting the development of fair trade, FLO's main activity is to formulate and revise the various fair trade specifications according to products – in other words, once again, global meta-standards.

RETAILER STANDARDS A third type of private voluntary standards should also be mentioned, one that is particularly important in the agri-food industry: "retailer" standards. They are unrelated to organic or fair trade standards by the way in which they are aimed at the mainstream and not at "niche" markets and they are not based on explicitly political objectives. Such standards, for example GlobalGAP and IFS[3], are utilised by stakeholders involved in large-scale retailing. GlobalGAP specifications vary according to product type. They include a set of basic specific requirements, as well as additional stricter criteria that are recommended but not mandatory, allowing the definition of several certification "levels".

GlobalGAP standards were created in 1997 under the name of EurepGAP (Euro-retailer produce working group for good agricultural practices) and were originally to be applied exclusively to food safety issues. The standards resulted from an initiative of the key stakeholders involved in European supermarkets, who participated in the Eurep working group, under strong British influence from companies such as Tesco and Sainsbury's. The use of GlobalGAP standards was later extended beyond issues of food safety, and their specifications now integrate sustainable development along with social, environmental and ethical concerns.

These private standards are not a priori obligatory and their application remains theoretically voluntary, but the fact that all major retailers demand such standards explains their very rapid extension and widespread implementation. In fact, for many purposes, these standards have replaced public standards (Jaffee and Henson, 2004). Moreover, because of the extent of their application, EurepGAP standards were often assumed to be obligatory by countries that exported to the European Union, and that to avoid confusion with mandatory public standards the EU pushed for the name to be changed to GlobalGAP, which was achieved in September 2007. In addition, due to the market power of retailer standards, they are more often used as reference for governments: they are sometimes used as explicit objectives or as a "basis" for national and international public policy. Thus, to assist producers with their upgrading of operations and practices, and to "facilitate their market access", courses are funded and delivered and grants are specifically targeted by public policies for the implementation of these private standards.

3. These two examples are European, but a similar situation exists in the United States or Australia for example, concerning similar standards.

MULTI-STAKEHOLDER STANDARDS FOR SUSTAINABLE COMMODITIES Finally, a fourth category of standards has emerged in the 2000s with explicit focus on "sustainability" or "responsibility". These standards are characterized by the governance structures that produce them, which involve specific so-called "multi-stakeholder" decision-making processes; and the related standardization organizations are often called "round tables". These organizations claim that their functioning is based on inclusive and participatory processes described as "democratic" because they allow all those involved or concerned with the commodity and sector in question (producers, financiers, importers, agri-food industries, upstream industries, exporters, supermarkets, social and environmental NGOs) to provide input and contribute to decisions. Round tables are based on equitable representation of the different stakeholders and on very precise and codified participatory procedures, and the mode of interaction within them is mainly based on the continuous search for consensus.

Initially applied to forests (FSC – Forest Stewardship Council, 1993) and then fisheries (MSC – Marine Stewardship Council, 1999), and also established for textiles and mining among others, sustainable multi-stakeholder voluntary standards have increased in the last decade for agricultural commodities originating from the tropics and traded on international markets. Examples include the Roundtable on Sustainable Palm Oil (RSPO, 2003) and on Responsible Soy (RTRS, 2005) and also for sugar cane (BSCI, 2006), biofuels (RSB, 2007), cotton (BCI, 2007), etc.

ISEAL, THE GLOBAL ASSOCIATION FOR SUSTAINABILITY STANDARDS This brief overview shows the coexistence of different standard types (organic agriculture, fair trade, retailer, sustainable commodity standards), each marked by a tendency towards international or transnational organization (IFOAM, FLO and WFTO, GlobalGAP). It should also be noted that this tendency towards organization on a transnational scale asserts itself despite the objective competition between different standards of

BOX 1 DO RETAILER STANDARDS WORK IN THE INTEREST OF PRODUCERS?

Retailer standards exemplify the power relationships that voluntary standards can exacerbate in each sector. Indeed, besides the fact that they provide a way for retailers to differentiate themselves from their competitors, the standards can be used to anticipate regulatory changes or as a form of risk management by its transference onto upstream operators (Henson, 2008). Moreover, the requirement for specific standards by retailers, who control virtually all of the market, constitutes a real barrier to entry into European markets and represent a significant risk of excluding small producers, including those from developing countries: if they are not of sufficient size and cannot afford additional certification costs – or rather the various certifications necessary for export – they will go out of business. So while GlobalGAP standards claim to work "in the interests of producers", several studies highlight the fact that they are of most benefit to large supermarket chains, increasing their regulatory capacity while reducing their direct supervision and support, thus minimizing their liability in case of problems, with the costs being carried forward to producers and processors (Hatanaka et al., 2005; Henson, 2008). For some critics, such standards and public programmes are likely to lead to "a reinvention of European colonial food relations" (Campbell, 2005), by encouraging farmers in developing countries to export, rather than to provide for the local market.

the same type: rather than competing over the exact content of specifications or on the practical feasibility of corresponding audits, or seeking to obtain the maximum amount of potential customers for a product or region, it is primarily in the interest of standards organizations to cooperate to establish themselves as credible and effective, and therefore legitimate, regulators. In a similar way, and further strengthening the trend towards convergence, since the mid-2000s there has been a gradual establishment of the ISEAL (International Social and Environmental Accreditation and Labelling) Alliance, an organization that aims to combine different standard types.

In many ways, ISEAL is a functional equivalent of the ISO (International Standards Organization) specializing in the field of voluntary environmental and social standards, which aims to unite and harmonize the different standard types. Launched in 1999 by members of four pioneering voluntary standards organizations (FSC, IFOAM, FLO and MSC) to address the risks of confusion and of a declining credibility of standards, its formal status as a non-profit organization was established in 2002. Its membership now comprises 12 standardization (and accreditation) bodies and seven associate members. This membership includes standards for fair trade (FLO, Utz certified) as well as GlobalGAP or standards such as the Common Code for the Coffee Community or the Better Sugar Cane Initiative for example, as well as organizations related to organic farming (IFOAM until 2010, IOAS).

ISEAL presents itself as "the global association for social and environmental standards", a "community of practice" which aims to "*create a world where environmental sustainability and social justice are the normal conditions of business*". The main activity of ISEAL is to develop "best practices" for the producers of sectorial standards (considered as clients), which are used as procedural rules for the production of "good" voluntary standards in environmental and social fields. These criteria are contained in meta-standards (codes) that aim to guide members in the development, improvement and strengthening of standards. There are four main codes: a "Code of Ethics", a "Code of Good Practice for Setting Social and Environmental Standards," a "Code of Good Practice for Assessing the Impacts of Social and Environmental Standards Systems" and a "Code for Assuring Compliance with Social and Environmental Standards". Another important activity of ISEAL is to promote private voluntary standards to states, international organizations, academic communities, private firms, etc. as effective regulatory instruments.

PRIVATE STANDARDS AND PUBLIC STANDARDS: A DISTINCTION THAT IS LESS THAN CLEAR...

Despite their great diversity, the standards discussed above have much in common. They are based on the same principle: regulation through the market. In theory, through the purchasing of certified products, consumers encourage the establishment of more socially fair and environmentally friendly production conditions. In this regard, the regulatory dimension is secondary: it is the market that guides the type of productive practices. The implementation and monitoring of standards is based almost exclusively on private actors, including certifiers and independent

FIGURE 1 THE PRODUCTION OF PRIVATE STANDARDS, A VERY COMPETITIVE HISTORY

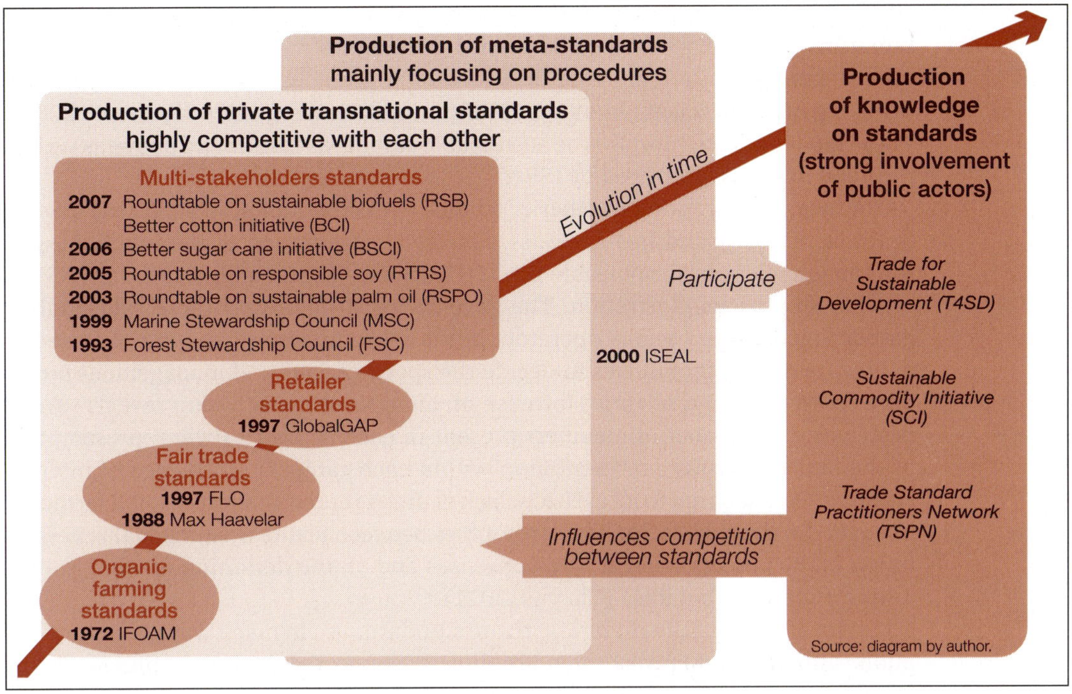

Source: diagram by author.

organizations accredited by the standard holder. As a service provider, certification agencies are paid by producers to monitor their practices for compliance with the specifications of a particular standard. Agencies conduct audits of producer practices and ultimately decide whether or not a corresponding label is to be granted. Third party certification creates considerable economic activity and certifiers have a vested interest in the proliferation of voluntary standards (Djama et al., 2011). Public or private, voluntary standards are a form of privatization of the regulation process.

Another striking feature that emerges from a study of the proliferation of private standards is that they all claim to aim for the common good, to produce a "public good". Thus, we can see in the willingness shown by standards organizations to emphasize the "democratic" nature of their decision-making processes an aspiration to compete with, if not to replace, public decisions in terms of legitimacy. Rhetoric highlighting the benefits of private management of environmental and social issues stems from two academic schools of thoughts: the first being political science, where it draws on the most normative work of the school of global governance, which emphasizes the inability, considered to be both structural and functional, of public authorities to deal with the complex issues related to globalization and the benefits expected from an increased involvement of private actors in the framework of public-private partnerships (Reinicke, 1998); and the second is management science, where it derives from the work on the concept of corporate social responsibility (CSR), which would give companies crucial comparative advantages in a competitive environment (Porter and Kramer, 2006). At the meeting point between these two areas, a vast literature has developed on collaborative governance methods that propose ideal procedures to ensure legitimate decisions on a given issue, based on procedures involving multiple stakeholders (Van Huijstee and Glasbergen, 2008; Zadek, 2008). These models directly inspire the sustainable standards mentioned above as well as the good practices promoted by ISEAL.

However, an increasing number of studies stress the limitations of these devices in terms of equity and inclusiveness. Empirical observation of the functioning of the round tables on "Responsible Soy" (RTRS) and "Sustainable Palm Oil" (RSPO) provide an obvious illustration. The governance structures of these round tables gather together the various operators in the relevant sector, who are expected to deliberate among themselves to decide the specifications and management procedures of the certified sectors. However, an analysis of the situation reveals several problems: the number of members present in each category (over-representation of the industry), their representation within each category (influence of international NGOs compared to local NGOs, lack of direct representation of small farmers), unequal distribution of discursive resources between actors (i.e. unequal representation in the process of forming a consensus and on the definition of what is, or is not, considered debatable) (Cheyns, 2011).

Moreover, beyond the rhetoric, private standards are in many ways related to the public authorities. Many of them are directly supported, including financially, by governments. The International Financial Corporation (IFC), for example, which is the branch of the World Bank Group dedicated to financing the private sector, is a direct member of several round tables. Many of its fund allocation programmes

BOX 2 ARE STANDARDS ORGANIZATIONS INCLUSIVE?

The various standards organizations do not all operate in the same way and some appear more truly inclusive than others. For example, in the RSPO there are seven categories of stakeholders among which five represent the different stages of the value chain (production, industry, finance, etc.) and two are NGO categories respectively representing social and environmental interests. In the RTRS, social and environmental NGOs are grouped into one "civil society" category, ahead of a "producer" category and an "industry" category. There is therefore a noticeable difference between these two round tables and the Forest Stewardship Council (FSC), which is their historical model. The FSC is divided into three chambers, "economic", "environmental" and "social", each with an equal percentage of voting rights. Compared to the FSC, the RSPO and RTRS associations therefore allocate much greater power to economic interests at the expense of NGOs in terms of the formal allocation of seats: 75% of votes on the RSPO board and 66.6% for the RTRS, compared with only 33.3% in the FSC. In addition, in the FSC, 50% of the votes within each chamber are distributed to representatives of the South and 50% to its members of the North. Such a division is not made in the RSPO, or in the RTRS, or in any other sustainable standardization initiative for agricultural commodities. Despite characteristics that appear more equitable, actors of other round tables often regard the provisions of the FSC as overly cumbersome and inefficient in terms of decision-making, some even referring to it as a "psychotic democracy" (Bartley and Smith, 2010).

require beneficiary companies to participate in the round tables. Various states are also involved. In all sectors, there is particularly active bilateral cooperation from the US, Canada, Britain, Germany and Sweden. While in agriculture, the Netherlands, Germany and Switzerland play leading roles. Dutch cooperative agencies have developed programmes such as the Sustainable Trade Initiative to encourage sustainability in international trade, which particularly involve the establishment of transnational private voluntary standards. Since the beginning of the 2000s, Germany has also been very active in the support of sustainable standards and has directly participated in the launch of several round tables, including the 4C Association (Common Code for the Coffee Community), in association with cooperation from Switzerland. In parallel, national cooperation agencies are also taking an active role in the defence of voluntary standards as regulatory instruments. Conferences, such as in Berlin in 2006 and 2008, are frequently organized to bring together the sustainable standards community as actors and to promote these regulatory tools, to "expand and deepen the positive effects" of voluntary standards and to "make them a mainstream phenomenon". The term paradigm shift has even been used. Finally, there is a proliferation of institutionalized transnational networks that combine public and private authorities on the issue of private voluntary standards. Three such examples are given below:

The Trade Standards Practitioners Network (TSPN), launched by the German Cooperation agency, USAID and the World Bank, brings together various organizations and institutions involved in standard setting and capacity building in related fields. The TSPN consists of three main working groups. The first is aimed at policy makers in developing countries, including through a guide that aims to help them "*understand the catalytic function of voluntary standards for development and their role in the success of export strategies*". The second aims to "*maximize the positive impacts of the actions of cooperation and technical assistance related to standards (...) and to ensure the sustainability of these impacts*". It aims to define a shared understanding of the impacts of standards, and to provide the tools to measure them. This type of evaluation relates to particular techniques, which is a highly political area: measuring the impact of standards ultimately enables them to be classified as "good" or "bad" instruments; and also, according to the criteria used, the value of a standard can be calculated in completely different ways. The third working group of the TSPN, which is mainly led by the IFC, is specifically aimed at the private sector in developing countries to increase the commitment to voluntary standards initiatives by its actors (international companies in these countries that source from international markets, exporters in developing countries, national and international companies that provide, on a commercial basis, services in connection with the voluntary standards – certification agencies, auditing offices and training providers). This refers to an obvious issue of legitimacy and to a recurring problem for the different standards systems: they can only be developed if the economic actors in the field adopt them. The promoters of these systems therefore develop methods of persuasion in this direction.

Launched in 2003 at the initiative of the Canadian think tank IISD and the United Nations Commission on Trade and Development (UNCTAD), the Sustainable

Commodity Initiative (SCI) promotes sustainable standards as a potential driver for a paradigm shift: "*voluntary supply-chain approaches have the potential to establish a new paradigm for commodity production and trade*." The projects relating to the SCI bring together dozens of partner organizations of all types (companies, cooperation and development agencies, many research organizations, international organizations, consulting firms, TSPN, standards organizations, ISEAL, certifiers, etc.) that "collectively support initiatives to improve the sustainability of international trade and global markets for raw materials". SCI faces the challenge of assembling and disseminating as much information as possible on sustainable standards systems and to produce specific knowledge to refine and improve them in terms of impact assessment, capacity building and training, financing and the reporting of standards information. All of these works also share a clearly and explicitly political goal: to influence public authorities so that public policies support and promote sustainable standards.

Finally, the Trade for Sustainable Development (T4SD) project is supported by the International Trade Centre, an organization formed jointly by the UNCTAD and WTO. It aims to establish a comprehensive database of all existing sustainable standards systems, a training programme, and a research observatory on the subject. While T4SD defines itself as "a UN sponsored neutral information repository", it provides many examples that illustrate the political dimension of the production of knowledge on sustainable standards and the competitive relationships that sometimes cause standards organizations to oppose each other. For example, during a meeting in Amsterdam organised by ISEAL, the representative of the German FSC explained her reluctance vis-à-vis the creation of the T4SD database and her annoyance that ISEAL was encouraging its members to work with the ITC: she was sceptical about the way in which the ITC measured the impacts and the criteria that they might employ, which could disadvantage her organization in favour of its main rival PEFC. Indeed, since the PEFC certifies more timber than the FSC, but based on less demanding specifications, if performance standards were measured on the basis of the volume of wood certified for example, this would make the FSC appear less efficient and therefore less attractive to loggers, and could lead to a loss of market share.

This proliferation of transnational networks that combine public and private authorities – which are often interlinked – attests to the undeniable support for private voluntary standards globally. Furthermore, although the actors involved tend to stress that these networks primarily address the "need for dialogue" and the "sharing of experiences" that the proliferation of standards has created, such a phenomenon highlights concerns from public authorities that they have been caught slightly off guard by the explosion in these instruments, and their willingness to refocus on the issue and take control of their development. ■

REFERENCES

BARTLEY T. and SMITH S.N., 2010, "Communities of practice as cause and consequence of transnational governance", In: Djelic M.L. and Quack S. (eds.), *Transnational Communities. Shaping Global Economic Governance*, Cambridge University Press: 347-374.

BUSCH L., 2011, *Standards: Recipes for Reality*, The MIT Press. 384 p.

CAMPBELL H., 2005, "The Rise and Rise of EurepGap: European (Re)Invention of Colonial Food Relations?", International Journal of Sociology of Food and Agriculture, 13 (2): 6-19.

CHEYNS, E., 2011, "Multi-stakeholder initiatives for sustainable agriculture: The limits of the 'inclusiveness' paradigm", In: Ponte et al., Governing through standards: Origins, drivers and limits. London: Palgrave.

DAVIRON B., VAGNERON I, 2011, From Commoditisation to De-commoditisation... and Back Again: Discussing the Role of Sustainability Standards for Agricultural Products. *Development Policy Review*, 29(1): 91-113.

DJAMA M., FOUILLEUX E., VAGNERON I., 2011, "Standard Setting, Certifying and Benchmarking: A Governmentality Approach to Sustainability Standards in the Agro-Food Sector" *In*: Ponte, S., Vestergaard, J.& Gibbon, P. (eds.), *Governing through standards: Origins, drivers and limits*. London: Palgrave:187-209.

FOUILLEUX E., 2010, Les standards volontaires, instruments montants des politiques agricoles et alimentaires. Entre internationalisation et privatisation, in: Hervieu et al., *Les Mondes Agricoles en Politique*, Presses de Science Po, Paris: 372-396.

HENSON S., 2008, The Role of Public and Private Standards in Regulating International Food Markets, *Journal of International Agricultural Trade and Development*, 4(1):63-81.

JAFFEE S., HENSON S., Standards and Agri-food Export from Developing Countries. Rebalancing the Debate, *Policy Research Working Paper Series*, 3348, Washington (D. C.), World Bank, 2004.

RAYNOLDS L., 2004, The Globalization of Organic Agro-Food Networks, *World Development*, 32(5):725-743.

RAYNOLDS L., MURRAY D.L., WILKINSON J., 2007, Fair Trade. The challenges of transforming globalization, Routeledge.

PORTER M.E., KRAMER M.R., 2006, "Strategy and society: the link between competitive advantage and corporate social responsibility", *Harvard Business Review*, 84(12):78-92.

REINICKE W.H., 1998, *Global public policy: governing without government?*, Washington: Brookings Institution Press.

VAN HUIJSTEE M., GLASBERGEN P., 2008, "The Practice of Stakeholder Dialogue between Multinationals and NGOs", *Corporate Social Responsibility and Environment Management*, 15:298-310.

ZADEK S., 2008, Global Collaborative Governance: there is no alternative", *Corporate Governance*, 8(4):374-388.

QUALITY LABELS AS A TOOL TO ADD VALUE TO MEDITERRANEAN OLIVE OIL

Annarita Antonelli,
Mediterranean Agronomic Institute, Bari, Italy
Sébastien Abis,
International Centre for Advanced Mediterranean Agronomic Studies, Montpellier, France

The evolution of the agri-food sector has resulted in profound changes in agricultural policies, development models and consumption patterns. These changes are based on two phenomena: the emergence of a new consumer culture with individuals increasingly influenced in their purchases by the intangible characteristics of goods and the assertion of two different types of agriculture, one with a purely productive function, the other oriented towards environmental protection. In this context, the qualitative differentiation of products, particularly typical products, which are *"expressions of production areas"*, now plays a strategic role.

This is particularly true for the Mediterranean and its olive oil sector (CIHEAM, 2011). Olive cultivation contributes to the fight against erosion, enhances agricultural land and establishes populations in areas with difficult access. This activity generates income, creates employment and contributes to the export trade of countries. Olive oil is therefore a flagship product of the Mediterranean area for historical, sociocultural and economic reasons.

Faced with competition that is accelerating and increasing consumer demands for quality, the olive oil sector has gradually modernized with an emphasis on diversification of products according to various criteria relating to history, the specification of production area and the search for new export opportunities.

PDOs: A MECHANISM TO SAFEGUARD, DIFFERENTIATE AND DEVELOP

The system of Protected Designation of Origin (PDO) and Protected Geographical Indication (PGI) was established in 1992 by the European Union (Regulation (EEC) 2081/92) to replace national systems. In a context of a growing liberalization in agricultural trade, this system provides a guarantee to consumers of characteristics and properties offering added value to products, as well as compliance with basic standards through the certification systems of quality assurance, and the protection of typical products relating to their origin. Regulation 2081/92 has the objective *"to promote the diversification of agricultural production to achieve a better balance between supply and demand in the market"* through insisting on the territorial component.

The main objective of the quality certification process is to obtain income differentiation through the existence of a collective brand which certifies compliance with standards that set conditions of quality and typicality for oil. These differentiation strategies are a key competitive choice for the preservation of small farms and food companies located in rural areas, generally disadvantaged in the competitive struggle for undifferentiated products, the market of which is governed by dominant price and cost strategies. These approaches are also an effective way to counter the growing market presence of products from new

FIGURE 1 OLIVE OIL: TOWARDS THE LABELLING OF A MEDITERRANEAN TRADITION?

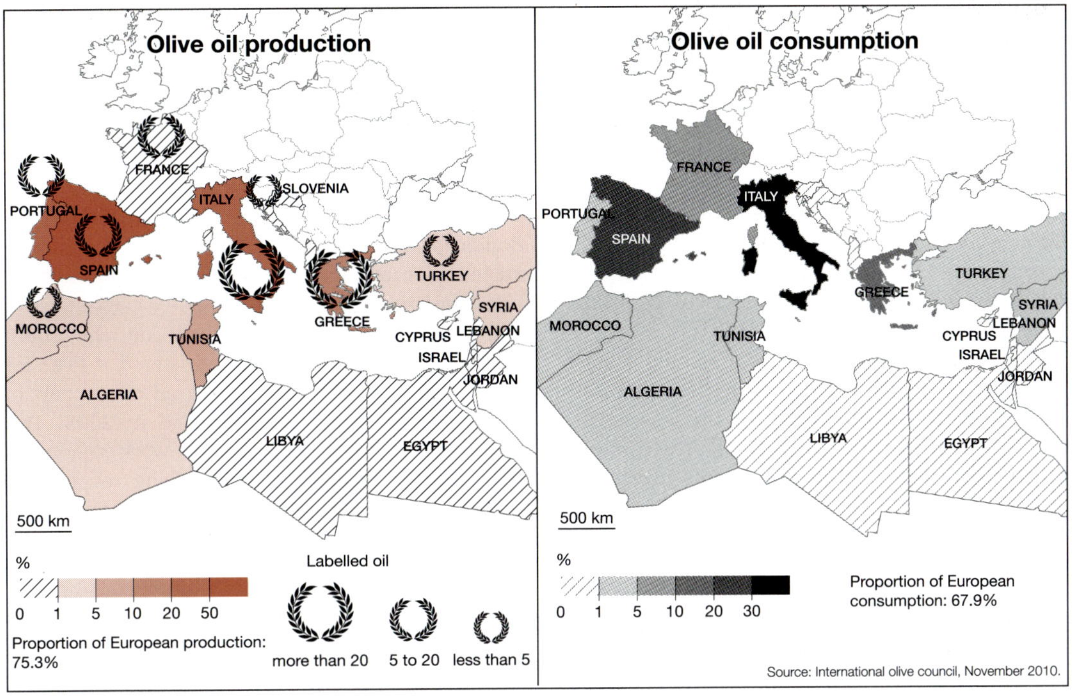

producing countries (United States, Australia, Chile and, in future, China).

At the end of the twentieth century, very few Mediterranean *terroirs* were implementing quality strategies to differentiate their oil, namely the olive-growing regions of central and northern Italy and the PACA region in France. In Spain, the designation of agricultural origin was developed in response to the olive crisis of 1970.

Today, new business opportunities arise because the image of quality olive oil carries an increasingly positive connotation among consumers in developed countries and the upper middle classes in emerging countries. Certain segments of consumers increasingly personalize their consumption according to the multifunctional characteristics of good quality oil (sensory, environmental, social).

In the main producing countries, olive oil has bridged the gap between an increasingly segmented demand, and an increasingly articulated supply (different oil types based on *terroirs* such as PDOs, organic and integrated production or according to variety or flavour etc.) that is not always ready to meet market requirements. Olive oil is an "*experience good*": experience is required to assess its quality; and even more a "*credence good*": as consumers are not able to know all the features of the product, even after purchase, they must therefore rely on a producer's reputation or the legal assurance of certification. The use of designation of origin is an effective strategy to provide a guarantee to consumers and gives the producer the opportunity to add value at sustainable costs (pooling resources with other producers in the territory).

In olive oil producing countries, there has been a proliferation of new olive oil PDOs: there were four in 1993 and 105 in 2010. In Spain, between 2000 to 2008, PDOs rose from 7 to 28, but Italian PDOs predominate (44 PDOs for oils and 1 PGI), a sign of a strong dynamism of the Italian olive-growing industry which finds in these tools an effective way to consolidate the sector by enhancing the quality and originality of their products. A similar trend is found in French olive growing, reflecting the variety of growing regions, despite its very low importance (0.2% of EU production).

In 2010, a study conducted by the International Olive Council (IOC) revealed that most of the world's origin-designated olive oils are registered in the EU (19 refer to table olives and 101 to olive oils). Among the non-EU member Mediterranean countries, only Morocco and Turkey have registered a few products, even though many oils have the potential to benefit from origin designation (IOC, 2010).

IS ORIGIN DESIGNATION EXPORTABLE TO THE MEDITERRANEAN'S SOUTH COAST?

Adding value to a product through quality is one of the typical issues discussed in the agricultural sector for Mediterranean countries, but the relevance of this approach is still debated. While the economic efficiency of such a strategy for some investors or groups of producers is undoubtable, the olive oil sector in these countries still has many limitations. Tunisia with 5.9% market share in 2008-2009 is the third largest exporter of olive oil in the world, after Spain and Italy, and has increased its efforts to promote its flagship product (which on average accounts for 40% of its national agri-food exports) and to improve its quality (through a quality label policy) from 2000. However, the exported quantity that is not actually sold in bulk remains insignificant: underlying the oft-cited example of Italian operators that purchase bulk oil in Tunisia and sell bottles stamped "Made in Italy" is the problem that the added commercial value of the product is not passed back to the Tunisian stakeholders. The issue is even more crucial now that, since 2009, European legislation requires the "non-community" origin of a product from a third country to be declared. The Tunisian objective is to increase the export rate of bottled olive oil from 1% to 10% by 2011, especially with the opening of Tunisian olive oil to new markets (United States, Canada and Japan). Following the creation of a fund dedicated to bottled olive oil (FOPROHOC), the Ministry of Industry and Technology asked the Technical Centre of Packing and Packaging (PACKTEC) in 2007 to implement a generic promotion campaign for bottled olive oil. But these approaches are still weakened by several handicaps (lack of bottling units, poor organization of stakeholders and lack of reputation of Tunisian olive oil at the international level).

Morocco is the only country on the south Mediterranean coast to have registered a PDO for olive oil ("Tyout Chiadma"), carried out as part of the Green Morocco Plan launched in 2008. This step resulted from participatory work conducted by a multidisciplinary team involving researchers, farmers, maâsra (oil mills) owners and developers, and represents a test case at the national level. Factors that encouraged the achievement of the project include: the organization of olive growers within an association; the special geographical situation of the irrigated area of Tyout, which contains over 12,000 ancient olive trees covering 100 hectares; the site's good technical management and monitoring abilities; the willingness of farmers; and the site's cultural history.

Current experience shows that quality labelling strategies to add value to products can only work as part of a wider ranging agriculture and rural development strategy, which is based on participation by all operators within the sector (producers, processors and exporters) and public stakeholders, in order to make the necessary innovations to ensure quality products and to obtain a higher added value. There is also a necessity to raise the awareness of the consumer. ■

REFERENCES

Angles, S., 2007, *Les Appellations d'Origine Protégée (AOP) d'huile d'olive de l'Union européenne*, *Méditerranée*, 109. URL: http://mediterranee.revues.org/index91.html.

Baldascini, A., Vandencandelare, E., 2007, *Les produits de qualité liée à l'origine et aux traditions en Méditerranée*, Report of the FAO Regional Seminar, Casablanca, 8-9 November 2007

CIHEAM, 2011, "Huile d'olive et oléiculture en Méditerranée", Lettre de veille Hiver 2011, n°16, Paris

International Olive Council, 2010, Acts of the International seminar on geographical indications in the olive oil and table olive sector, 21 - 22 October 2010, Reggio Calabria (Italy).

Karray B. (et al.), 2009, *Programmes et acquis de recherches pour la rénovation de la filière huile d'olive et l'amélioration de ses performances*, Institut de l'Olivier, Tunisia.

CHAPTER 14

Latin America, a rising continent in international agricultural trade, offers a wide variety of products and production models. Such diversity means that this region is a veritable laboratory of agricultural policies, allowing the analysis of the forms and the scales of sectoral regulations.

LATIN AMERICA AND THE CARIBBEAN: REBUILDING AGRICULTURAL POLICIES

Adrián Rodríguez, Mônica Rodrigues and Octavio Sotomayor,
ECLAC (United Nations Economic Commission for Latin America and the Caribbean), Santiago, Chile

In their joint report in 2009, the ECLAC, FAO and IICA[1] stressed that "*Latin America has great potential [...] to contribute to global food security, since it is one of the few parts of the world with sufficient availability of land and water to increase agricultural production*" (ECLAC-FAO-IICA, 2009).

However, significant differences exist between countries. Thus, as one of the largest states with the most potential, Brazil has an abundance of water across a significant proportion of its territory, and a vast availability of land with over 68.5 million hectares in 2008. With such resources, Brazil has become the world's largest exporter of sugar, coffee, soybeans, orange juice, poultry, beef, tobacco and ethanol, and seeks to achieve similar results for other produce. The agricultural sector is critical to the economy, which in 2007 accounted for 25.1% of GDP (including processing activities), 37% of employment and 36.4% of exports, with a continuous annual growth of almost 5% per year over the last decade. Argentina is another major world agricultural exporter, with 33 million hectares cultivated in 2008, including 16.3 million hectares of transgenic soybeans; certain regions have thus become areas of mono-production, resulting in social and environmental risks. The number of cattle in Argentina has reached 50 million which, along with the 205 million heads of cattle in Brazil, represents 20% of the global herd.

Conversely, land availability is very small in the Caribbean and agriculture plays a lesser role. Between these two extremes are countries such as Colombia, Ecuador, Mexico, Peru, Venezuela and the Central American countries, where agriculture still

1. The Inter-American Institute for Cooperation on Agriculture

FIGURE 1 AGRICULTURE IN LATIN AMERICA, A VARIETY OF PROFILES

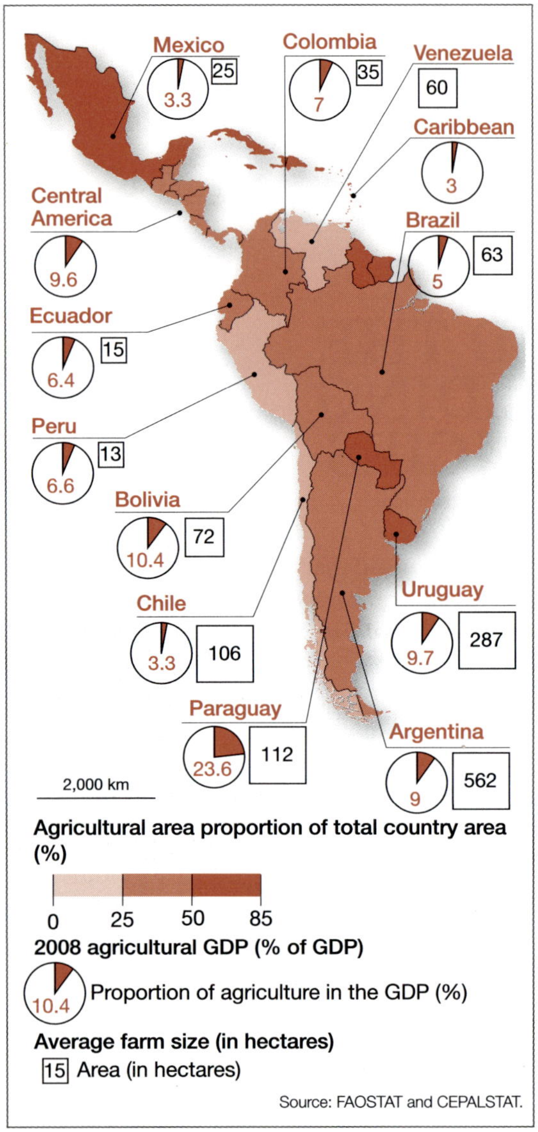

The powerful growth of the two Latin American agricultural giants, Brazil and Argentina, should not obscure the great diversity of agriculture on the continent. There is not one single but many agricultural profiles in terms of average farm size, the proportion of land area under cultivation and, finally, the share of GDP from agriculture.

plays an important economic role but their agricultural structures are highly fragmented.

The strong growth of the two Latin American agricultural giants, Brazil and Argentina, should not obscure the great diversity of agriculture on the continent. There is not one single agricultural profile, but many, in terms of average farm size, the proportion of land area under cultivation and, finally, the share of agriculture in the GDP.

Figure 1 distinguishes three main agricultural profiles. The first group includes countries with high land availability and where agriculture has a major economic influence (such as Bolivia). Farms may be vast or otherwise, as in Argentina or Brazil. A second group consists of countries like Peru, where smaller farms engender a more employment-intensive type of agriculture. The development of this group over the coming years is likely to be related to the production of products with high added value, unlike the Southern Cone countries that specialize in commodities. In this second group, production for auto consumption is very important due to strong smallholder and indigenous agriculture. Finally, agriculture in the Caribbean has a great potential for fruits and vegetables, both for the domestic market (including tourism) and for the very nearby US market. Beyond the wide variety of situations, the majority of countries in the region differ from other parts of the world by a significant potential for agricultural and forestry production (figure 2).

POLICIES HAVE A HISTORY

During the 1980s, the Latin American region initiated a series of institutional reforms which had the common objectives of opening-up trade, reducing the presence of the state, market deregulation and macroeconomic stabilization. Beyond the differences between countries, there are three main stages in this process (Sotomayor et al., 2011):

FIRST STAGE. STRUCTURAL ADJUSTMENT AND THE DISMANTLING OF PUBLIC SYSTEMS The goal at the time was the economic stabilization of countries

BOX 1 STRATEGIC DILEMMAS

This regional and national diversity poses a number of dilemmas:

To increase the productivity of all sectors, or focus on the most "viable" areas: we must go beyond the opposition between "subsistence/commercial" agriculture or "viable/non-viable" agriculture. Faced with the great contrasts that exist between different farming types, there are those that claim that only peasant family farming should be supported, since industrial agriculture does not need state support. Others focus on entrepreneurial agriculture, considering family farming as a social issue only, ignoring the productive and environmental potential of small farms. In fact, rather than choosing between the two sectors, what is needed is a comprehensive approach that exploits the synergies between the different forms of agriculture.

To export, or to supply the domestic market: domestic markets play a very important role, as do exports, which have an ability to improve competitiveness and stimulate growth. To achieve responsible globalization, that addresses the sectors affected by trade integration, public policies must both exploit the potential of each country, while simultaneously moderating the risks and costs associated with the process of removing barriers.

The provision of public and private goods: a number of public goods (statistical information, infrastructure, etc.) that are produced for all and not for one specific sector, are essential for improving sectoral competitiveness. In some cases the subsidisation of private goods (fertilizers, for example) may also be legitimate to enhance competitiveness, although they may be subject to appropriation by specific groups. Each country must make this policy choice based on its economic, political or environmental situation.

Biofuels or food production: the relatively recent development of biofuels as an alternative energy source has led to increasing competition for land between biofuels, livestock and food crops. This has disrupted the functioning of world markets, as demonstrated (at least partially) by the price shock of 2008. Several studies show that Latin America has sufficient land to meet this new demand, even if this resources has its limits: excessive land pressure can cause food or environmental imbalances.

A sectoral or system approach: the experience of Bovine Spongiform Encephalopathy (BSE) and other health and environmental crises, have led to a loss of confidence, that had been previously high, in progress and experts. Productive models and public policies are beginning to be debated outside the agricultural sector, involving multiple actors: ministries of health and the environment, environmental organizations and consumer groups, regional governments and local communities.

Vertical or participatory approaches: growing citizen movements are today profoundly rejecting the model of policy formulation and decision-making according to poorly consultative vertical systems. In several countries, efforts have been made to comprehensively engage farmers and citizens in policy formulation, through the exploration of new forms of public dialogue and innovative methods of consultation and participation.

Centralism or territorialism: in all countries of the region, public services related to agriculture are being restructured. For the implementation of the new principles and approaches to agriculture and the rural domain, the use of the territory as a unit of planning and management is important for encouraging improved competitiveness.

Evaluation and redesign or institutional inertia: institutional inertia is usually the consequence of organisations having a low capacity for change and weak learning capabilities. This can be addressed by the introduction of evaluation systems (for the relevance of the design and the implementation of strategies), thus giving greater "technical rationality" to public policy.

Source: Sotomayor, O et al., 2011

through macro- and microeconomic instruments. This process started in Chile in the mid-1970s, with the interruption of land reform. The size of state institutions was radically reduced, agricultural prices were liberalized and markets were opened up to the outside world. During this period of radical neoliberal policies, health institutions were however strengthened, and some instruments were created to encourage exports, reforestation and technical assistance to small farmers. In the

FIGURE1 FARMERS AT THE GLOBAL LEVEL

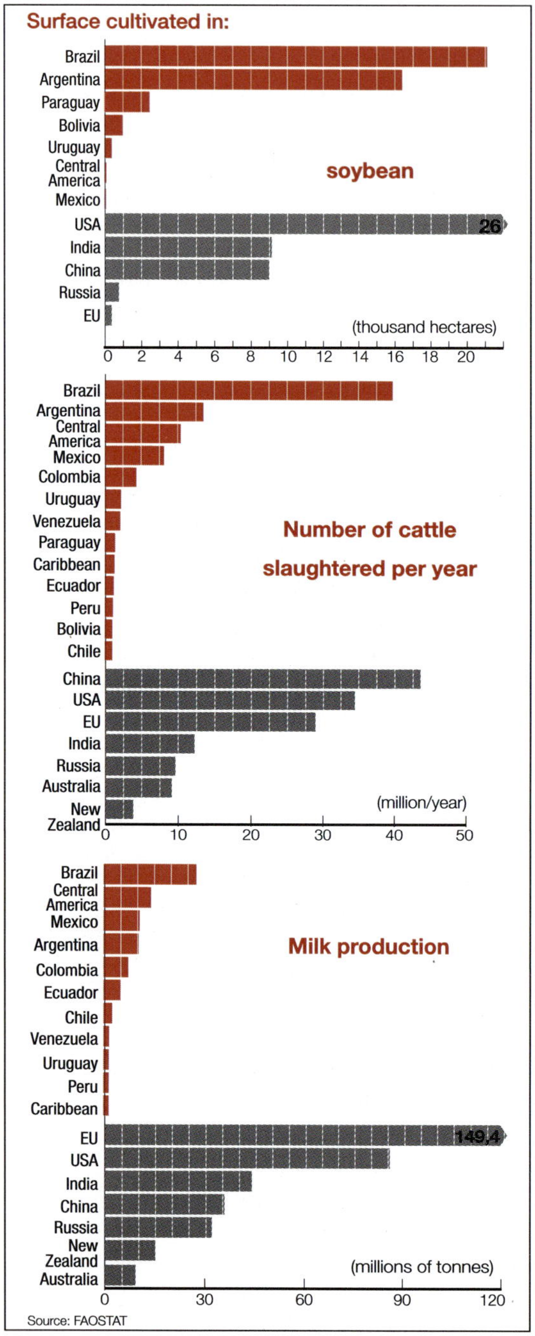

case of Mexico, this stage was characterized firstly by the 1986 accession to the General Agreement on Tariffs and Trade (GATT), and also by the closure of special credit lines for the agricultural sector, the reduction or closure of public institutes for trade or technical assistance, and the confinement of agricultural research to technical institutes and universities. Brazil, meanwhile, abandoned the policy of import substitution in the late 1980s, and embarked on a comprehensive plan of reforms, including macroeconomic stabilization and structural reforms such as the privatization of public enterprises, the deregulation of domestic markets, and the reduction in size of a number of public sector agricultural organisations. Regarding commercial matters, Brazil signed the Mercosur customs union, which reduced tariffs and eliminated many export licenses and non-tariff barriers to agricultural trade. These policies, with some specificities, were also applied in Argentina, Bolivia, Colombia, Ecuador, Peru and other Central American countries.

SECOND STAGE. THE IMPLEMENTATION OF CONTEMPORARY SYSTEMS OF AGRICULTURAL DEVELOPMENT

The objective of this stage was the establishment of a development system to transfer fiscal resources to producers to improve their productivity. In Chile, this period began in the mid-1980s with the creation of technology transfer programmes for medium and large producers, as well as programmes to support access to irrigation technologies. The arrival of democracy in 1990 strengthened the support for these programmes and substantially expanded the measures that targeted smallholder agriculture. Then, the negotiations and signing of the Mercosur agreement (1996) led to the creation of programmes to re-establish soil and technological innovation, and the expansion of programmes to encourage exports, among others. In Mexico, the signing of the North American Free Trade Agreement led to the creation of PROCAMPO (1993), which made direct financial payments to producers as compensation for the effects of trade liberalization. This policy continued in 1996 with the creation of the Programme of Marketing Support and the Alliance for the Countryside, which became the key development programmes and channels for the improvement of the sector's competitiveness.

During this period, Brazil made several adjustments to previously used systems of price fixing that had been made ineffective due to the failure of successive attempts to control inflation. In 1990, the markets for wheat and coffee (domestic and export) were deregulated. In the second half of the 1990s, markets for ethanol and sugar cane were also deregulated, and some improvements in the system of guaranteed minimum pricing were introduced. The acceleration of land reform was accompanied by the creation of the National Programme for the Strengthening of Family Agriculture (PRONAF), and policies which targeted the most disadvantaged people in rural areas, including: subsidized credit, agricultural training and advice, and the promotion of economic activities which add value to primary production.

Other countries in the region implemented major rural development programmes. The Integrated Rural Development Project launched in Colombia in the mid-1970s has been amended several times since 1983. The Provincial Agricultural Services Programme (PROSAP) was created in 1992 in Argentina, with the support of the World Bank and the Inter-American Development Bank (IDB), to channel the transfer of resources to the provinces to strengthen public services, improve basic infrastructure and water management. In 1998 these efforts were supplemented by the creation of the Small Farmer Development Project (PROINDER), which finances small-scale non-refundable investment to support poor producers. The same approach guided the territorial projects in Peru in the 1980s and 1990s, especially in the Andean region, as well as the platform of agricultural services (information and statistics, health, land ownership, irrigation and poverty) in the coastal areas where agriculture is entrepreneurial and export-focused.

STAGE THREE. INSTITUTIONAL RECONSTRUCTION/TRANSFORMATION During this stage that began in the early 2000s, the least developed countries sought to reconstruct sectoral institutions to ensure the smooth functioning of the sector. In the most advanced countries, the series of structural reforms ended at this stage and ways to optimize sectoral policy devices were sought through the improvement of existing development systems and the creation of new programmes and instruments.

In Chile, this stage was marked by the creation of Agricultural Insurance, a Commodity Exchange and other instruments such as technology hubs. In Mexico, the main results were the approval in 2001 of the Law of Sustainable Rural Development, which established guidance for rural development at national, regional, district and municipal levels. In Brazil, the creation in 2003 of the "Zero Hunger" programme and the National Council on Food and Nutrition Security strengthened policies that targeted the fight against hunger and poverty. In Argentina, PROSAP resources were increased, allowing the expansion of its coverage, the modification of existing projects (such as the Risk Project and Agricultural Insurance), and the creation of new policy instruments such as the Argentinian Food Quality Programme

(PROCAL), which was established in 2001. In Uruguay, the Fund for the Reconstruction and Development of Farms (FRFG) was created in 2002, followed by a series of programmes to promote economic initiatives, to reduce the technology gap and improve the training of producers and technicians. In Peru, this period was marked by the signing of new free trade agreements and the fine-tuning of integrated development programmes in the Andean region. Similar approaches were implemented in Ecuador, Colombia, Costa Rica and other countries in the region.

TOWARDS MORE EFFICIENT AGRICULTURAL POLICIES

STRATEGIC DIRECTION AND GOVERNABILITY: PROGRAMMES BY SECTOR AND BY INDUSTRY

The general trend in the region is for the improvement of devices for agricultural policy at several levels. At the macro-sectoral level, some countries have launched participative approaches with a long-term vision aiming at transversal political agreements, such as the Agri-food and Agribusiness Strategic Plan 2010-2016 in Argentina, and the Politics of State 2010-2021 for the Food Sector and Rural Development in Costa Rica (MAG, 2010). Other participatory approaches seek to integrate government programmes and international cooperation: such as for example the Rural Development Plan of Nicaragua (PRORURAL), which is designed to include an extensive consultation process with the private sector and international cooperation (MAGFOR, 2009). However, in some other countries, sectoral plans have been implemented without consultation, such as: the Sector Programme on Agricultural, Livestock and Fisheries, that is developed in Mexico at the beginning of each government (SAGARPA, 2007); the Family Agriculture Plan for the period 2011-2014 in El Salvador (MAG, 2011); and the Strategic Action Plan for the Agricultural Sector 2010-2014 in Panama (MIDA, 2010). A final option is to create regional agricultural policy systems, such as the Agricultural Policy in Central America 2008-2017 (PACA) launched by Belize, Costa Rica, El Salvador, Guatemala, Honduras, Nicaragua and Panama. Even if it is part of the economic integration process implemented through the Economic Integration System (SICA), the design of a common agricultural policy aims to provide the necessary strengthening of public and private regional institutions to support the application of this type of sectoral policy (CAC-SICA, 2007).

Another almost universal trend is the use of the "agro-chains" concept to manage the sectoral policies from the upstream to the downstream of specific industries, based on the principle that each industry is a specific universe with its own stakeholders, problems and issues. The improvement of the competitiveness of each national agricultural sector involves systematic efforts focused on the main branches or productive chains, which requires the coordination of all stakeholders, according to medium and long-term strategies.

THEMATIC DEVICES Based on these sectoral agendas and agro-chains, Latin American countries have deployed a set of specific agricultural policies (Sotomayor et al., 2011). Trade policy is a key component of the agricultural sector policy of each country, since it defines the way agriculture fits into the global economy: the role assigned to agriculture in the national development strategy, the type of production specialization, the form of insertion into globalization processes, the identification of regions that gain or lose, and a prospective long-term restructuration to adapt to globalization. Based on these choices, governments have applied different strategies for market access (free trade agreements, health agreements, trade promotion, etc.), as well as tariff policies and other internal mechanisms to protect trade. These choices have also determined the design of other sectoral regulations and production development policies to accelerate structural change.

Technological innovation policies: science and technology policies implemented in the region for the generation of new knowledge have recently undergone profound changes. There has been a shift from a linear model of technology supply, where the public sector played a key role through the identification of priorities and direct involvement in research and development, to a model that includes a range of public and private actors. These stakeholders operate on the basis of grants that generate a demand for technology (Argentina, Chile and Mexico) or on the coordination mechanisms between supply and demand (Brazil). These innovation systems are linked with long-established programmes for technical assistance to farmers. Today, several approaches co-exist for the transfer of technology: ranging from traditional agricultural development to social network approaches, through horizontal peer groups, productive chains, precision agriculture and specialized technical assistance programmes in the fight against poverty.

In terms of access to land, social inequalities remain and have made the problem of access to land, which has always been a major political problem in spite of agrarian structures, now more dynamic and fragmented. Since its return to democracy in 1985, Brazil has initiated active land reform (expropriations, direct procurement, allocation of public land and legal recognition of occupied land). Other countries have chosen to allocate state land, either through a programme of organized settlement or through the regularization of existing settlements on public lands. A third option, the "Land Fund", supported by the World Bank, offers credit for land purchases to small producers who do not have access to capital markets. Described as "market-assisted land reforms" they are criticized for their high costs and the limited number of beneficiaries.

Soil management: the early twentieth century saw the first soil management initiatives such as programmes against desertification and the first areas of state-protected land. In a second stage, research projects have been implemented in national agricultural research institutes, although these still receive very limited funding. In a third step, a number of schemes for soil management have been developed: increasing the complexity of themes and instruments used, the promotion of no-tillage agriculture as a new production paradigm and a greater coverage of public programmes (such as public investment in soil fertility).

Water and irrigation: in arid areas of the region (the Andean area of Argentina, central Chile, the valleys of Peru, northern Mexico, among others), irrigation has been a public policy issue since the early twentieth century when the state began to build the first water works (reservoirs, canals...). Generally of medium size, they reinforced the canal network of the colonial era. In the 1960s, the second wave of irrigation schemes commenced, with an added focused on large hydroelectric infrastructure, with the support of the IDB, World Bank and other financial institutions. In many countries, large reservoirs and other control structures were built and managed by the state through specialized agencies (irrigation commissions or districts). A third phase saw an improvement of infrastructure that was accompanied by an optimization of water management through cross-sectoral irrigation planning and innovative schemes for the management of water demand (water footprints, social control systems, tradable water rights markets, irrigation technologies...).

Biosecurity and safety: international regulations for plant and animal health have become increasingly strict and now include issues of safety (human health) and of environmental and animal welfare. The health issue is at the heart of trade negotiations, which sometimes generates insurmountable barriers. For this reason, on top of the customary activities of services for the protection of plant and animal heritage, are added new tasks relating to the international dimension of the agricultural sector. This is despite the fact that services for plant and animal health in the region are often operating on reduced budgets and are restricted to purely reactive and emergency interventions. Safety policies face the same challenges, since the intensification of productive systems, aimed at increasing production, has augmented the use of fertilizers, pesticides, veterinary drugs and other substances that may remain in the food chain. In addition, the growing importance of preventive health in terms of consumer perceptions has led governments to impose increasingly important health standards.

The first quality policies in the region were related to the identification of goods (flour, oils and others) and later gave rise to health regulations for food. To these were added various voluntary standards of quality, aiming at differentiating certain products from standard ones. Industrial property rights have recently emerged: the region already has 42 Appellations of Origin, 16 Geographical Indications and 30 Collective Labels. In addition, other quality labels also exist, such as organic farming - in virtually all countries, high-quality labels in Mexico and Argentina and Fair Trade, which has developed over several years through the work of NGOs and small producer organizations in the region.

Funding: The region has few financial systems in rural areas, despite recent efforts to expand their reach through the financial institutions of ministries of agriculture and development banks, which target small businesses, such as in Brazil. In Mexico and Argentina, public banks are developing a two-tier system in which they transfer funds to private banks and other financial intermediaries, so that they can provide credit to micro-entrepreneurs. The development of specific methodologies for family farmers with lower resources is in line with the "micro-financial revolution", which has been driven since the 1970s by non-financial structures (NGOs, cooperatives and mutuals), who develop financial products for customers that are traditionally ignored. Currently, there seems to be a consensus that the market is the best way of dealing with the steadily growing number of customers. It requires, however, policies, procedures and technologies to overcome or compensate for the failures of major financial markets, with appropriate incentives and new institutional structures (coordination between development banks and other public entities, support for investment projects and technical support to manage the credit demand from producers).

Price stabilization systems: In Latin America and the Caribbean, several countries had previously implemented major public programmes for trade and price controls. Most were dismantled during the structural reforms of the 1980s due to their high cost to the taxpayer and their distortive effects on the economy. Some countries such as Brazil and Mexico, however, have maintained their systems but moved towards public policies that are less interventionist and less expensive for the state. More generally, due to growing concerns about the volatility of international prices for agricultural products, three main types of policy schemes have been developed: (i) grants to finance state-induced business operations (or direct procurement) and to co-finance the purchase of price insurance premiums that are contracted by producers in commodity exchanges (Brazil, Mexico and Colombia); (ii) income stabilization funds for specific industries, co-financed by the producers and the state (Colombia and Costa Rica); and (iii) price range policies for certain products (Argentina, Chile, Colombia, Ecuador, Peru and Venezuela).

Climate insurance: agricultural insurance is traditionally underdeveloped in the region, particularly due to the complexities involved in the sale of intangible assets. Several governments are now making efforts to expand the use of these instruments, using state subsidies towards premiums. Agricultural insurance for climate risk management is gradually spreading (Argentina, Brazil, Mexico...).

Environmental policy: faced with global challenges such as poverty and climate change, and to improve the competitiveness of the agri-food sector, several countries in the region are encouraging a real transition in agricultural policy to fully integrate environmental challenges. This integration is complex, consisting of multiple levels and different technical issues of environmental policy. Environmental issues are specific to agro-ecological conditions, types of agriculture and the natural ecosystems that are specific to each region in the different countries. However, in most countries the Ministries of Agriculture are active in the management of public policies related to environmental issues, particularly those relating to mitigation and adaptation to climate change.

CONCLUSION

The heterogeneity of the agrarian structure remains within Latin American agriculture: among the diverse types of productive enterprises, some smallholder microenterprises, which form the majority in number, still have very low productivity levels, leading to very low incomes and high levels of poverty and exclusion. To reduce these asymmetries, the process of trade liberalization must be accompanied by safeguards that are both innovative and as inexpensive as possible, given the variations in world prices (of products and services, exchange rates and interest rates). These processes must be accompanied by a strong development system aimed at improving sectoral competitiveness, sustainability and social inclusion. While some countries such as Argentina, Brazil, Chile, Colombia, Costa Rica, Peru and Mexico have been able to implement devices such as these, other countries in the region suffer from high levels of institutional weakness of their Ministries of Agriculture, as a result of state dismantling in the 1980s and 1990s due to restrictions on fiscal resources and/or the mechanical application of the Washington Consensus recommendations. The region therefore displays a wide variety of interventional approaches. Some countries have a system of basic policies (agricultural research, health and trade), while others have devices for intermediary development, based on instruments such as credit, technical assistance, irrigation, soil management, and polices for trade, health, quality and the environment. Finally, some countries have advanced policy systems, including all of the instruments discussed above, in addition to R&D programmes, climate insurance and mechanisms of price stabilization at the level of the major productive chains. However, there remains much to be done, and a new wave of reforms is needed. This requires the region to be considered as a major area of innovation and experimentation, in which the successful experience of one country can be applied to another, with the necessary adaptation. Much of the world arsenal of potential public policies is reflected in the wide range of agricultural policies in the region. Latin America and the Caribbean are therefore generally as advanced as other parts of the world: the regional system has many innovators, and many successful and interesting experiences. ■

REFERENCES

CAC, SICA, 2007, Política Agrícola Centroamericana 2008-2017. Una agricultura competitiva e integrada para un mundo global. Document for national and regional consultation. April 2007.

CEPAL, 2011, International seminar "Politicas para la agricultura en América Latina y el Caribe: competitividad, sostenibilidad e inclusión social". Santiago de Chile, 6-7 December 2011.

CEPAL, FAO, IICA, 2009, Perspectivas de la Agricultura y del Desarrollo Rural en las Américas: Una Mirada hacia América Latina y el Caribe. San José de Costa Rica.

MAG, 2010, Política de Estado para el Sector Agroalimentario y el Desarrollo Rural Costarricense 2010-2021. Draft version, San José, September. www.mag.go.cr/Politica/index.html

MAG, 2011, Plan de Agricultura Familiar. Período 2011-2014. Power Point Presentation, 11 February, El Salvador.

MAGFOR, 2009, Plan Sectorial PRORURAL Incluyente 2010 -2014. July, Managua, Nicaragua.

MIDA, 2010, Plan de Acción Estratégico del Sector Agropecuario 2010-2014. October, Panama.

SAGARPA, 2007, Programa Sectorial de Desarrollo Agropecuario y Pesquero 2007-2012, Mexico, DF.

Sotomayor, O., Rodríguez A. and Rodrigues M., 2011, Competitividad, sostenibilidad e inclusión social en la agricultura: nuevas direcciones en el diseño de políticas en América Latina y el Caribe. CEPAL, Santiago de Chile.

LEARNING SUSTAINABILITY: 50 YEARS OF EUROPEAN AND AMERICAN AGRICULTURAL POLICIES

Antonin Vergez,
Centre International de Recherche Agronomique pour le Développement, Paris, France
Simon Liu,
United States Department of Agriculture, Washington, United States of America

For over half a century, the United States (US) and the European Union (EU) have been implementing agricultural policies. In a world that is increasingly shaped by market forces, the importance and longevity of such policies is exceptional. Sometimes praised as an embodiment of market-compatible regulation, agricultural policies are also much criticized (by environmentalists, taxpayers, consumers, member or federal states, third-party countries and even farmers although they are also beneficiaries). Despite objections, agricultural policies remain today. How have they changed over the last 40 years? Have sustainability issues (environmental, social and economic) been integrated into their design and implementation, and if so, to what extent, how often and under what political processes?

In the US, agricultural laws, known as "Farm Bills", are renewed approximately every five years. Since the first Farm Bill of 1933, they have been organized into "titles".

In 1964 and 1977, concerns about sustainability were gradually incorporated into "Food Stamps" titles, then into the "Conservation"[1] (of land, resources and natural areas) title in 1985, and "Bioenergy" in 2002. The term sustainable agriculture was first defined in the 1990 Farm Bill, as: an integrated system of plant and animal production practices having a site-specific application that will, over the long term, 1) satisfy human food and fiber needs, 2) enhance environmental quality and the natural resource base upon which the agricultural economy depends, 3) make the most efficient use of nonrenewable resources and on-farm resources and integrate, where appropriate, natural biological cycles and controls, 4) sustain the economic viability of farm operations and 5) enhance the quality of life for farmers and society as a whole.

In the 1980s and 1990s, progress was made on the issues of organic agriculture, research and development on alternatives to synthetic inputs, and policies on resource and habitat conservation. Policies to limit watershed pollution were implemented in the 1980s: the Department of Agriculture offered technical and financial assistance to farmers willing to modify their practices, through annual payments over a five year period along with federal and local government cost sharing (Chuang et al., 1998). Reports published by the National Research Council fuelled discussions on sustainability (e.g. a 1993 report on pesticide residues and the health of children and infants).

Unlike the EU's Common Agricultural Policy

1. Programmes related to the environment belong to the "Conservation" title.

(CAP), the Farm Bill includes a social component related to the significant amount[2] of food aid for poor people living in the US. Similarly to CAP, the Farm Bill has had to continually adapt to WTO rules, as well as to domestic critics (who point out the high cost to the US taxpayer), by modifying its aid to become less distorting and more efficient. Economics, in its normative aspect, has played an important role (Bureau, 2007) in the successive reforms of agricultural policies on both sides of the Atlantic.

In Europe, the 1957 Treaty of Rome fixed the CAP objectives on: increasing production and productivity, improving farmer income and improving the food supply to European consumers at reasonable prices. The EAGGF (European Agriculture Guidance and Guarantee Fund), the financial arm of the CAP, was deployed in 1962 in the six member states of the time. It served as a catalyst for the modernization of production structures and enabled market intervention through a system of guaranteed prices and CMOs[3]. While the CAP achieved its intended goals, in the 1980s it became a victim of its own success. High guaranteed prices encouraged overproduction, allowing food stocks to increase exponentially (creating the notorious "mountains" of meat, cereals, butter...). Even though export subsidies enabled surpluses to be sold, the United Kingdom and some third-party countries, whose producers felt they were being penalized, denounced this system as inconsistent and expensive. NGOs and some farming unions also stressed the negative impacts of this policy on the environment

The CAP, which was considered unsustainable, was reformed in 1992, 1999, 2003 and 2008. The aims of these reforms were i) to restore the balance between supply and demand in the EU through prices by (re)connecting producers to market signals, and ii) to improve the way in which "societal" expectations, rather than exclusively agricultural ones, are taken into account. To limit overproduction and budget drift, a paradigm shift was proposed in 1992 involving the lowering of intervention prices and in turn the compensation of farmers through direct income support. In 1999, the "multifunctionality of agriculture" was highlighted, with the establishment of the CAP's "second pillar", comprising rural development aid and agri-environmental measures (animal welfare, biodiversity and soil conservation, etc.). In 2003 and 2008, direct "first pillar" payments, previously related to production, were decoupled from particular crops to be replaced by single payment entitlements without a production requirement; while environmental concerns over cross-compliance[4] were also incorporated and the relative share of the second pillar was increased.

Over a period of 40 years, while the concepts have been renewed, the aims and the incentives have changed and the distortive potential of aid has decreased. At the same time, the volume of financial transfers in the EU and the US has, in general, been maintained. The range of stakeholders able to have an influence on agricultural policies continues to grow and the balance of power has shifted slightly. Sectoral lobbies that were initially very strong in the design and implementation of agricultural policies are now facing a challenge to their hegemony: a fact evidenced by the WTO negotiations, which have accelerated the disappearance of export refunds[5].

Gradually, agriculture is being shaped through channels other than agricultural policies. Distributors in the agri-food sector keep one step ahead of legislation, meeting the demand for "greener" products by forming direct contracts with farmers, or even entire sectors. There are a number of cities that innovate and complement the purely technical solutions through integrated territorial management approaches. Barraqué and Pincetl (2010) show that to stop "*water quality degradation caused by agricultural intensification in drinking water catchment areas*", cities or private stakeholders can sign cooperative arrangements with farmers to encourage them to modify

2. 68% of the 284 billion dollar budget of the current Farm Bill (2008-2012) is allocated for nutrition which includes food stamps (coupons redeemable for food in stores).

3. Common Market Organisations. These are now almost completely dismantled. For 30 years they framed production to varying degrees, depending on the product, guaranteeing prices that were higher than world prices and stabilizing supply.

4. A farmer's non-compliance with good agri-environmental practices (crop rotation, minimum coverage, etc.) or of European directives (environment, animal welfare, etc.) can lead to a reduction in direct payments.

5. The disappearance of this aid was announced following the WTO negotiations in Hong Kong in 2005.

FIGURE 1 40 YEARS OF EU-US AGRICULTURAL POLICY

FARM BILL UNITED STATES					
1933 ▪ *Agricultural Adjustment Act, a programme to support agricultural prices (maize, wheat and cotton)*	**1964** ▪ *Food Stamp Act, amended in 1977 (simplification of the eligibility of households) and 1994 (improved access to retail stores)*	**1981** ▪ *Agriculture and Food Act.* **1985** ▪ *Food Security Act*	**1990** ▪ *Food, Agriculture, Conservation, and Trade Act* **1996** ▪ Federal Agriculture Improvement and Reform Act	**2002** ▪ *Farm Security and Rural Investment Act*	**2008** ▪ *Food, Conservation, and Energy Act*
COMMON AGRICULTURAL POLICY – EUROPEAN UNION					
	1957-58 ▪ Treaty of Rome and the Stresa resolution set out the objectives of the CAP **1962** ▪ Deployment of the CAP in 6 countries	**1980s** ▪ Overproduction: criticism from countries both within the CAP and outside it **1984** ▪ *"I want my money back": Margaret Thatcher negotiates Britain's EU budget rebate*	**1992** ▪ CAP changes from price support to direct income support **1999** ▪ Agenda 2000 recognizes the multifunctionality of agriculture and created the Second Pillar: rural development and agri-environmental measures.	**2003** ▪ Decoupling of direct support: single payment entitlements without having to produce	**2008** ▪ "Modulation" of aid in favour of the Second Pillar, eco-conditionality of aid from the First Pillar ▪ Ratification of the Lisbon Treaty: the European Parliament makes joint decisions on agriculture **2013** ▪ Total elimination of export subsidies
CONTEXT					
		1980 ▪ First Federal Coordinator for organic farming **1985** ▪ Research and training programme on LISA ("Low-Input/Sustainable Agriculture") ▪ President Reagan: "America would be nothing without its farmers: they are the backbone of this country and everything we do to help support our country and its future." **1986-1994** *WTO: the Uruguay Round, which included agriculture in the negotiations*	**1989** ▪ President Bush's policy of "zero net loss of wetlands" **1980s and 1990s** ▪ Freshwater preservation programmes ▪ Research and training programmes on sustainable agriculture and on alternatives to synthetic inputs **1991** ▪ Nitrates Directive, the first European environmental directive specifically targeting agricultural activities	**2000s** ▪ Programmes to improve irrigation infrastructure, tree planting and crop diversification to improve water quality in watersheds **2000** ▪ Water Framework Directive, sets the objective of good water quality (chemical and ecological) by 2015 **2001 - ?** *WTO Doha Round seeking to improve the access of developing countries to agricultural markets*	**2008** ▪ Increased funding of environmental programmes (wetlands, farmland and pastures), supply of school meals with fresh products and the promotion of organic farming ▪ EU: Adoption of the "climate and energy package" that will affect agriculture, including the production of sustainable biofuels

European Union | United States | World Trade Organization

their practices. For example, the city authorities in New York and Munich pay farmers for practices that protect their natural water reservoirs, and in France, Perrier Vittel directly contractorizes 37 farmers on 3500 hectares. These new and heterogeneous stakeholders promote their own sustainability criteria, but they remain few in number and their power is limited.

The evolution of institutional mechanisms could change the balance of power. In the US, agricultural lobbies seek to influence Congress and the President, and must now take into account environmental and social NGOs (Orden et al., 2008). A change of the majority party in Congress in 2010, the budget deficits and the instability of agricultural markets will all have a strong influence on the 2012 Farm Bill. For some, the fiscal imbalance in favour of the CAP's first pillar (in 2010 the CAP accounted for 42% of the EU budget, 75% of which went to the first pillar) illustrates the power of agricultural trade unionism in the EU. The commencement of the Treaty of Lisbon in 2010 could change this situation: agriculture now, as for other sectors, is subject to the decisions of both the EU Council and its Parliament. The latter, which is the only EU body directly elected by European citizens, could aim to make payments for the production of public goods via an increased role of the second pillar, rather than by directly supporting farmer income.

Finally, regulations that are not directly agricultural, such as the incorporation of renewable energy in transport (the 2008 EU "Climate and Energy" package stipulates 10% in 2020, while the US has ambitious goals for bioethanol production from maize), have and will have increasing effects on agriculture and markets.

In the US and the EU, a growing, but still low, number of farmers and ranchers are moving towards more sustainable practices that are both integrated and innovative. However, much research remains to be done to fully evaluate these systems and their sustainability. For future agricultural policies (Farm Bill 2012 and the CAP in 2014 to 2020), the challenge will be to further integrate sustainability concerns, including: local and global environments, equity within and between nations and, more "conventionally", issues regarding farmer incomes and consumer prices. In a multi-level governance context, the role of the EU and the US federal government could be to coordinate agricultural policies (participation and involvement at various stages to ensure a legitimate diversity of goals while limiting any conflicts and providing effective instruments) rather than to fully control it. ■

REFERENCES

BARRAQUÉ B. and PINCETL S., 2010, The city returns to its hinterlands for natural resources (chapter 7). In A Planet for life, Cities: Steering towards sustainability.

BUREAU J-C., 2007, La politique agricole commune. *Ed.* La Découverte, coll. Repères, 121 p.

CHATELLIER V. and GUYOMARD H., 2011, Réformer la PAC dans un contexte budgétaire difficile. Problèmes économiques, n°3 022. La documentation française.

CHUANG L., CORCORAN E., and HENDERSON A., 1998, Toward a Sustainable Agricultural Policy – Evolution in the Conservation Titles of the 1985, 1990, and 1996 Farm Bills. Evaluating natural resource use in agriculture, 1998, p. 85-113.

ORDEN D., BLANDFORD D. and JOSLING T., 2008, Determinants of Farm Policies in the United States, 1996-2008. International Agricultural Trade Research Consortium.

CAP 2013, AN OPPORTUNITY TO RETHINK AGRICULTURAL POLICY?

Henri Nallet,
Fondation Jean Jaurès, Paris, France

In what way does CAP 2013 offer an opportunity to rethink agricultural policy?
The current reform offers Europeans an opportunity to redefine what they understand by the Common Agricultural Policy (CAP), or in other words what type of agriculture and products, what place in international trade and what regulatory systems they want.

The time is ripe for two reasons. First, European agriculture, like the rest of Europe's economy, is in crisis – a crisis that does not simply involve adjusting to the market but which concerns the actual model of European agricultural development followed over the last 40 years. This can be explained on the one hand by strong competition from other types of farming (e.g. in the area of large-scale production) and on the other because this model has damaging effects on ecosystems and contributes to climate change. New development models thus need to be developed. Second, and paradoxically, unlike all previous reforms, the current CAP reform is not being undertaken under strong international constraints (surpluses or trade negotiations). We should thus take advantage of this situation to get the work done calmly and thoroughly.

The most difficult exercise in the negotiations will nonetheless be to reach a socially acceptable agreement on agricultural subsidies, especially as the negotiations are being held in a context of enormous strain on public finances.

Will discussions be focused on the budgetary aspect?
The budgetary aspect will of course be decisive, quite simply because a number of countries, and not the least important, will be applying austerity programs to their population. Take the case of France: in the talks on the CAP, which has been very largely beneficial for France over the last fifty years, our country will, for the first time, be in the position of net contributor to the Community budget. This situation will force France to be prudent in its demands, and Italy and Spain will find themselves in a similar situation.

Achieving policy reform within the 27-member Union is a not simple matter. If there is tension in the budget talks because a blocking minority decides that other items cannot be discussed until agreement is reached on the allocation of subsidy volumes, then the reform will be in jeopardy. The art of negotiating is to keep the talks on the table, but the crisis in the euro zone will have a tremendous impact on the CAP negotiations and bring to light totally different conceptions of the European Union.

Is the Commission's proposal to green part of Pillar 1 subsidies broadly consistent with your recommendations?
Not entirely perhaps. In substance, I think that the inspiration is the same, that's to say, what justification for farming subsidies can we put forward to

FIGURE 1 DISTRIBUTION OF THE CAP BUDGET

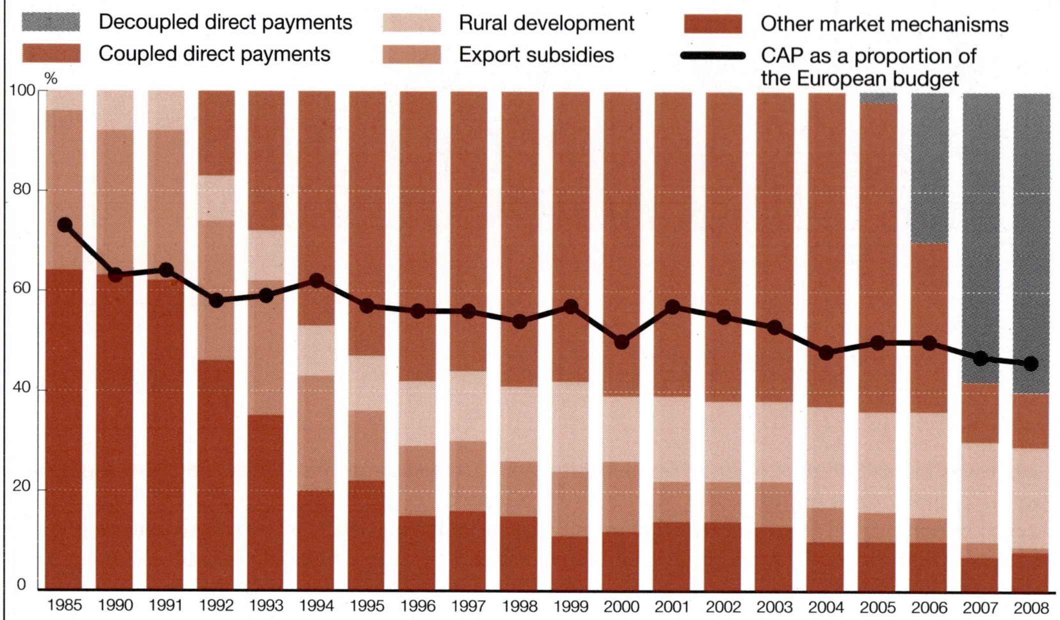

Source: European Parliament, 2010:21, *Lyon Report*.

taxpayers. Farmers produce goods that are useful to the population, but they also provide public goods valuable to the community. Land use management, wildlife and vegetation management, etc. Many of these public goods are not remunerated, which justifies government subsidies. This is especially true as we know that we are entering a period of heightened tension over the appropriation of land, which remains a scarce resource, either because there are other use requirements (urban uses) within our communities or because climate change and land degradation will create a great deal of pressure internationally on agricultural land.

This new justification for public aid appears in the Commission's proposals even though there is no explicit reference to public goods. Yet, the current formulation used by the Commission is awkward as it presents this in terms of blanket agri-environmental measures, and farmers have very bad memories of these as they lead to highly complicated bureaucracy. It will be up to the negotiations to improve these proposals and the Commission seems ready for this.

Are these proposals capable of triggering the transition towards the sustainable agriculture that you are hoping to see?

The proposals are heading in this direction, but we need to go much further, in my opinion, than simply mainstreaming Pillar 2 agri-environmental measures. We need to foster the transformation of all European agriculture so that it becomes more sustainable, by using less water, fertilizer and pesticides and implementing new tillage techniques. We need to mainstream sound agricultural and ecologically intensive practices, as Michel Griffon calls them, etc. Many farmers, cooperatives and agricultural schools are already following this direction. These players do not wait for ministerial circulars in order to change, as we can already see a host of organic farming initiatives, the preservation of local animal breeds and species, zero-till seeding, etc. The role of the CAP should be to provide a framework and encourage this movement, with simple and effective measures.

Other aspects support greater sustainability. On the one hand, the fact that the CAP recognizes agricultural diversity. We should no longer oppose one form of agriculture against another. I believe that sustainability means a horizontal approach. It involves the ability to combine a production system integrating the specific characteristics of soils, climate, etc. Those, such as myself, who held positions of responsibility during the 1960-80s developed a very vertical type of agriculture, a one-size-fits-all model applied everywhere in whatever context. This must be abandoned. On the other hand, aid needs to be regionalized. Great headway would be made if at least Pillar 2, and even part of Pillar 1, were managed at regional level. This would mean much greater support from farmers, as well as the possibility of combining Community, national and regional structures with agricultural organizations.

What needs to be done is thus to assume the cost of public goods, recognize diversity and regionalize aid.

Finally, the last aspect is to keep the tools as simple as possible. We have known what needs to be done for several years, but have found it difficult to move forward as some farmers viewed these measures as punitive. They were criticized, accused, monitored, asked to do paperwork and subjected to a bureaucratic regime. I think this approach is mistaken, and support should be given to everything that relies on simple criteria, that is efficient and can be controlled. Naturally, there is the risk of fraud, but we know how to set up effective controls.

Alongside the environmental stakes, what about employment in the agricultural sector? Is the question of employment being discussed today?

With the exception of some Central and Eastern European countries, farming populations are now at a very low level. For some places, the question arises not of how to maintain populations, but of how to maintain production, with the distinct possibility of land being left uncultivated. This is particularly true in difficult areas, mountainous zones. On this count, European policy has been positive. Some production, such as livestock farming in mountainous regions, has been kept going artificially with respect to the market through direct subsidies. When we ask ourselves what kind of agriculture we want, we should ask whether it needs to cover the whole territory. I think it does, and that in difficult regions specific subsidies should enable young farmers to set up in normal conditions. This does not pose any technical problems. It is a political issue.

In countries like Poland, Romania, Bulgaria, this shift should be accompanied. There is a natural shift, related to aging, towards a decrease in the number of very small farms. Today, we do not have the financial means that we had in the 1960s, when there were funds for training young farmers and early retirement annuity schemes. We must – while respecting diversity – move towards a certain competitive convergence.

In your book, you speak of new prospects for European agriculture and the CAP. What about this reform?

The reform has nothing on the Euro-Mediterranean outlook of the CAP, and yet this is a key issue. It is not just a matter of developing agriculture in these countries, or simply a question of a timetable for commercialization. For Europeans, it means being more present in the agricultural development of countries on the southern shores of the Mediterranean: through equity investments, associations, direct investment. We should be discussing with them what resources the European Union could make available so they can develop their own agriculture. Not with a view to sharing production opportunities "you do this, I do that", but in order to create local employment and foster agricultural production for their domestic markets and for export to the European Union. It is now within reach of the 27-member EU to be more open to Southern countries in the area of agriculture. This makes good business sense, despite the reticence of many Europeans. The more we buy the goods they produce best, with regard to their comparative, mainly climate-related advantages, the more we will be in a position to sell them the products we produce better, such as cereals, livestock, processed agrifood products.

Will any other work streams come under discussion?

The question of reciprocity is a major issue that the reform does not address. Do we consider that European agriculture should always be open, or on the contrary that it deserves some degree of protection when its competitors do not operate in the same conditions or do not comply with the same requirements? Some large agricultural countries may not only be our competitors but also severely endanger our own productions: Brazilian soya, Argentine beef, Canadian wheat. This is a difficult but necessary debate, particularly as the absence of a Doha Round agreement re-opens the question of liberalization in the agricultural sector. The Commission has not wanted to tackle these sensitive and controversial topics. This is a mistake as the questions related to protecting our agricultural and industrial production systems have been raised and if political leaders do not address them, their citizens will force them to do so. ■

For too long, agricultural, energy and development policies have lost sight of the need for food security. Yet, as a global public good, food security should be a structural objective of global governance. But reform of the latter is not sufficient and we should be fully aware of the role to be played by national policies, and of the growing importance of private stakeholders.

FOOD SECURITY AS A GLOBAL PUBLIC GOOD

Olivier De Schutter,
United Nations Special Rapporteur on the Right to Food

To understand food security as a global public good and the need, therefore, for international coordination, we must begin our analysis at the food price crisis of 2007-2008. This crisis arose from a series of immediate causes[1]. Some causes are related to the supply of agricultural raw materials. In some regions, crop yields in 2005 and 2006 were low due to weather phenomena, causing an initial price rise. While this could have stimulated production, increasing energy prices, both for agricultural production and for the transportation of inputs and harvested crops, restricted the ability of farmers to respond to market signals. In addition, growth in agricultural productivity has slowed significantly since the 2000s: productivity per hectare has probably reached a ceiling in many regions of developed countries, and remained stagnant in some developing countries (DCs) because of the difficulties faced by producers in accessing credit and public support, and the weakness of rural infrastructure. The increase in energy prices has also meant an increase in the prices of synthetic fertilizers, pesticides, transport and processing of crops, thus creating a gap between food price trends on international markets and farmer incomes: when commodity prices started to rise, the benefits that should in principle have been directed towards producers, were largely absorbed by increased production costs.

1. For an analysis of key factors, cf. the analysis I presented in my official capacity as Special Rapporteur on the Right to Food, 2nd May 2008, available on: http://www.srfood.org/index.php/en/documents-issued/other-documents-issued.

Causes stemming from the demand side include rising energy prices, which led to a growing demand for biofuels - resulting in increased prices for maize, then soybean and palm oil - and to an increased pressure on arable land due to the fact that crop production for food had to compete with fodder and bioenergy production[2]. The gap between the growth in demand and the supply capacity of agricultural raw materials gradually widened. Food stocks, which were already in global decline since the late 1990s, have from 2006 shrunk to dangerous levels. Against this already tense background, speculation played a role both in the physical markets, particularly for rice (Slayton, 2009), as well as in the derivatives markets: all this was enough to generate feelings of new shortages, which in spring 2008 resulted in a market panic that created the sharp price rises that we all still recall.

These causes are well known and it is not necessary to dwell further on them. What must be emphasized, however, is the fact that the crisis highlighted both the inconsistencies in global food security governance – or rather the cost of its absence – and the need for a reinforced international coordination to address the issue. Indeed, underlying the immediate causes of the agricultural commodity price increases are more structural developments, three of which are particularly important to consider.

THERE REMAINS A MAJOR OBSTACLE: THE POOREST STATES LACK THE BUDGETARY RESOURCES TO FUND THE POLICIES THEY WOULD LIKE TO IMPLEMENT.

Firstly, for nearly forty years, there has been inadequate investment in subsistence agriculture in developing countries. In Latin America and several countries in South Asia, the 1960s and 1970s were characterised by the launch of a "green revolution", gambling on a more capitalized form of agriculture that involved massive irrigation, mechanization of agricultural production, a reliance on external inputs – chemical fertilizers and pesticides – and the distribution of high-yielding seeds developed in public agricultural research centres (Borlaug, 2000; Holt-Gimenez and Patel, 2009; Shiva, 1991). A high degree of international cooperation has characterized this development because the implementation of public research, for example in Mexico and the Philippines, has been of benefit to many countries throughout the continents concerned. In Sub-Saharan Africa, however, this technological revolution did not occur over the same timeframe: farmers have been largely left to fend for themselves, aside from limited support from some governments through centralized public procurement to promote industrialization and to ensure access to cheap food for their populations. These very different trajectories have one common aspect: rural poverty and inequality have increased because those responsible for agricultural development have generally favoured producers with the most land and capital, and the best transport infrastructure; while small farmers in remote areas or working on more marginal lands have not received due attention. In a context dominated by a fear of shortages along with

2. 70% of the increase in maize production between 2004 and 2007 was due to ethanol production (Christiaensen, 2009)

rapid population growth, there has been a general reliance on the enhancement of agricultural production efficiency, without giving the same degree of consideration to issues of access and distributive justice. The quasi-generalized dismantling of public intervention systems in the agricultural sector during the 1980s and 1990s, largely due to structural adjustment programmes imposed on indebted DCs, has further accentuated this trend: where liberalization is dominant, only the most competitive producers can emerge, confining many others to subsistence agriculture or forcing them to leave the sector to swell the slums of huge cities.

Through the advances of the "green revolution", world agricultural production has increased over the last fifty years. And it has done so everywhere, although in Sub-Saharan Africa production has remained stagnant relative to the population (due to high population growth) and the only production gains there have been achieved by an extension of cultivated areas rather than by increases in yields per hectare. This progress in the availability of agricultural raw materials has, however, been accompanied by an increase in rural poverty because the situation for family smallholders, upon which a majority of rural people depend, has not improved. Such farmers are often facing increased pressure on land and water resources due to competition from highly capitalized agriculture that is more competitive on markets. Therefore, there has been an increase in hunger at the same time as production was being raised. Among other factors, this is a failure of governance, in that public policies were designed in a technocratic manner, based on an incomplete diagnosis of the causes of hunger, and were not aligned with the objective of improving food security. Investment in agriculture has been excessively geared towards the promotion of export crops while giving insufficient attention to the social and environmental impacts of technological advances, and the chosen indicators of success (production increase per capita, growth of competitiveness and increased export revenue) have guided the development trajectory towards aims other than food security (De Schutter, 2011).

Not only had agricultural policies shifted the focus from the objective of food security; the liberalization of agricultural trade and trade policies has also diverted governments from pursuing this goal. This is a second major structural cause of the crisis. The macro-economic prescriptions of the 1980s have prompted governments of many developing countries to lower import duties on foodstuffs. This suited their short-term interest in promoting access to affordable food for urban populations. But small local producers that were unable to compete on international markets or that grew staple food crops not meant for export, have had to compete on their own markets with products that are heavily subsidized by OECD countries: the result was that a large number of small farmers selling their products on local markets came to fail[3]. In 1994, the conclusion of the Marrakesh Agreement that established the World Trade Organization (WTO), only confirmed what had, at that time, already become a

3. Moreover, this has led to development cooperation agencies gradually moving away from agriculture in general, as they did not view the sector as one that had any future potential for DCs (World Bank, 2007)

FIGURE 1 FOOD PRICES DO NOT INCREASE ALONE

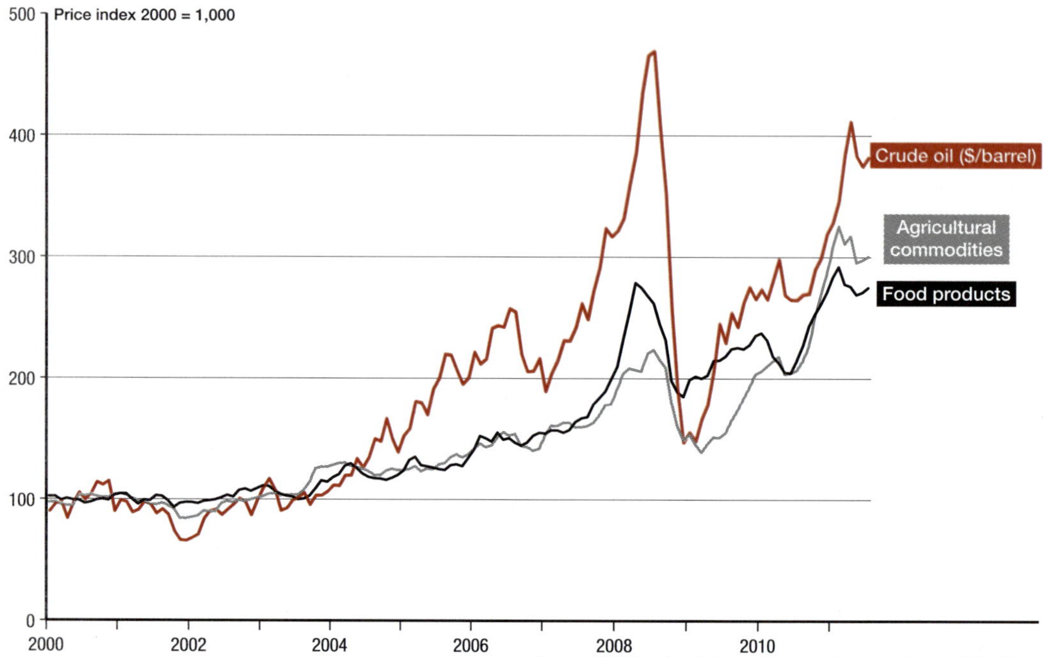

Source: UNCTAD, Monthly global commodity price indices, UNCTADstat.

Food prices have risen steadily since 2000, with an acceleration in this increase from 2008. Prices of agricultural commodities have followed a very similar trend. Both have reflected the increase in prices of other basic resources such as oil, which influences the cost of many chemical inputs as well as the transport of food products.

reality: the gradual reduction of tariff rates on agricultural products that encourages developing countries to both specialize in the production of agricultural commodities, and to depend to a greater extent on the import of processed food products. The international division of labour codified in the agreement is not necessarily favourable towards the least advanced countries. Above all, it has ambiguous impacts on food security. While some efficiency gains have been achieved, the dependence of these poor countries on imports leaves them vulnerable to price rises on international markets. Poverty in rural areas has increased, and with it, automatically, the number of people lacking access to affordable food.

Finally, the third development to consider is that energy policies were not geared towards the need for food security. In the 1940s, the food systems of OECD countries developed a strong dependence on fossil fuels. Other contributing factors were the mechanization of production, the use of mineral fertilizers and pesticides from the chemical industry, and the lengthening of supply chains - increasing transport and storage costs due to the longer distances involved. The way that the "green revolution" has spread throughout developing countries has extended this dependency in several important areas. More recently, the decision to focus on biofuels, ethanol or biodiesel products from plants (maize, sugar cane, beet and jatropha for example) has accentuated the fusion between the energy and agriculture markets with inevitable consequences: when oil and gas prices rise, the prices of agricultural products follow suit and food prices may therefore be adversely affected[4] (Figure 1).

In other words, the food price crisis of 2007-2008 highlighted not only the vulnerability of a number of poor countries that are net importers of foodstuffs and vulnerable to price increases, but also the fact that agricultural trade and energy policies were not in alignment, or were even detrimental, to the objective of food security. Even food aid deployed over the last thirty years has sometimes produced perverse effects, destabilizing the already fragile local markets by competing with producers in recipient regions (FAO, 2006). The observation of these inconsistencies has given rise to the idea of a better international coordination of the sectoral policies that impact upon food security at the national level (Christiaensen, 2009).

THE REFORM OF GLOBAL GOVERNANCE FOR FOOD SECURITY

When food prices soared in spring 2008, UN agencies responded by forming the High-Level Task Force on the Global Food Security Crisis (HLTF), under the chairmanship of the Secretary General of the United Nations. Under the auspices of the UN and FAO, the HLTF gathers together representatives of specialized UN agencies (including the International Fund for Agricultural Development (IFAD) and UNICEF), as well as the Bretton Woods institutions and the WTO.

The HLTF is primarily intended to enable governments to identify a "range of options" that they are encouraged to explore in order to address the issue of food security, in particular in relation to price increases: these options are set out in a "Comprehensive Framework for Action", highlighting the different routes available for States to take, and the associated consequences. This represented an initial coordination attempt, launched as early as April 2008. However, the HLTF remains a consultation structure amongst international agencies; neither governments nor civil society are represented. While a better alignment of sectoral policies that impact on food security is essential, it seems vital that these stakeholders are integrated into any coordinative structure.

There was a subsequent attempt by the Committee on World Food Security (CFS) to integrate these stakeholders. Officially, the CFS was simply an FAO committee, established in 1974 following the first World Summit on Food Security, to make proposals to the FAO Council on the basis of a policy review relating to global food security. However, the thirty-fifth CFS annual session was seized as an opportunity to instigate its ambitious reform, which was endorsed by the FAO Council (CFS, 2009). The reforms are intended to make the CFS a forum for dialogue between all stakeholders, aiming to contribute to food security and improved nutrition, and to establish the CFS as the main international and intergovernmental platform for this purpose.

The CFS, it deserves underscoring, is not a forum for donors: its purpose is not to encourage the governments of rich countries to make financial commitments

4. However, the debate on this issue is not closed.

to support agriculture in DCs. It should instead be considered as a place to forge consensus on certain issues of common interest for the entire world, at the junction between the spheres of several international agencies, and concerning both public and private investment. However, there is a question mark regarding the gap between the professed ambitions and the ability to deliver the changes considered necessary.

THE BLIND SPOT OF THE REFORMS: THE ROLE OF PRIVATE INVESTMENT

Indeed, there remains a major obstacle to achieving consistency and to making the much-needed transition towards more sustainable agricultural and food systems: the poorest States lack the budgetary resources to fund the policies they would like to implement.

This results in these poor countries being highly dependent on the donors. However, in recent years, the tendency of donors to impose various requirements on recipients, without any discussion between them, has been increasingly criticized because of the burden it imposes on the administrative capacity of partner governments, and because of the resulting inefficiencies in the flow of aid (Easterly, 2006). On 2 March 2005, the Paris Declaration on development aid effectiveness was adopted under the auspices of the OECD to improve the quality of aid. It was approved by 122 governments, the European Commission and 27 organizations. The commitments contained therein are based on five principles aiming at shifting the direction of aid policies: national ownership, alignment with national-level priorities, harmonization between donors, result-focused management and mutual accountability (OECD, 2007).

But even more aid shall not suffice to bridge the financing gap. The least developed countries also depend on the arrival of private investment for the development of storage or communication infrastructure, or to support the provision of private goods – fertilizers, high-yielding seeds or pesticides – to farmers. These countries therefore face a dilemma: to reject such investment would mean an end to the ability to fund long overdue programmes for agricultural recovery; but to accept is to risk not being able to direct funds towards the most urgent needs, for example to support small farmers located in remote areas or working on the most marginal land – precisely those that need the most help.

It is in these terms that we can consider the debate on large-scale agricultural investment, which since the end of 2008 has attracted much attention (De Schutter, 2011b). These investments take the form of purchases or long-term leases of large tracts of arable land in DCs, both by private operators (agri-food companies and investment funds) and by the governments of countries that are cash-rich but are relatively poor in natural resources (and concerned about ensuring long-term food security for their own populations by developing agricultural production capacity elsewhere). Given that such investments entail the development of export crops on large plantations that are capital-intensive rather than labour-intensive and contribute very little to the local economy, but that utilize land that may be the most fertile and best located, these investments appear unfavourable to local food security. However, these areas of land would probably remain underexploited, and average surface productivity low, if not "developed" through the acceptance of foreign investors.

Contract farming creates a similar dilemma. Here agri-food buyers make long-term agreements with producers to ensure a stable supply of agricultural raw materials. In exchange for a commitment to supply the buyer and to meet their requirements in terms of volumes, delivery times and standards, the producers gain guaranteed market access; they receive some technical assistance; and they are often able to buy the necessary inputs at less than market prices since the buyer is able to purchase such inputs in bulk in order to redistribute them to contracted producers. Such agreements seem to be in the interest of both parties. They exempt governments from having to provide small farmers with the services offered by the buyer, such as inputs and technical assistance. At the same time, if governments do not play an active role in guiding these investments, the most marginalized small farmers are at risk of being excluded. These farmers may be located far from transportation infrastructure or have plots that are too small (not allowing them to invest in land with adequate economies of scale) and are sometimes illiterate or poorly trained: for all these reasons, potential investors may be deterred from supplying these farmers due to the transaction costs being too high to allow long-term agreements to be made (Runsten & Key, 1999). Given the situation, should we discourage potential investors that would be unlikely to promote an increase in income for poor farmers? Or should such investors be encouraged, but on the proviso that they conduct their affairs within a framework that makes a contribution towards development strategies agreed at the national level in the host country? Can a relationship of subsidiarity be envisaged between the respective roles of the private investor and the State?

AGRA AIMS TO REPRODUCE IN AFRICA THE TECHNOLOGICAL REVOLUTION IN AGRICULTURE THAT OCCURRED IN LATIN AMERICA AND SOUTHEAST ASIA IN THE 1960S, WHILE LEARNING FROM THE NEGATIVE ENVIRONMENTAL CONSEQUENCES OF THE FIRST GREEN REVOLUTION.

The emergence in 2006 of the Alliance for a Green Revolution in Africa (AGRA), further illustrates this difficulty (Toenniessen et al, 2008; Dano, 2007). Created with the support of the Gates and Rockefeller private foundations (which contributed 100 and 50 million USD respectively at the outset), AGRA aims to reproduce in Africa the technological revolution in agriculture that occurred in Latin America and Southeast Asia in the 1960s, while learning from the negative environmental consequences of the first green revolution (Conway, 1997). To achieve this goal, AGRA focuses on breadbasket regions, in which it aims to support farmers by providing inputs such as high-yielding seed varieties: between 2009 to 2012, 40% of AGRA's support will go to Ghana, Mali, Mozambique and Tanzania, countries that are identified as suitable according to the World Bank's "Doing Business" index, i.e. countries that have created an environment that is favourable for private investment (AGRA, 2009)[5]. The objective is to create markets that will eventually become viable. In the thirteen countries where it operates today, the Alliance wants to encourage the emergence of "farmer-entrepreneurs" to develop agriculture that is sufficiently capitalized in order to encourage private companies to supply inputs and to invest in regional-level distribution.

The AGRA project sets out to remove the barriers faced by smallholders in the acquisition of inputs (their purchasing power being too low to attract input providers, especially taking into account the distribution costs associated with their geographical dispersal) and those that limit their market access (barriers related to a lack of adequate infrastructure and to insufficient information on prices or the buyer expectations). This should create linkages between farmers – in their capacity as input buyers and suppliers of agricultural products - and the markets: this improved market access is defined as the key to allowing farmers to escape from subsistence agriculture. The main AGRA programmes converge towards this goal, supporting 40 African seed companies and 10,000 distributors (Program on African Seed Systems); encouraging the creation of rural markets, harvest exchanges and storage units (Market Access Program); encouraging better soil maintenance, starting with a mapping of African soils (SoilHealth Program); and encouraging African governments to strengthen their support for agriculture and to subsidize input purchases (Policy and Partnership Program). In addition, AGRA collaborates with several partners. Thus, on 15 April 2009, it announced the creation of an Agricultural Investment Fund for Africa, set up jointly by AGRA and the African Development Bank (ADB) and IFAD, with the support of the French Development Agency, aiming at raising 500 million euros in support of agroindustry and agricultural cooperatives in Africa.

Even a superficial review of these projects and their evaluation would go far beyond the scope of this article. However, it should be noted that one of the challenges awaiting African governments is, once again, that of consistency. AGRA's available budgets, that attract the interest and support of the private sector, are superior to those at the disposal of African governments to revive their agriculture. Under these conditions, how can national ownership of programmes be ensured? Are AGRA's priorities meeting those laid down by national States? If AGRA is able to benefit farmers that have the potential to transform into "entrepreneurs", and that can gradually make the transition to a more capitalized agriculture that attracts private investors, what will happen to smaller farmers for whom this type of development is not viable? Will the pressures on land and water increase, and will competition between farmers eventually impact negatively upon those that are most in need of help - the poorest farmers working in the least favourable agro-ecological areas, that are unable to benefit from the introduction of new technologies?

5. In addition, 40% of AGRA's support will go to nine other countries (Burkina Faso, Ethiopia, Kenya, Malawi, Niger, Nigeria, Rwanda, Uganda and Zimbabwe), and an additional 13% to awareness-raising operations throughout the entire African continent, and 7% will go to operating costs.

THE ROLE OF NATIONAL STRATEGIES

A key issue is therefore that of the inter-sectoral coordination and consistency between investment and development aid on one hand, and the setting of national priorities on the other. In this regard, the declaration adopted by the G20 agriculture ministers during the meeting held under the French Presidency on 22 and 23 June 2011, recognizes that investments in agriculture "should be made in support of country-owned investment plans" (paragraph 19). Similarly, the declaration adopted in Rome during the World Summit on Food Security on 16-18 November 2009, identifies the first "Principles of Rome", which is to "invest in country-owned plans, aimed at channelling resources to well-designed and result-based programmes and partnerships"[6]. The L'Aquila Initiative on Food Security, adopted at the G8 meeting on 8-10 July 2009, also insists on this dimension of aligning investments with national strategies.

This concern for national ownership is also at the heart of the approach taken by agricultural development policies, which stems from the right to food. Following the call made at the 1996 World Food Summit for the clarification of normative requirements on the right to food, the Committee on Economic, Social and Cultural Rights (CESCR) – the body of experts responsible for monitoring compliance with the International Covenant on Economic, Social and Cultural Rights – dedicated a general comment on the right to food, which clarified its implications for governments (CESCR, 1999). From 2002 to 2004, governments within the FAO negotiated a text containing a set of recommendations on the implementation of the right to food: this text, titled the Voluntary guidelines to support the progressive realisation of the right to adequate food in the context of national food security, was adopted unanimously by FAO Council members on 23 November 2004.

Like the CESCR's General Comment No. 12, the guidelines on the right to food called for the adoption of national strategies defining which measures should be taken, by whom, within which timeframe and following which procedures. Such national strategies or action plans are designed to ensure that adequate resources are mobilized. They aim to improve coordination across different ministerial departments to ensure that the multiple causes of hunger (which are interlinked) will be addressed. These strategies require accountability: by assigning roles to each party and by defining responsibilities, they facilitate the work of civil society organisations and independent authorities – such as human rights national institutions or jurisdictions – which will be able to demand from public authorities that they justify their choices in the light of the agreed roadmap. These action plans sometimes allow ombudsmen or equivalent institutions to respond to the passivity of authorities. They also promote collective learning: if indicators are used to measure progress, policies that fail to produce results can be corrected at an early stage. Finally, because

6. World leaders reaffirmed that "the responsibility for food safety rests with countries and any programme aiming at meeting the challenge of food security must be formulated, developed, supported and led by countries and built on a consultation with all key stakeholders" (FAO, 2009).

these strategies must be designed to be participatory and inclusive, they encourage a democratization in decision-making, and they guarantee that decisions will be made in light of actual needs, as expressed by those concerned.

Governments must ensure that their agricultural policies and their decisions regarding international trade, energy policies, the fight against climate change, food aid and development cooperation, converge towards the aim of improving food security. They must also retake control of agricultural investment by ensuring that priorities are not defined by investors seeking to acquire the best land, or by private foundations convinced that the solution to poverty and a lack of rural development, that afflicts such a large number of smallholders involved in family farming, is a transition to highly capitalized agriculture based on farmers that are connected to markets. The reform of the governance of food security has a key role to play in this respect, to promote this convergence and to ensure this ownership. It is urgent that within DCs, especially those where agriculture has stalled and must now be replaced at the heart of development policies, governments must be willing to transform their modes of governance and develop these national strategies by accepting the control of public opinion and the sharing of power with farmer organisations. Reinvesting in agriculture is not only ensuring that agricultural productivity will increase in areas where it has remained low and virtually unchanged for forty years, it is also winning the fight against rural poverty, and therefore fighting against hunger and malnutrition. Furthermore, through the increase in the standard of living of rural populations whose purchasing power is currently insufficient to constitute a market for local suppliers of goods and services, it promotes the development of all economic sectors in very poor countries, based on agricultural regeneration (Adelman, 1984).

Governance reform within States must go hand in hand with improvements to global governance. The reform of the CFS, if it leads to the adoption of a robust global strategic framework containing a specific reconstruction agenda, is promising in this regard. This framework should inspire the work of various international agencies gathered within the CFS, and it should guide the relationships between donor governments and their Southern partners. Gradually, the global strategic framework must tend towards the improvement of the international environment so that poor countries will see their efforts rewarded. It is this objective that we must not fail to achieve. ■

REFERENCES

Adelman, Irma, 1984, "Beyond Export-Led Growth", World Development, vol. 12(9): 937-949.

Alliance for a Green Revolution in Africa, 2009, Strategy for an African Green Revolution?

Borlaug, Norman E., 2000, The Green Revolution Revisited and the Road Ahead, Special 30th Anniversary Lecture, Norwegian Nobel Institute, Oslo.

CFS, 2009, Reform of the committee on world food security, Doc. CFS:2009/2Rev. 2 (Oct. 2009).

Christiaensen, Luc 2009 "Revisiting the Global Food Architecture. Lessons from the 2008 Food Crisis", Review of Business and Economics, 2009/3, pp. 345-361, note 10.

Comity in Economic, Social and Cultural Rights, General comment n°12: the right to adequate food (1999), doc. ONU E/C.12/1999/5.

Conway, Gordon, 1997, The Doubly Green Revolution: Food for All in the 21st Century,. Penguin Books, London, 1997.

Dano, Elenita, 2007, Unmasking the New Green Revolution in Africa: Motives, Players and Dynamics, Third World Network, Penang.

De Schutter, Olivier, 2011, "International Trade in Agriculture and the Right to Food", in O. De Schutter and K. Cordes (eds), Accounting for Hunger: The Right to Food in the Era of Globalization, Hart Publ., Oxford, chapter 5.

De Schutter, Olivier, 2011b, "The Green Rush: The Race for Farmland and the Rights of Land Users", Harvard International Law Journal, vol. 52(2) (2011), pp. 503-559.

Easterly, William, 2006, The White Man's Burden: Why the West's Efforts to Aid the Rest have done so much Ill and so little Good, Oxford, Oxford Univ. Press.

FAO, 2006, The State of Food and Agriculture 2006 – Food aid for food security?, Rome, FAO.

FAO, 2008, The State of Food and Agriculture 2008. Biofuels: prospects, risks and opportunities, Rome, FAO.

FAO, 2009, document FAO WSFS 2009/2, available on http://www.fao.org/wsfs/wsfs-list-documents/en/

Holt-Gimenez, Eric et Patel, Raj, 2009 Food Rebellions! Crisis and the Hunger of Justice, Pambazuka Press, Food First Books, and Grassroots International, 2009, pp. 26-37

Key, N. et Runsten, D., 1999, "Contract Farming, Smallholders, and Rural Development in Latin America: The Organization of Agroprocessing Firms and the Scale of Outgrower Production," World Development 27(2) (1999): 381-401

OECD, 2007, Aid Effectiveness: 2006 Survey on Monitoring the Paris Declaration. Overview of the Results, Paris, OECD

Toenniessen, G., Adesina, A., Devries, J., 2008, "Building an Alliance for a Green Revolution in Africa", Annals of the New York Academy of Sciences, vol. 1136 (2008), 233-242.

Slayton, T., 2009, Rice Crisis Forensics: How Asian Governments Carelessly Set the World Rice Market on Fire, Center for Global Development Working Paper 163, Washington D.C., Center for Global Development

Shiva, Vandana, 1991, The Violence of the Green Revolution: Third World Agriculture, Ecology, and Politics, London, Zed Books, 1991.

Word Bank, 2007, World Development Report 2008: Agriculture for development, World Bank, New York.

World Food Programme, 2008, Food aid flows.

REFORM OF THE COMMITTEE ON FOOD SECURITY: AN OPPORTUNITY FOR GLOBAL GOVERNANCE?

Sélim Louafi,
Centre International de Recherche Agronomique pour le Développement, Montpellier, France

The food price crises of 2008 and 2011 have put the issue of food security high on the international agenda. Faced with these cyclical price rises, several international institutions were called upon and mobilized in response to the emergency. These organizations cover a broad spectrum of institutional arrangements: UN intergovernmental agencies (FAO, IFAD, WFP); international organizations outside the UN framework (World Bank, CGIAR); intergovernmental treaties or conventions on particular aspects of food security (such as the International Plant Protection Convention); international initiatives and programmes such as the UN Special Rapporteur on the right to food initiative and the G8.

This global picture is completed by a number of actions carried out by organizations or programmes operating on a transnational scale (such as non-governmental organizations, multinational corporations or regional cooperation programmes such as CADDP in Africa).

However, according to the mandate, theme or sector of each institution or initiative, the responses remain partial. This leads to the fragmentation of agendas within institutions and to initiatives with different dynamics but with occasionally overlapping functions and aims.

In a context where the strong leadership of an institution, state or group of states is lacking, this fragmentation of the food security global governance landscape could be problematic.

THE CHALLENGES OF THE COMMITTEE ON FOOD SECURITY

It is in this context of fragmentation that the challenges facing the reform of the Committee on Food Security (CFS) must be understood. The CFS was established after the 1974 World Food Conference in the perspective of another "global food crisis". It was intended as a forum for analysis and for monitoring policies dedicated to all aspects of food security throughout the world. However, the Committee failed to engage other UN agencies and civil society in the process, and was hence confined to a technical role, mainly focusing on agricultural aspects. The Committee considered food crises in terms of availability, and regarded increases in agricultural production as the main solution for food security. Although the FAO had developed, thanks to extra-budgetary funding, programmes and other initiatives seeking to build links with the socio-economic (poverty, health) and environmental (climate change, biodiversity) issues of food security, and such projects had only a very slight effect on the activities of the Committee, whose influence on the international food security agenda remained very limited until recently.

The revitalization of the CFS is part of the radical reforms that the FAO has engaged in since 2005. The process, which is unprecedented for a United Nations agency, represents a takeover by Member States of an organization considered to be losing legitimacy (McCalla, 2007; Lele, 2009). The CFS reform, driven by Member States, was approved in October 2009. It emerged in the form of a re-evaluation of the Committee's legal status that enabled it to communicate its decisions not only to the FAO Conference, but also to the United Nations General Assembly through the Economic and Social Council. The objective of CFS revitalization is to make it "(...) the main international and intergovernmental open platform, gathering a broad range of stakeholders who are committed to working together in a coordinated way and in support of processes instigated by countries for the elimination of hunger and to ensure food and nutrition security for the whole of humanity"[1]. The Committee therefore claims to act explicitly to improve the quality of international coordination, asserting itself as the legitimate intergovernmental platform par excellence, in the mould of the Conferences of the Parties on climate change or biodiversity. In other words, a common platform for all stakeholders acting on a global scale, who are keen to benefit from a comprehensive framework that can support their actions to address the challenge of world hunger. This aim is based mainly on two mechanisms: an inclusive and participatory process and an increased mobilization of expertise.

CFS ENLARGEMENT AND EXPERTISE

The transformation of the relative role of states in the management of world affairs has become a truism in the literature on international relations (see for example Stone, 2008). The construction of international rules entails the involvement of an increasing number of stakeholders, or groups of stakeholders, who do not necessarily (or do not only) apply their influence through the channels of national representation. Rather than a weakening, this represents a transformation in the role of states. The issue of food security is no exception to this development and the creation of an advisory group within the CFS reflects this change. The advisory group – comprising representatives of non-governmental organizations, professional organizations, the private sector and foundations, "sister" and "associate" groups connected to international organizations, and international financial organizations – provides a procedural legitimacy that intergovernmental coordination alone no longer delivers. In particular, the advisory group enables stakeholder participation to go beyond the annual CFS plenary sessions, allowing in particular their involvement during the inter-sessional phases. Such participation plays an undeniable role in increasing the transparency of decision-making while ensuring a wider representation of interests at the table. In addition, it increases the mobilized knowledge base (broadening the scope further than the agricultural field alone and its focus on increased production) and maximizes the opportunity to increase social learning through frequent member interaction. It is too early to measure the impact of such inclusiveness on the coordination process or, even more so, on the problems to be addressed. However, the degree to which the various advisory group stakeholders have become engaged provides a good way to assess the impact of the changes. It is within the group of non-governmental organizations that the most dramatic changes can be seen (see ETC, 2009; IPC, 2010). Traditionally excluded from political discussions, at the request of the CFS they have established their own coordination mechanism to strengthen their legitimacy and thus have more impact in debates. For example, during discussions on the issue of land grabbing, NGOs pushed for the international community to consider the voluntary guidelines developed by the four UN agencies (FAO, IFAD, UNCTAD and World Bank) as insufficient, regarding these guidelines as too timid. In consequence, an open working group to develop principles for the responsible investment of the CFS was put in place during its last session in October 2010. This working group should benefit in particular from an inventory of all existing initiatives on the subject, a report requested from the High level panel of experts (see below), and also from the establishment of a wide consultation

1. See document CFS: 2009/2 Rev 2, § 4, FAO, Rome

process to provide feedback, data and new recommendations. Without trying to predict whether this will lead to something truly innovative and likely to effectively regulate the behaviour of stakeholders investing in farmland throughout the world, it is however reasonable to assume that such a process would not have been possible without the presence and involvement of the advisory group members.

The second pillar of the CFS revitalization is related to the mobilization of scientific and technical expertise. Given that international coordination is frequently hampered by non-consensual opinions and by major controversies over possible solutions, the establishment of a High level expert panel (influenced in part by the Intergovernmental Panel on Climate Change – IPCC – model) is expected to provide the international policy-making process with data and recommendations that are as objective as possible. The panel comprises a steering committee of 15 members selected for their experience and international reputation. The committee oversees the establishment of ad hoc teams to deal specifically with given problems, under the responsibility of one or more members of the steering committee. While the IPCC's main aim was to involve the hard science community in the field of politics, the creation of the Expert Panel on Food Security should especially help the emergence of debates and solutions that are usually "filtered" by the interest – real or imagined – of existing institutions or states. The first two topics commissioned by the panel of experts were: the volatility of agricultural prices; and land tenure and responsible investment. The reception of these initial two studies during the thirty-seventh session of the CFS in October 2011, given that they are highly controversial and already heavily debated by various international bodies, will provide a good indicator of the value brought by independent expertise that is less subject to political or institutional influences. It will also reveal the extent of the role actually given to such expertise in building the political debate and the development of innovative solutions.

CONCLUSION: IS FOOD SECURITY BECOMING A GLOBAL ISSUE?

A procedural legitimacy, therefore, through the inclusion of concerned stakeholders and a better understanding of the substantive nature of discussions by using a specialized analytical approach, should lead, at least on paper, to the enhancement and increased effectiveness of political debate and international coordination. However, there are still many obstacles to tackle before the CFS becomes a real platform for cooperation on food security.

The first barrier relates to institutional aspects that may undermine the expectations of CFS reform. Remaining embedded within the FAO, there is a risk that the CFS will be consistently fixated, consciously or not, on the issue of food security in its agricultural dimension. Nutritional, commercial, environmental or social (poverty-related or health) components remain underrepresented despite the broadening of the advisory group or the constitution of the expert panel. It will be some time before the CFS, which exists within an organization that specializes in agricultural and food issues, can claim legitimacy in these fields that are not directly included in its mandate.

Other obstacles relate to the nature of the problem itself: while the global nature of environmental issues is clear, this is not the case for food security, which remains a problem that is essentially envisaged at the national level. If we add the lack of consensus regarding the relative importance attributed to the various dimensions of food security, it is particularly difficult to agree on what should be the focus of coordination and, even more so, on the tools and policy options likely to add value in comparison with a situation where global collective action is low or absent.

In this context, it is clear that at this stage no one (state, group of states or international institution) has sufficient legitimacy to impose a single model of collective international action. The current fragmentation may therefore continue. Undeniably, the CFS already delivers benefits by providing a unique platform where all stakeholders can, at the very least, exchange ideas, learn and discuss current initiatives; it will, however, take time and require obstacles to be addressed before the forum becomes one where food security is truly conceived as a global issue and where solutions can be developed with the consent of all parties. The continued investment of states (beyond periods of global food crises); the construction of synergies with existing global initiatives (such as the G20 or the Secretary General's high-level task force on the global food security crisis); assigning a major role to independent expertise and enabling the FAO to rise to the challenge; are all key determinants for the CFS to have a sustained influence on the international food security agenda, and to make a real impact on dealing with hunger. ■

REFERENCES

ETC Group, 2009, Report Card from Reform School: The Committee for World Food Security Passes a Test, but...

IPC, 2010, the reformed committee on World food security, a briefing paper for civil society

Lele, U., 2009, Global Food and Agricultural Institutions: The Cosmology of International Development Assistance, Book Review Article, *Development Policy Review*, *27*(6), 771-784.

McCalla, A., 2007, FAO in the Changing Global Landscape. *Giannini Foundation of Agricultural Economics*, (07-006).

Maxwell, S., 1996, Food security: a post-modern perspective. *Food Policy*, *21* (2), 155–170. Elsevier.

Stone, D., 2008, 'Global Public Policy, Transnational Policy Communities and their Networks', Journal of Policy Sciences, 36 (10), pp.19–38.